Visit classzone.com and get connected

Online resources provide instruction, practice, and learning support correlated to your text.

- **Misconceptions database** provides solutions for common student misconceptions about science and their world.
- **Professional development links**, including SciLinks, offer additional teaching resources.
- **Animations and visualizations** help improve comprehension.
- **Math Tutorial** helps strengthen students' math skills.
- **Flashcards** help students review vocabulary.
- **State test practice** prepares students for assessments.

You have immediate access to *ClassZone's* teacher resources.

MCDTCOWDMSSZ

Use this code to create your own username and password.

Also visit *ClassZone* to learn more about these innovative and updated online resources.

- eEdition Plus Online
- eTest Plus Online
- EasyPlanner Plus Online
- Content Review Online

Now it all clicks!™

McDougal Littell

TEACHER'S EDITION

McDougal Littell Science

Matter and Energy

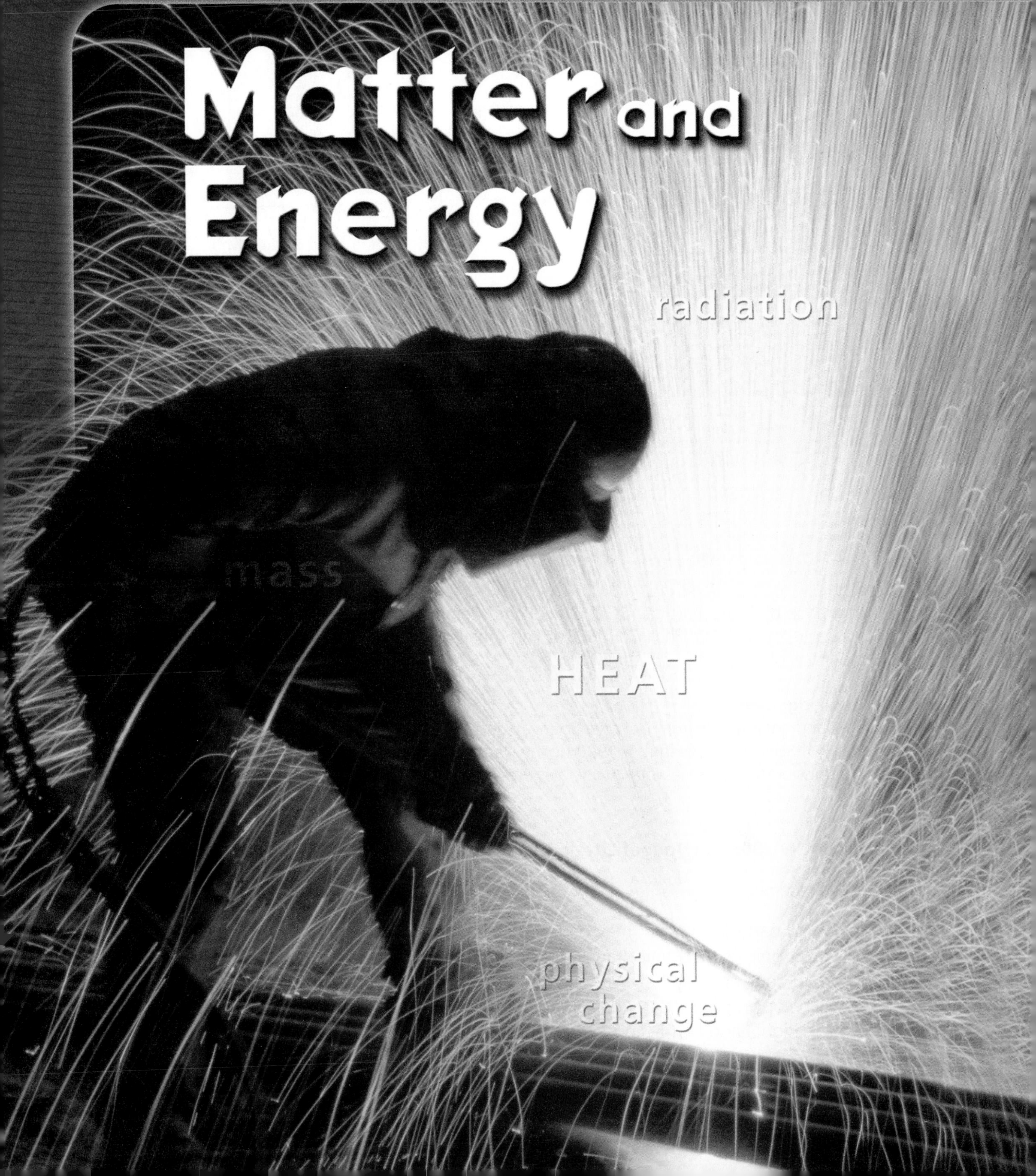

radiation

mass

HEAT

physical change

Credits
5B Illustration by Stephen Durke; **5C** © Omni Photo Communications, Inc./Index Stock; **37C** © David Young-Wolff/PhotoEdit; **67B** © Left Lane Productions/Corbis; **67C** © Joe Sohm/Visions of America, LLC/PictureQuest.

Acknowledgements
Excerpts and adaptations from National Science Education Standards by the National Academy of Sciences. Copyright © 1996 by the National Academy of Sciences. Reprinted with permission from the National Academies Press, Washington, D.C.

ISBN: 0-618-33445-9 1 2 3 4 5 6 7 8 VJM 08 07 06 05 04

Internet Web Site: http://www.mcdougallittell.com

McDougal Littell Science

Effective Science Instruction Tailored for Middle School Learners

Matter and Energy Teacher's Edition Contents

Consultants and Reviewers

Science Consultants

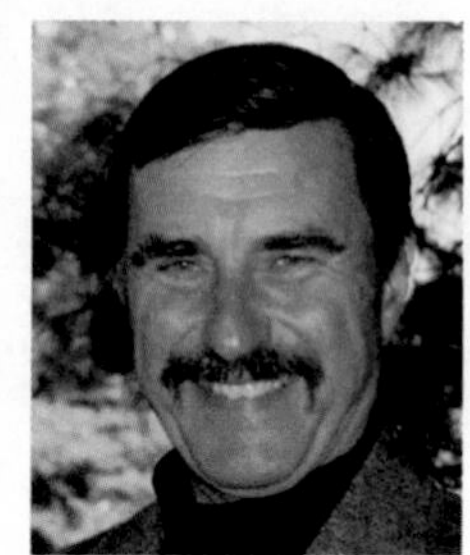

Chief Science Consultant

James Trefil, Ph.D. is the Clarence J. Robinson Professor of Physics at George Mason University. He is the author or co-author of more than 25 books, including *Science Matters* and *The Nature of Science*. Dr. Trefil is a member of the American Association for the Advancement of Science's Committee on the Public Understanding of Science and Technology. He is also a fellow of the World Economic Forum and a frequent contributor to *Smithsonian* magazine.

Rita Ann Calvo, Ph.D. is Senior Lecturer in Molecular Biology and Genetics at Cornell University, where for 12 years she also directed the Cornell Institute for Biology Teachers. Dr. Calvo is the 1999 recipient of the College and University Teaching Award from the National Association of Biology Teachers.

Kenneth Cutler, M.S. is the Education Coordinator for the Julius L. Chambers Biomedical Biotechnology Research Institute at North Carolina Central University. A former middle school and high school science teacher, he received a 1999 Presidential Award for Excellence in Science Teaching.

Instructional Design Consultants

Douglas Carnine, Ph.D. is Professor of Education and Director of the National Center for Improving the Tools of Educators at the University of Oregon. He is the author of seven books and over 100 other scholarly publications, primarily in the areas of instructional design and effective instructional strategies and tools for diverse learners. Dr. Carnine also serves as a member of the National Institute for Literacy Advisory Board.

Linda Carnine, Ph.D. consults with school districts on curriculum development and effective instruction for students struggling academically. A former teacher and school administrator, Dr. Carnine also co-authored a popular remedial reading program.

Sam Miller, Ph.D. is a middle school science teacher and the Teacher Development Liaison for the Eugene, Oregon, Public Schools. He is the author of curricula for teaching science, mathematics, computer skills, and language arts.

Donald Steely, Ph.D. serves as principal investigator at the Oregon Center for Applied Science (ORCAS) on federal grants for science and language arts programs. His background also includes teaching and authoring of print and multimedia programs in science, mathematics, history, and spelling.

Vicky Vachon, Ph.D. consults with school districts throughout the United States and Canada on improving overall academic achievement with a focus on literacy. She is also co-author of a widely used program for remedial readers.

Content Reviewers

John Beaver, Ph.D.
Ecology
Professor, Director of Science Education Center
College of Education and Human Services
Western Illinois University
Macomb, IL

Donald J. DeCoste, Ph.D.
Matter and Energy, Chemical Interactions
Chemistry Instructor
University of Illinois
Urbana-Champaign, IL

Dorothy Ann Fallows, Ph.D., MSc
Diversity of Living Things, Microbiology
Partners in Health
Boston, MA

Michael Foote, Ph.D.
The Changing Earth, Life Over Time
Associate Professor
Department of the Geophysical Sciences
The University of Chicago
Chicago, IL

Lucy Fortson, Ph.D.
Space Science
Director of Astronomy
Adler Planetarium and Astronomy Museum
Chicago, IL

Elizabeth Godrick, Ph.D.
Human Biology
Professor, CAS Biology
Boston University
Boston, MA

Isabelle Sacramento Grilo, M.S.
The Changing Earth
Lecturer, Department of the Geological Sciences
Montana State University
Bozeman, MT

David Harbster, MSc
Diversity of Living Things
Professor of Biology
Paradise Valley Community College
Phoenix, AZ

Richard D. Norris, Ph.D.
Earth's Waters
Professor of Paleobiology
Scripps Institution of Oceanography
University of California, San Diego
La Jolla, CA

Donald B. Peck, M.S.
Motion and Forces; Waves, Sound, and Light; Electricity and Magnetism
Director of the Center for Science Education (retired)
Fairleigh Dickinson University
Madison, NJ

Javier Penalosa, Ph.D.
Diversity of Living Things, Plants
Associate Professor, Biology Department
Buffalo State College
Buffalo, NY

Raymond T. Pierrehumbert, Ph.D.
Earth's Atmosphere
Professor in Geophysical Sciences (Atmospheric Science)
The University of Chicago
Chicago, IL

Brian J. Skinner, Ph.D.
Earth's Surface
Eugene Higgins Professor of Geology and Geophysics
Yale University
New Haven, CT

Nancy E. Spaulding, M.S.
Earth's Surface, The Changing Earth, Earth's Waters
Earth Science Teacher (retired)
Elmira Free Academy
Elmira, NY

Steven S. Zumdahl, Ph.D.
Matter and Energy, Chemical Interactions
Professor Emeritus of Chemistry
University of Illinois
Urbana-Champaign, IL

Susan L. Zumdahl, M.S.
Matter and Energy, Chemical Interactions
Chemistry Education Specialist
University of Illinois
Urbana-Champaign, IL

Safety Consultant

Juliana Texley, Ph.D.
Former K–12 Science Teacher and School Superintendent
Boca Raton, FL

English Language Advisor

Judy Lewis, M.A.
Director, State and Federal Programs for reading proficiency and high risk populations
Rancho Cordova, CA

Research-Based Solutions for Your Classroom

The distinguished program consultant team and a thorough, research-based planning and development process assure that *McDougal Littell Science* supports all students in learning science concepts, acquiring inquiry skills, and thinking scientifically.

Standards-Based Instruction

Concepts and skills were selected based on careful analysis of national and state standards.

- National Science Education Standards
- Project 2061 Benchmarks for Science Literacy
- Comprehensive database of state science standards

Standards and Benchmarks

Each chapter in **Matter and Energy** covers some of the learning goals that are described in the *National Science Education Standards* (NSES) and the Project 2061 *Benchmarks for Science Literacy*. Selected content and skill standards are shown below in shortened form. The following National Science Education Standards are covered on pages xii–xxvii, in Frontiers in Science, and in Timelines in Science, as well as in chapter features and laboratory investigations: Understandings About Scientific Inquiry (A.9), Understandings About Science and Technology (E.6), Science and Technology in Society (F.5), Science as a Human Endeavor (G.1), Nature of Science (G.2), and History of Science (G.3).

Content Standards

1 Introduction to Matter

National Science Education Standards

B.1.c | There are more than 100 known elements that combine to produce compounds.

Project 2061 Benchmarks

4.D.1 | All matter is made up of atoms.

4.D.3 | Atoms and molecules are always in motion.
- In solids, they vibrate.
- In liquids, they slide past one another.
- In gases, they move freely.

2 Properties of Matter

National Science Education Standards

B.1.a
- A substance has characteristic properties.
- A mixture often can be separated into the original substances using these properties.

Project 2061 Benchmarks

4.D.2 | Equal volumes of different substances usually have different weights.

8.B.1 | The choice of materials for a job depends on their properties.

3 Energy

National Science Education Standards

B.3.a | Energy is a property of substances that is often associated with
- heat
- light
- electricity
- mechanical motion
- sound
- atomic nuclei
- chemical compounds

Energy is transferred in many ways.

Project 2061 Benchmarks

4.E.1 | Energy cannot be created or destroyed, but it can be changed from one form to another.

4.E.2 | Most of what goes on in the universe involves energy transformations.

4.E.4 | Energy has many different forms, including
- heat
- chemical
- mechanical
- gravitational

x Unit: **Matter and Energy**

CHAPTER 1

Introduction to Matter

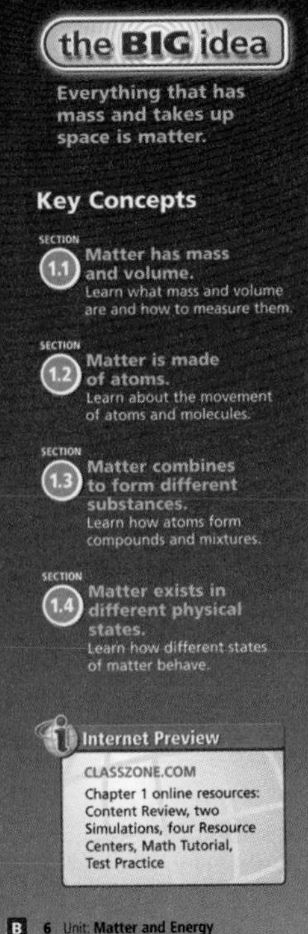
the BIG idea

Everything that has mass and takes up space is matter.

Key Concepts

SECTION 1.1 **Matter has mass and volume.** Learn what mass and volume are and how to measure them.

SECTION 1.2 **Matter is made of atoms.** Learn about the movement of atoms and molecules.

SECTION 1.3 **Matter combines to form different substances.** Learn how atoms form compounds and mixtures.

SECTION 1.4 **Matter exists in different physical states.** Learn how different states of matter behave.

Internet Preview

CLASSZONE.COM

Chapter 1 online resources: Content Review, two Simulations, four Resource Centers, Math Tutorial, Test Practice

What matter can you identify in this photograph?

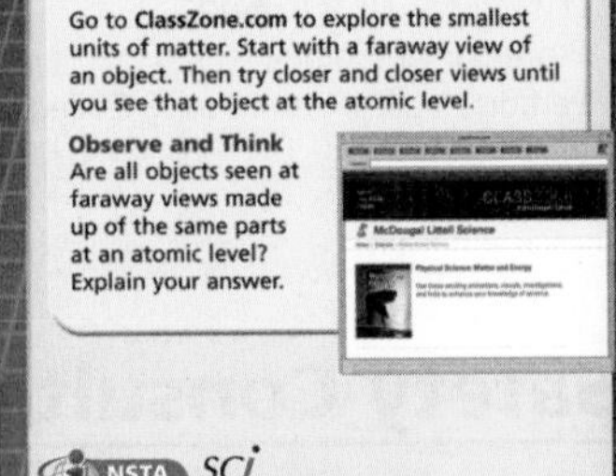
Go to **ClassZone.com** to explore the smallest units of matter. Start with a faraway view of an object. Then try closer and closer views until you see that object at the atomic level.

Observe and Think Are all objects seen at faraway views made up of the same parts at an atomic level? Explain your answer.

NSTA scilinks.org SCILINKS

Solids, Liquids, and Gases Code: MDL061

B 6 Unit: **Matter and Energy**

Chapter 1: **Introduction to Matter** 7 B

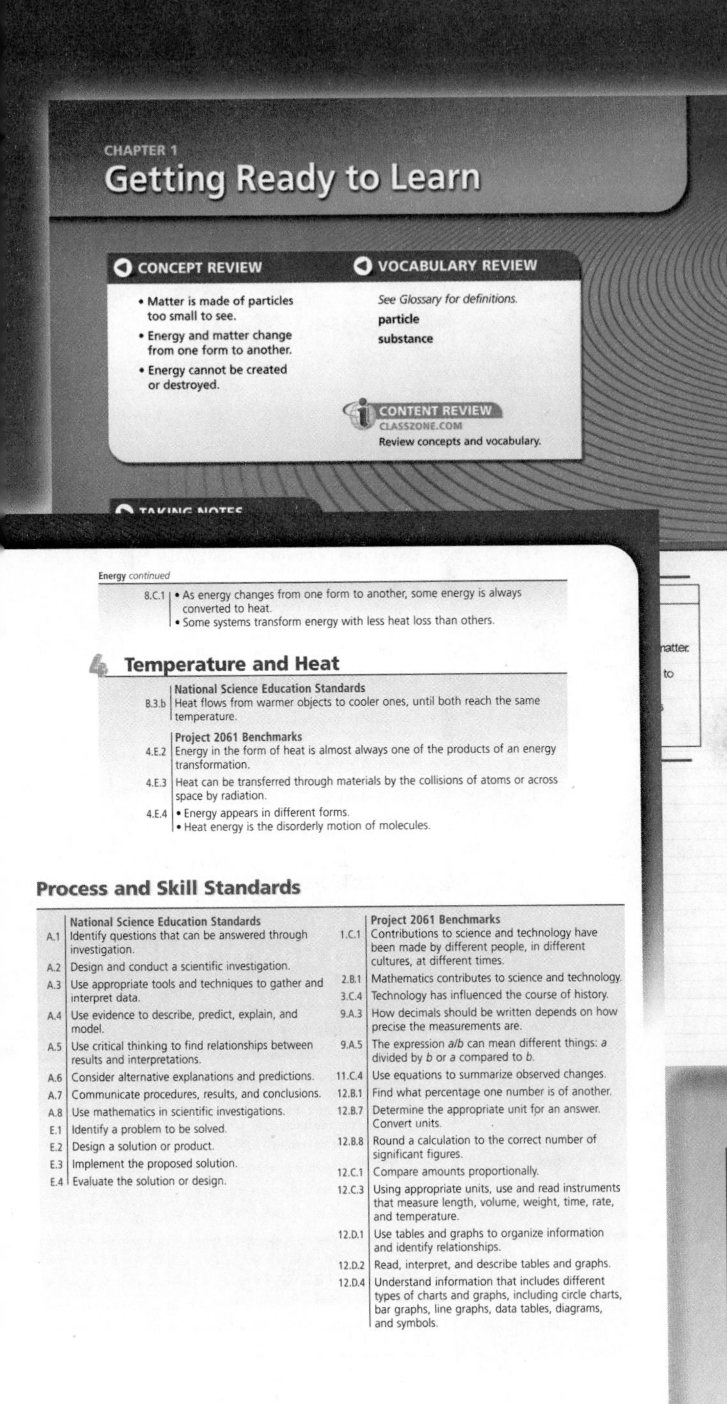

CHAPTER 1

Getting Ready to Learn

CONCEPT REVIEW

- Matter is made of particles too small to see.
- Energy and matter change from one form to another.
- Energy cannot be created or destroyed.

VOCABULARY REVIEW

See Glossary for definitions.

particle

substance

CONTENT REVIEW CLASSZONE.COM

Review concepts and vocabulary.

Energy *continued*

8.C.1	• As energy changes from one form to another, some energy is always converted to heat. • Some systems transform energy with less heat loss than others.

4 Temperature and Heat

	National Science Education Standards
B.3.b	Heat flows from warmer objects to cooler ones, until both reach the same temperature.
	Project 2061 Benchmarks
4.E.2	Energy in the form of heat is almost always one of the products of an energy transformation.
4.E.3	Heat can be transferred through materials by the collisions of atoms or across space by radiation.
4.E.4	• Energy appears in different forms. • Heat energy is the disorderly motion of molecules.

Process and Skill Standards

	National Science Education Standards
A.1	Identify questions that can be answered through investigation.
A.2	Design and conduct a scientific investigation.
A.3	Use appropriate tools and techniques to gather and interpret data.
A.4	Use evidence to describe, predict, explain, and model.
A.5	Use critical thinking to find relationships between results and interpretations.
A.6	Consider alternative explanations and predictions.
A.7	Communicate procedures, results, and conclusions.
A.8	Use mathematics in scientific investigations.
E.1	Identify a problem to be solved.
E.2	Design a solution or product.
E.3	Implement the proposed solution.
E.4	Evaluate the solution or design.

	Project 2061 Benchmarks
1.C.1	Contributions to science and technology have been made by different people, in different cultures, at different times.
2.B.1	Mathematics contributes to science and technology.
3.C.4	Technology has influenced the course of history.
9.A.3	How decimals should be written depends on how precise the measurements are.
9.A.5	The expression *a/b* can mean different things: *a* divided by *b* or *a* compared to *b*.
11.C.4	Use equations to summarize observed changes.
12.B.1	Find what percentage one number is of another.
12.B.7	Determine the appropriate unit for an answer. Convert units.
12.B.8	Round a calculation to the correct number of significant figures.
12.C.1	Compare amounts proportionally.
12.C.3	Using appropriate units, use and read instruments that measure length, volume, weight, time, rate, and temperature.
12.D.1	Use tables and graphs to organize information and identify relationships.
12.D.2	Read, interpret, and describe tables and graphs.
12.D.4	Understand information that includes different types of charts and graphs, including circle charts, bar graphs, line graphs, data tables, diagrams, and symbols.

Standards and Benchmarks xi

Effective Instructional Strategies

McDougal Littell Science incorporates strategies that research shows are effective in improving student achievement. These strategies include

- Notetaking and nonlinguistic representations (Marzano, Pickering, and Pollock)
- A focus on big ideas (Kameenui and Carnine)
- Background knowledge and active involvement (Project CRISS)

Robert J. Marzano, Debra J. Pickering, and Jane E. Pollock, *Classroom Instruction that Works; Research-Based Strategies for Increasing Student Achievement* (ASCD, 2001)

Edward J. Kameenui and Douglas Carnine, *Effective Teaching Strategies that Accommodate Diverse Learners* (Pearson, 2002)

Project CRISS (Creating Independence through Student Owned Strategies)

All objects are made of matter.

Suppose your class takes a field trip to a museum. During the course of the day you see mammoth bones, sparkling crystals, hot-air balloons, and an astronaut's space suit. All of these things are matter.

Matter is what makes up all of the objects and living organisms in the universe. As you will see, **matter** is anything that has mass and takes up space. Your body is matter. The air that you breathe and the water that you drink are also matter. Matter makes up the materials around you. Matter is made of particles called atoms, which are too small to see. You will learn more about atoms in the next section.

Not everything is matter. Light and sound, for example, are not matter. Light does not take up space or have mass in the same way that a table does. Although air is made of atoms, a sound traveling through air is not.

VOCABULARY Make four square diagrams for *matter* and for *mass* in your notebook to help you understand their relationship.

CHECK YOUR READING What is matter? How can you tell if something is matter?

Chapter 1: Introduction to Matter 9 B

Comprehensive Research, Review, and Field Testing

An ongoing program of research and review guided the development of *McDougal Littell Science.*

- Program plans based on extensive data from classroom visits, research surveys, teacher panels, and focus groups
- All pupil edition activities and labs classroom-tested by middle school teachers and students
- All chapters reviewed for clarity and scientific accuracy by the Content Reviewers listed on page T5
- Selected chapters field-tested in the classroom to assess student learning, ease of use, and student interest

Content Organized Around Big Ideas

Each chapter develops a big idea of science, helping students to place key concepts in context.

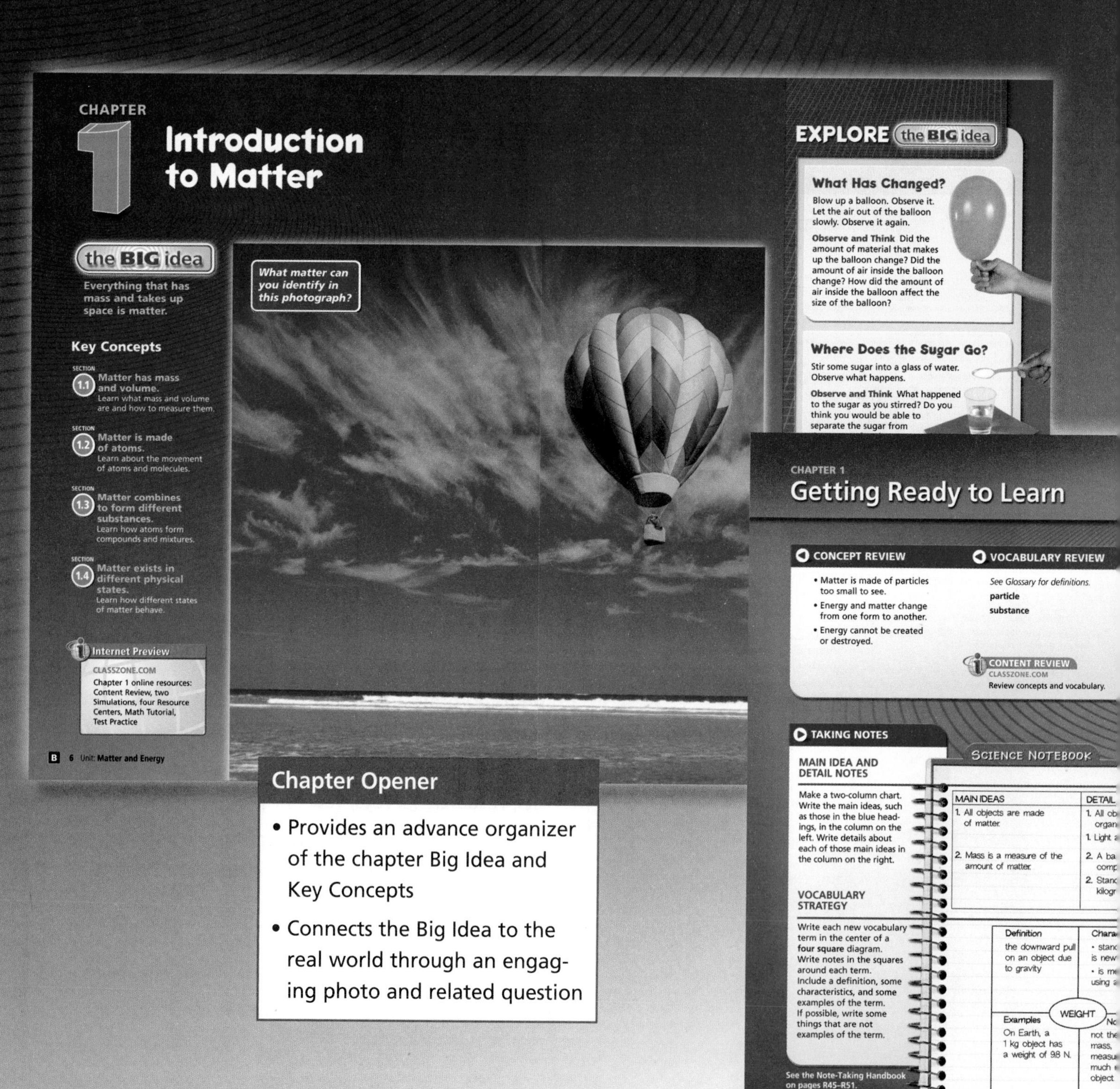

Chapter Opener

- Provides an advance organizer of the chapter Big Idea and Key Concepts
- Connects the Big Idea to the real world through an engaging photo and related question

Visual Summary

- Summarizes Key Concepts using both text and visuals
- Reinforces the connection of Key Concepts to the Big Idea

1 Chapter Review

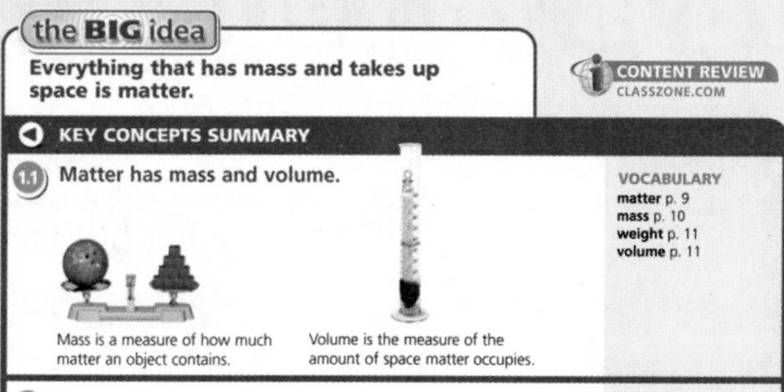

CONTENT REVIEW CLASSZONE.COM

1.2 Matter is made of atoms.

An atom is the smallest basic unit of matter. Two or more atoms bonded together form a molecule. Atoms and molecules are always in motion.

VOCABULARY
atom p. 16
molecule p. 18

1.3 Matter combines to form different substances.

Matter can be pure, such as an element (gold), or a compound (water).

Matter can be a mixture. Mixtures contain two or

VOCABULARY
element p. 22
compound p. 23
mixture p. 23

1.4 Matter exists in different

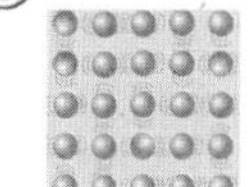

Solids have a fixed volume and a fixed shape. Liquids have ... ume but no

Reviewing Vocabulary

Copy and complete the chart below. If the right column is blank, give a brief description or definition. If the left column is blank, give the correct term.

Term	Description
1.	the downward pull of gravity on an object
2. liquid	
3.	the smallest basic unit of matter
4. solid	
5.	state of matter with no fixed volume and no fixed shape
6.	a combination of different substances that remain individual substances
7. matter	
8.	a measure of how much matter an object contains
9. element	
10.	a particle made of two or more atoms bonded together
11. compound	

Reviewing Key Concepts

Section Opener

- Highlights the Key Concept
- Connects new learning to prior knowledge
- Previews important vocabulary

KEY CONCEPT

Matter has mass and volume.

BEFORE, you learned

- Scientists study the world by asking questions and collecting data
- Scientists use tools such as microscopes, thermometers, and computers

NOW, you will learn

- What matter is
- How to measure the mass of matter
- How to measure the volume of matter

VOCABULARY
matter p. 9
mass p. 10
weight p. 11
volume p. 11

EXPLORE Similar Objects

How can two similar objects differ?

PROCEDURE

1. Look at the two balls but do not pick them up. Compare their sizes and shapes. Record your observations.
2. Pick up each ball. Compare the way the balls feel in your hands. Record your observations.

MATERIALS
2 balls of different sizes

WHAT DO YOU THINK?
How would your observations be different if the larger ball were made of foam?

All objects are made of matter.

Suppose your class takes a field trip to a museum. During the course of the day you see mammoth bones, sparkling crystals, hot-air balloons, and an astronaut's space suit. All of these things are matter.

Matter is what makes up all of the objects and living organisms in the universe. As you will see, **matter** is anything that has mass and takes up space. Your body is matter. The air that you breathe and the water that you drink are also matter. Matter makes up the materials around you. Matter is made of particles called atoms, which are too small to see. You will learn more about atoms in the next section.

Not everything is matter. Light and sound, for example, are not matter. Light does not take up space or have mass in the same way that a table does. Although air is made of atoms, a sound traveling through air is not.

VOCABULARY
Make four square diagrams for *matter* and for *mass* in your notebook to help you understand their relationship.

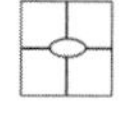

CHECK YOUR READING What is matter? How can you tell if something is matter?

Short Answer *Answer each of the following questions in a sentence or two.*

19. Describe the movement of particles in a solid, a liquid, and a gas.
20. In bright sunlight, dust particles in the air appear to dart about. What causes this effect?
21. Why is the volume of a rectangular object measured in cubic units?
22. Describe how the molecules in the air behave when you pump air into a bicycle tire.

Thinking Critically

23. CLASSIFY Write the headings *Matter* and *Not Matter* on your paper. Place each of these terms in the correct category: wood, water, metal, air, light, sound.
24. INFER If you could break up a carbon dioxide molecule, would you still have carbon dioxide? Explain your answer.
25. MODEL In what ways is sand in a bowl like a liquid? In what ways is it different?
26. INFER If you cut a hole in a basketball, what happens to the gas inside?
27. COMPARE AND CONTRAST Create a Venn diagram that shows how mixtures and compounds are alike and different.
28. ANALYZE If you place a solid rubber ball into a box, why doesn't the ball change its shape to fit the container?
29. CALCULATE What is the volume of an aquarium that is 120 cm long, 60 cm wide, and 100 cm high?
30. CALCULATE A truck whose bed is 2.5 m long, 1.5 m wide, and 1 m high is delivering sand for a sand-sculpture competition. How many trips must the truck make to deliver 7 cubic meters of sand?

Use the information in the photograph below to answer the next three questions.

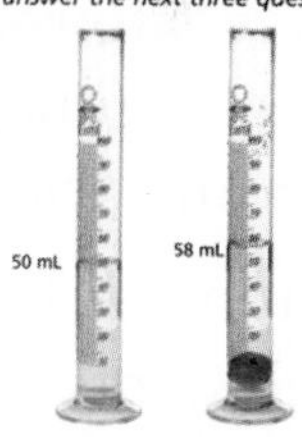

31. INFER One way to find the volume of a marble is by displacement. To determine a marble's volume, add 50 mL of water to a graduated cylinder and place the marble in the cylinder. Why does the water level change when you put the marble in the cylinder?
32. CALCULATE What is the volume of the marble?
33. PREDICT If you carefully removed the marble and let all of the water on it drain back into the cylinder, what would the volume of the water be? Explain.

the BIG idea

34. SYNTHESIZE Look back at the photograph on pages 6–7. Describe the picture in terms of states of matter.
35. WRITE Make a list of all the matter in a two-meter radius around you. Classify each as a solid, liquid, or gas.

UNIT PROJECTS

If you are doing a unit project, make a folder for your project. Include in your folder a list of the resources you will need, the date on which the project is due, and a schedule to track your progress. Begin gathering data.

The Big Idea Questions

- Help students connect their new learning back to the Big Idea
- Prompt students to synthesize and apply the Big Idea and Key Concepts

Many Ways to Learn

Because students learn in so many ways, *McDougal Littell Science* gives them a variety of experiences with important concepts and skills. Text, visuals, activities, and technology all focus on Big Ideas and Key Concepts.

Integrated Technology

- Interaction with Key Concepts through Simulations and Visualizations
- Easy access to relevant Web resources through Resource Centers and SciLinks
- Opportunities for review through Content Review and Math Tutorials

Visuals that Teach

- Information-rich visuals directly connected to the text
- Thoughtful pairing of diagrams and real-world photos
- Reading Visuals questions to support student learning

Kinetic energy and potential energy are the two general types of energy.

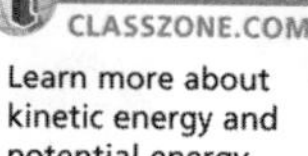

Learn more about kinetic energy and potential energy.

All of the forms of energy can be described in terms of two general types of energy—kinetic energy and potential energy. Anything that is moving, such as a car that is being driven or an atom in the air, has kinetic energy. All matter also has potential energy, or energy that is stored and can be released at a later time.

Kinetic Energy

READING TIP

Kinetic means "related to motion."

The energy of motion is called **kinetic energy.** It depends on both an object's mass and the speed at which the object is moving.

All objects are made of matter, and matter has mass. The more matter an object contains, the greater its mass. If you held a bowling ball in one hand and a soccer ball in the other, you could feel that the bowling ball has more mass than the soccer ball.

- **Kinetic energy increases as mass increases.** If the bowling ball and the soccer ball were moving at the same speed, the bowling ball would have more kinetic energy because of its greater mass.
- **Kinetic energy increases as speed increases.** If two identical bowling balls were rolling along at different speeds, the faster one would have more kinetic energy because of its greater speed. The speed skater in the photographs below has more kinetic energy when he is racing than he does when he is moving slowly.

High Speed

This skater has a large amount of kinetic energy when moving at a high speed.

Low Speed

When the same skater is moving more slowly, he has less kinetic energy.

READING VISUALS **APPLY** How could a skater with less mass than another skater have more kinetic energy?

B 74 Unit: Matter and Energy

Potential Energy

Suppose you are holding a soccer ball in your hands. Even if the ball is not moving, it has energy because it has the potential to fall. **Potential energy** is the stored energy that an object has due to its position or chemical composition. The ball's position above the ground gives it potential energy.

The most obvious form of potential energy is potential energy that results from gravity. Gravity is the force that pulls objects toward Earth's surface. The giant boulder on the right has potential energy because of its position above the ground. The mass of the boulder and its height above the ground determine how much potential energy it has due to gravity.

Because the boulder could fall, it has potential energy from gravity.

It is easy to know whether an object has kinetic energy because the object is moving. It is not so easy to know how much and what form of potential energy an object has, because objects can have potential energy from several sources. For example, in addition to potential energy from gravity, substances contain potential energy due to their chemical composition—the atoms they contain.

How can you tell kinetic energy and potential energy apart?

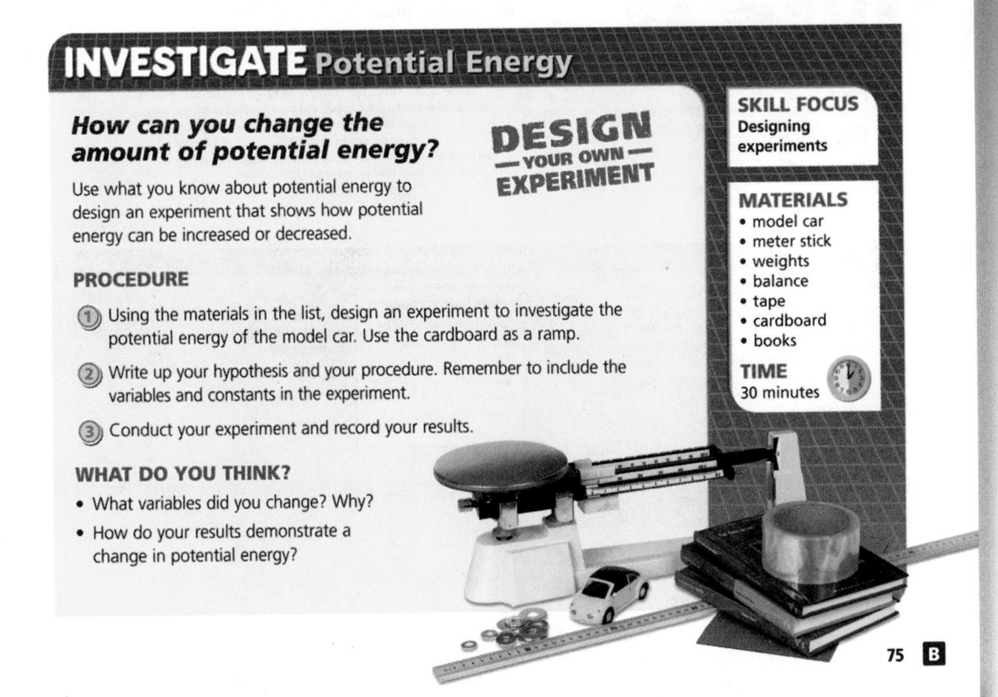

INVESTIGATE Potential Energy

How can you change the amount of potential energy?

DESIGN — YOUR OWN — EXPERIMENT

Use what you know about potential energy to design an experiment that shows how potential energy can be increased or decreased.

PROCEDURE

1. Using the materials in the list, design an experiment to investigate the potential energy of the model car. Use the cardboard as a ramp.
2. Write up your hypothesis and your procedure. Remember to include the variables and constants in the experiment.
3. Conduct your experiment and record your results.

WHAT DO YOU THINK?

- What variables did you change? Why?
- How do your results demonstrate a change in potential energy?

SKILL FOCUS
Designing experiments

MATERIALS
- model car
- meter stick
- weights
- balance
- tape
- cardboard
- books

TIME
30 minutes

75 B

Considerate Text

- Clear structure of meaningful headings
- Information clearly connected to main ideas
- Student-friendly writing style

Hands-on Learning

- Activities that reinforce Key Concepts
- Skill Focus for important inquiry and process skills
- Multiple activities in every chapter, from quick Explores to full-period Chapter Investigations

Differentiated Instruction

A full spectrum of resources for differentiating instruction supports you in reaching the wide range of learners in your classroom.

1.1 INSTRUCT

Develop Critical Thinking

APPLY Tell students that enough mass cubes are not available for the entire class. Have students apply what they know about measuring mass to invent standard units of mass based on common objects. Ask them to find the mass of objects based on the invented standard units.

Teach from Visuals

To help students interpret the photographs and illustrations comparing the size and mass of a bowling ball and a basketball, ask:

- How do the sizes of the bowling ball and the basketball in the photographs compare? *They are about the same.*
- How can you tell from the illustrations which ball has more mass? *More standard cubes are needed to balance the pan with the bowling ball than the pan with the basketball.*

Ongoing Assessment

Describe how to measure the mass of matter.

Ask: How would you find the mass of an object? *Compare its mass with the mass of standard units on a balance.*

CHECK YOUR READING *Answer: Mass is a measure of how much matter an object contains.*

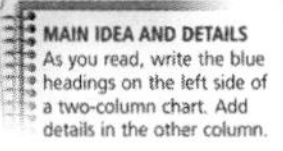

MAIN IDEA AND DETAILS As you read, write the blue headings on the left side of a two-column chart. Add details in the other column.

Mass is a measure of the amount of matter.

Different objects contain different amounts of matter. **Mass** is a measure of how much matter an object contains. A metal teaspoon, for example, contains more matter than a plastic teaspoon. Therefore, a metal teaspoon has a greater mass than a plastic teaspoon. An elephant has more mass than a mouse.

CHECK YOUR READING How are matter and mass related?

Measuring Mass

A When you measure mass, you compare the mass of the object with a standard amount, or unit, of mass. The standard unit of mass is the kilogram (kg). A large grapefruit has a mass of about one-half kilogram. Smaller masses are often measured in grams (g). There are 1000 grams in a kilogram. A penny has a mass of between two and three grams.

How can you compare the masses of two objects? One way is to use a pan balance, as shown below. If two objects balance each other on a pan balance, then they contain the same amount of matter. If a basketball balances a metal block, for example, then the basketball and the block have the same mass. Beam balances work in a similar way, but instead of comparing the masses of two objects, you compare the mass of an object with a standard mass on the beam.

A bowling ball and a basketball are about the same size, but a bowling ball has more mass.

Measuring Weight

B When you hold an object such as a backpack full of books, you feel it pulling down on your hands. This is because Earth's gravity pulls the backpack toward the ground. Gravity is the force that pulls two masses toward each other. In this example, the two masses are Earth and the backpack. **Weight** is the downward pull on an object due to gravity. If the pull of the backpack is strong, you would say that the backpack weighs a lot.

Weight is measured by using a scale, such as a spring scale like the one shown on the right, that tells how hard an object is pushing or pulling on it. The standard scientific unit for weight is the newton (N). A common unit for weight is the pound (lb).

Gravity is pulling down on both the girl and the backpack. The heavier the backpack is, the stronger the pull of gravity is on it.

Mass and weight are closely related, but they are not the same. Mass describes the amount of matter an object has, and weight describes how strongly gravity is pulling on that matter. On Earth, a one-kilogram object has a weight of 9.8 newtons (2.2 lb). When a person says that one kilogram is equal to 2.2 pounds, he or she is really saying that one kilogram has a weight of 2.2 pounds on Earth. On the Moon, however, gravity is one-sixth as strong as it is on Earth. On the Moon, the one-kilogram object would have a weight of 1.6 newtons (0.36 lb). The amount of matter in the object, or its mass, is the same on Earth as it is on the Moon, but the pull of gravity is different.

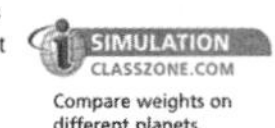

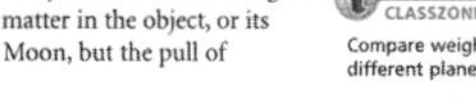

SIMULATION CLASSZONE.COM Compare weights on different planets.

CHECK YOUR READING What is the difference between mass and weight?

Volume is a measure of the space matter occupies.

Matter takes up space. A bricklayer stacks bricks on top of each other to build a wall. No two bricks can occupy the same place because the matter in each brick takes up space.

C The amount of space that matter in an object occupies is called the object's **volume.** The bowling ball and the basketball shown on page 10 take up approximately the same amount of space. Therefore, the two balls have about the same volume. Although the basketball is hollow, it is not empty. Air fills up the space inside the basketball. Air and other gases take up space and have volume.

DIFFERENTIATE INSTRUCTION

More Reading Support

A To measure mass, what do you compare the object's mass to? *standard units of mass*

English Learners Phrasal verbs such as "makes up" and "takes up" are used throughout this chapter (for example, on p. 9). Make sure students understand not to read "up" and other adverbs or prepositions in phrasal verbs as literal directions. If students are still confused, offer synonyms for phrasal verbs. For instance, "makes up" means composes, and "takes up" means occupies. English learners may also lack background knowledge of a field trip (p. 9).

DIFFERENTIATE INSTRUCTION

More Reading Support

B Which is affected by gravity, mass or weight? *weight*

C What do you call the amount of space an object takes up? *volume*

Below Level Have students make a table that compares mass and weight. Ask them to include a definition of each, the standard units used to measure them, and the tools used to measure them.

Teacher's Edition

- More Reading Support for below-level readers
- Strategies for below-level and advanced learners, English learners, and inclusion students

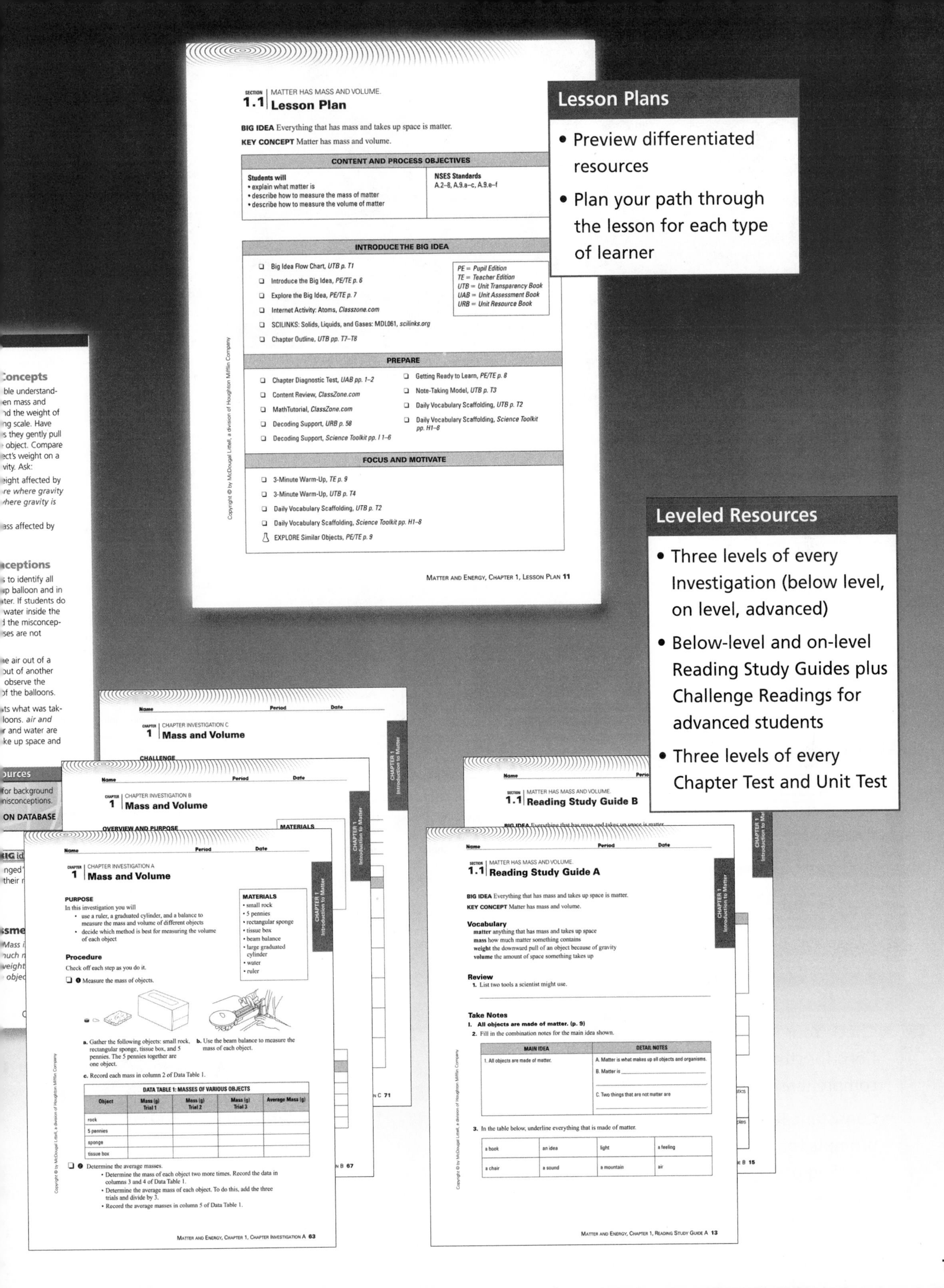

SECTION 1.1 | MATTER HAS MASS AND VOLUME.
Lesson Plan

BIG IDEA Everything that has mass and takes up space is matter.

KEY CONCEPT Matter has mass and volume.

CONTENT AND PROCESS OBJECTIVES	
Students will • explain what matter is • describe how to measure the mass of matter • describe how to measure the volume of matter	**NSES Standards** A.2–8, A.9.a–c, A.9.e–f

INTRODUCE THE BIG IDEA

- ❑ Big Idea Flow Chart, *UTB p. T1*
- ❑ Introduce the Big Idea, *PE/TE p. 6*
- ❑ Explore the Big Idea, *PE/TE p. 7*
- ❑ Internet Activity: Atoms, *Classzone.com*
- ❑ SCILINKS: Solids, Liquids, and Gases: MDL061, *scilinks.org*
- ❑ Chapter Outline, *UTB pp. T7–T8*

PE = Pupil Edition
TE = Teacher Edition
UTB = Unit Transparency Book
UAB = Unit Assessment Book
URB = Unit Resource Book

PREPARE

- ❑ Chapter Diagnostic Test, *UAB pp. 1–2*
- ❑ Content Review, *ClassZone.com*
- ❑ MathTutorial, *ClassZone.com*
- ❑ Decoding Support, *URB p. 58*
- ❑ Decoding Support, *Science Toolkit pp. I 1–6*
- ❑ Getting Ready to Learn, *PE/TE p. 8*
- ❑ Note-Taking Model, *UTB p. T3*
- ❑ Daily Vocabulary Scaffolding, *UTB p. T2*
- ❑ Daily Vocabulary Scaffolding, *Science Toolkit pp. H1–8*

FOCUS AND MOTIVATE

- ❑ 3-Minute Warm-Up, *TE p. 9*
- ❑ 3-Minute Warm-Up, *UTB p. T4*
- ❑ Daily Vocabulary Scaffolding, *UTB p. T2*
- ❑ Daily Vocabulary Scaffolding, *Science Toolkit pp. H1–8*
- EXPLORE Similar Objects, *PE/TE p. 9*

MATTER AND ENERGY, CHAPTER 1, LESSON PLAN 11

Lesson Plans

- Preview differentiated resources
- Plan your path through the lesson for each type of learner

Leveled Resources

- Three levels of every Investigation (below level, on level, advanced)
- Below-level and on-level Reading Study Guides plus Challenge Readings for advanced students
- Three levels of every Chapter Test and Unit Test

CHAPTER 1 | CHAPTER INVESTIGATION C
Mass and Volume
CHALLENGE
CHAPTER 1 | CHAPTER INVESTIGATION B
Mass and Volume
OVERVIEW AND PURPOSE
MATERIALS
CHAPTER 1 | CHAPTER INVESTIGATION A
Mass and Volume
PURPOSE
In this investigation you will
use a ruler, a graduated cylinder, and a balance to measure the mass and volume of different objects
decide which method is best for measuring the volume of each object
MATERIALS
small rock
5 pennies
rectangular sponge
tissue box
beam balance
large graduated cylinder
water
ruler
Procedure
Check off each step as you do it.
Measure the mass of objects.
MATTER AND ENERGY, CHAPTER 1, CHAPTER INVESTIGATION A 63
SECTION 1.1 | MATTER HAS MASS AND VOLUME.
Reading Study Guide B
SECTION 1.1 | MATTER HAS MASS AND VOLUME.
Reading Study Guide A
BIG IDEA Everything that has mass and takes up space is matter.
KEY CONCEPT Matter has mass and volume.
Vocabulary
matter anything that has mass and takes up space
mass how much matter something contains
weight the downward pull of an object because of gravity
volume the amount of space something takes up
Review
1. List two tools a scientist might use.
Take Notes
I. All objects are made of matter. (p. 9)
2. Fill in the combination notes for the main idea shown.
3. In the table below, underline everything that is made of matter.
a book
an idea
light
a feeling
a chair
a sound
a mountain
air
MATTER AND ENERGY, CHAPTER 1, READING STUDY GUIDE A 13

Effective Assessment

McDougal Littell Science incorporates a comprehensive set of resources for assessing student knowledge and performance before, during, and after instruction.

Diagnostic Tests

- Assessment of students' prior knowledge
- Readiness check for concepts and skills in the upcoming chapter

Measuring Volume by Displacement

Although a box has a regular shape, a rock does not. There is no simple formula for calculating the volume of something with an irregular shape. Instead, you can make use of the fact that two objects cannot be in the same place at the same time. This method of measuring is called displacement.

1. Add water to a graduated cylinder. Note the volume of the water by reading the water level on the cylinder.
2. Submerge the irregular object in the water. Because the object and the water cannot share the same space, the water is displaced, or moved upward. Note the new volume of the water with the object in it.
3. Subtract the volume of the water before you added the object from the volume of the water and the object together. The result is the volume of the object. The object displaces a volume of water equal to the volume of the object.

You measure the volume of a liquid by measuring how much space it takes up in a container. The volume of a liquid usually is measured in liters (L) or milliliters (mL). One liter is equal to 1000 milliliters. Milliliters and cubic centimeters are equivalent. This can be written as $1 \text{ mL} = 1 \text{ cm}^3$. If you had a box with a volume of one cubic centimeter and you filled it with water, you would have one milliliter of water.

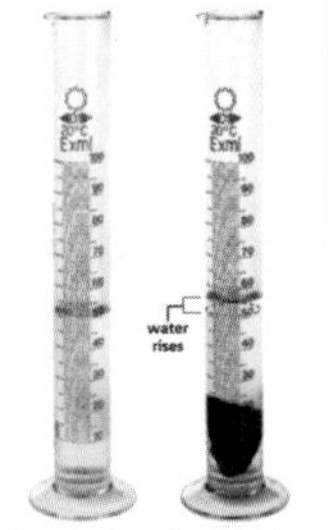

water rises

Measure the volume of water without the rock.

Measure the volume of water with the rock in it.

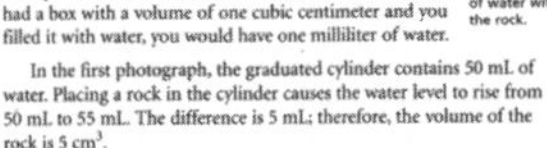

In the first photograph, the graduated cylinder contains 50 mL of water. Placing a rock in the cylinder causes the water level to rise from 50 mL to 55 mL. The difference is 5 mL; therefore, the volume of the rock is 5 cm^3.

1.1 Review

KEY CONCEPTS

1. Give three examples of matter.
2. What do weight and mass measure?
3. How can you measure the volume of an object that has an irregular shape?

CRITICAL THINKING

4. **Calculate** What is the volume of a box that is 12 cm long, 6 cm wide, and 4 cm high?
5. **Synthesize** What is the relationship between the units of measurement for the volume of a liquid and of a solid object?

CHALLENGE

6. **Infer** Why might a small increase in the dimensions of an object cause a large change in its volume?

Chapter 1: **Introduction to Matter** 13 B

ANSWERS

1. Sample answer: air, water, a rock

2. Weight measures the downward pull on an object due to gravity. Mass measures how much matter an object contains.

3. Submerge the object in water and subtract the volume of the water before the object was added from the volume of the water and the object together.

4. 12 cm · 6 cm · 4 cm = 288 cm³

5. One cubic centimeter of solid is equivalent to one milliliter of liquid.

6. Each of the three dimensions increases, so, for example, doubling the dimensions increases the volume by a multiple of eight.

Ongoing Assessment

Describe how to measure the volume of matter.

Ask: How would you find the volume of a box? of a small rock? *Multiply the box's length by its width and height. Put the small stone in a graduated cylinder of water and determine how much water the rock displaces.*

Teacher Demo

Place a clear glass that is full to its brim with water in a shallow tray. Ask students to predict what will happen if you slowly drop a stone into the glass. Ask them to explain their predictions.

Reinforce the BIG idea

Have students relate the section to the Big Idea.

R Reinforcing Key Concepts, p. 20

1.1 ASSESS & RETEACH

Assess

A Section 1.1 Quiz, p. 3

Reteach

Write the heading *Matter* on the board. Ask a volunteer to define *matter* and write the definition on the board. Add these two subheads under *Matter: Mass* and *Volume*. Ask a volunteer to define *mass* and *volume* and write the definitions on the board. Ask students to describe how each is measured, which tools are used for the measurements, and which units are used.

Technology Resources

Have students visit **ClassZone.com** for reteaching of Key Concepts.

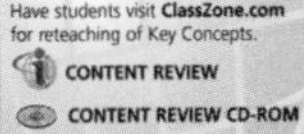

CONTENT REVIEW

CONTENT REVIEW CD-ROM

Chapter 1 13 B

Ongoing Assessment

- Check Your Reading questions for student self-check of comprehension
- Consistent Teacher Edition prompts for assessing understanding of Key Concepts

Reviewing Vocabulary

Copy and complete the chart below. If the right column is blank, give a brief description or definition. If the left column is blank, give the correct term.

Term	Description
1.	the downward pull of gravity on an object
2. liquid	
3.	the smallest basic unit of matter
4. solid	
5.	state of matter with no fixed volume and no fixed shape
6.	a combination of different substances that remain individual substances
7. matter	
8.	a measure of how much matter an object contains
9. element	
10.	a particle made of two or more atoms bonded together
11. compound	

Reviewing Key Concepts

Multiple Choice *Choose the letter of the best answer.*

12. The standard unit for measuring mass is the
- **a.** kilogram
- **b.** gram per cubic centimeter
- **c.** milliliter
- **d.** milliliter per cubic centimeter

13. A unit for measuring the volume of a liquid is the
- **a.** kilogram
- **b.** gram per cubic centimeter
- **c.** milliliter
- **d.** milliliter per cubic centimeter

14. The weight of an object is measured by using a scale that
- **a.** compares the mass of the object with a standard unit of mass
- **b.** shows the amount of space the object occupies
- **c.** indicates how much water is displaced by the object
- **d.** tells how hard the object is pushing or pulling on it

15. To find the volume of a rectangular box,
- **a.** divide the length by the height
- **b.** multiply the length, width, and height
- **c.** subtract the mass from the weight
- **d.** multiply one atom's mass by the total

16. Compounds can be separated only by
- **a.** breaking the atoms into smaller pieces
- **b.** breaking the bonds between the atoms
- **c.** using a magnet to attract certain atoms
- **d.** evaporating the liquid that contains the atoms

17. Whether a substance is a solid, a liquid, or a gas depends on how close its atoms are to one another and
- **a.** the volume of each atom
- **b.** how much matter the atoms have
- **c.** how free the atoms are to move
- **d.** the size of the container

18. A liquid has
- **a.** a fixed volume and a fixed shape
- **b.** no fixed volume and a fixed shape
- **c.** a fixed volume and no fixed shape
- **d.** no fixed volume and no fixed shape

...uction to Matter 35

Section and Chapter Reviews

- Focus on Key Concepts and critical thinking skills
- A full range of question types and levels of thinking

Leveled Chapter and Unit Tests

- Three levels of test for every chapter and unit
- Same Big Ideas, Key Concepts, and essential skills assessed on all levels

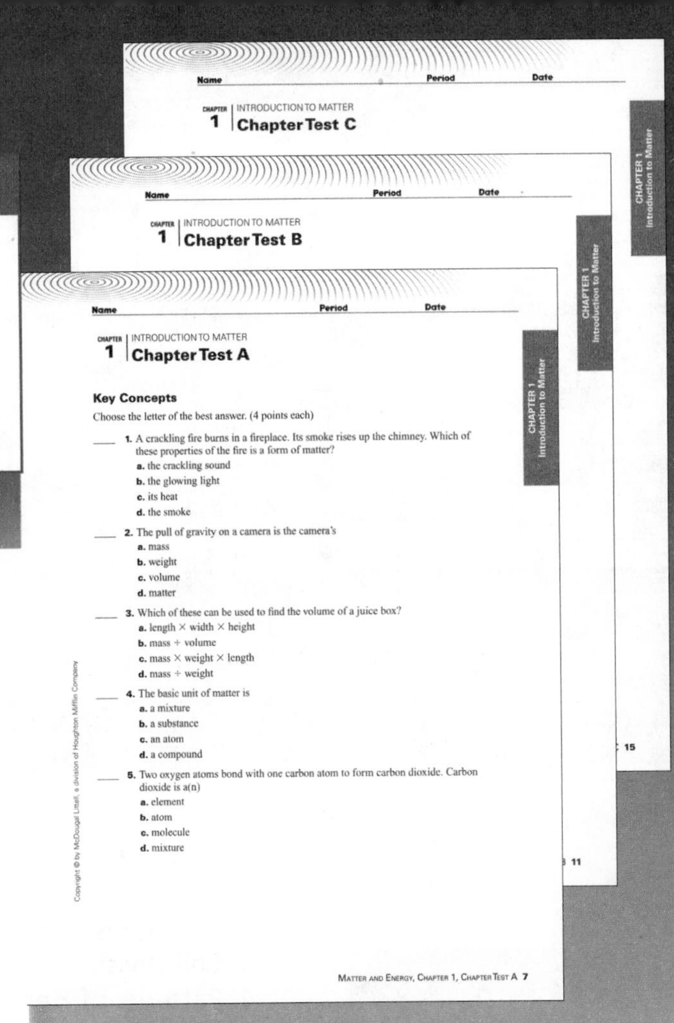

Name Period Date

CHAPTER 1 | INTRODUCTION TO MATTER
Chapter Test C

Name Period Date

CHAPTER 1 | INTRODUCTION TO MATTER
Chapter Test B

Name Period Date

CHAPTER 1 | INTRODUCTION TO MATTER
Chapter Test A

Key Concepts

Choose the letter of the best answer. (4 points each)

____ 1. A crackling fire burns in a fireplace. Its smoke rises up the chimney. Which of these properties of the fire is a form of matter?
a. the crackling sound
b. the glowing light
c. its heat
d. the smoke

____ 2. The pull of gravity on a camera is the camera's
a. mass
b. weight
c. volume
d. matter

____ 3. Which of these can be used to find the volume of a juice box?
a. length × width × height
b. mass ÷ volume
c. mass × weight × length
d. mass ÷ weight

____ 4. The basic unit of matter is
a. a mixture
b. a substance
c. an atom
d. a compound

____ 5. Two oxygen atoms bond with one carbon atom to form carbon dioxide. Carbon dioxide is a(n)
a. element
b. atom
c. molecule
d. mixture

CHAPTER 1 Introduction to Matter

MATTER AND ENERGY, CHAPTER 1, CHAPTER TEST A 7

11

15

Short Answer *Answer each of the following questions in a sentence or two.*

19. Describe the movement of particles in a solid, a liquid, and a gas.
20. In bright sunlight, dust particles in the air appear to dart about. What causes this effect?
21. Why is the volume of a rectangular object measured in cubic units?
22. Describe how the molecules in the air behave when you pump air into a bicycle tire.

Thinking Critically

23. CLASSIFY Write the headings *Matter* and *Not Matter* on your paper. Place each of these terms in the correct category: wood, water, metal, air, light, sound.
24. INFER If you could break up a carbon dioxide molecule, would you still have carbon dioxide? Explain your answer.
25. MODEL In what ways is sand in a bowl like a liquid? In what ways is it different?
26. INFER If you cut a hole in a basketball, what happens to the gas inside?
27. COMPARE AND CONTRAST Create a Venn diagram that shows how mixtures and compounds are alike and different.
28. ANALYZE If you place a solid rubber ball into a box, why doesn't the ball change its shape to fit the container?
29. CALCULATE What is the volume of an aquarium that is 120 cm long, 60 cm wide, and 100 cm high?
30. CALCULATE A truck whose bed is 2.5 m long, 1.5 m wide, and 1 m high is delivering sand for a sand-sculpture competition. How many trips must the truck make to deliver 7 cubic meters of sand?

Use the information in the photograph below to answer the next three questions.

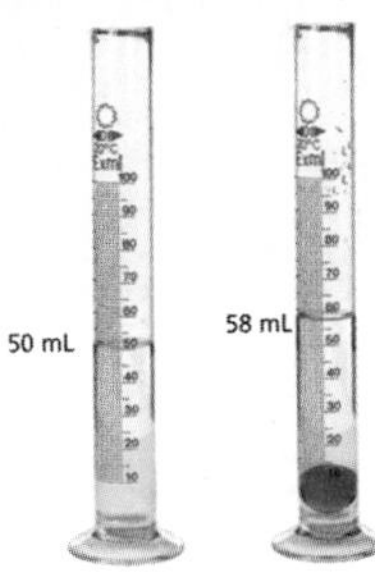

31. INFER One way to find the volume of a marble is by displacement. To determine a marble's volume, add 50 mL of water to a graduated cylinder and place the marble in the cylinder. Why does the water level change when you put the marble in the cylinder?
32. CALCULATE What is the volume of the marble?
33. PREDICT If you carefully removed the marble and let all of the water on it drain back into the cylinder, what would the volume of the water be? Explain.

the BIG idea

34. SYNTHESIZE Look back at the photograph on pages 6–7. Describe the picture in terms of states of matter.
35. WRITE Make a list of all the matter in a two-meter radius around you. Classify each as a solid, liquid, or gas.

UNIT PROJECTS

36 Unit: **Matter and Energy**

Rubrics

- Rubrics in Teacher Edition for all extended response questions
- Rubrics for all Unit Projects
- Alternative Assessment with rubric for each chapter
- A wide range of additional rubrics in the Science Toolkit

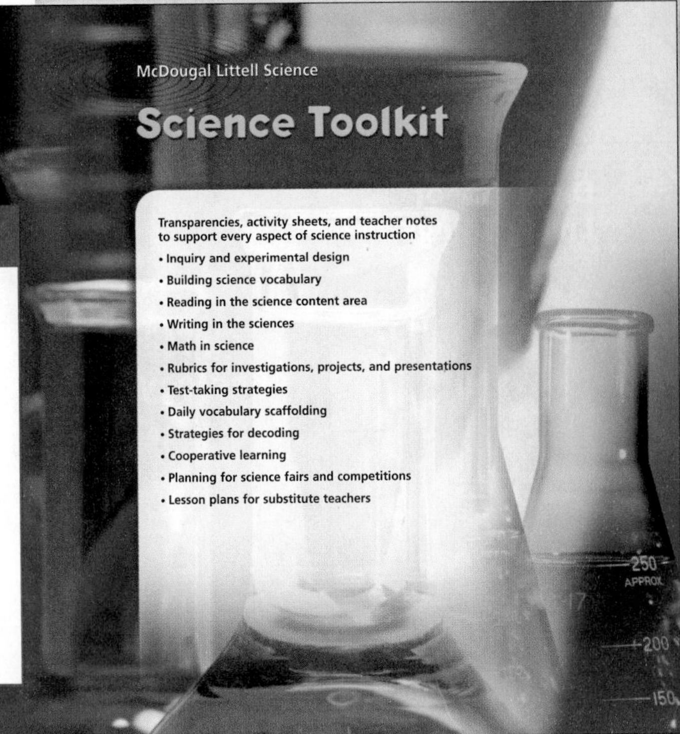

McDougal Littell Science
Science Toolkit
Transparencies, activity sheets, and teacher notes to support every aspect of science instruction
• Inquiry and experimental design
• Building science vocabulary
• Reading in the science content area
• Writing in the sciences
• Math in science
• Rubrics for investigations, projects, and presentations
• Test-taking strategies
• Daily vocabulary scaffolding
• Strategies for decoding
• Cooperative learning
• Planning for science fairs and competitions
• Lesson plans for substitute teachers
250
APPROX.
200
150

McDougal Littell Science Modular Series

McDougal Littell Science lets you choose the titles that match your curriculum. Each module in this flexible 15-book series takes an in-depth look at a specific area of life, earth, or physical science.

- Flexibility to match your curriculum
- Convenience of smaller books
- Complete Student Resource Handbooks in every module

Life Science Titles

A ▶ Cells and Heredity
1. The Cell
2. How Cells Function
3. Cell Division
4. Patterns of Heredity
5. DNA and Modern Genetics

B ▶ Life Over Time
1. The History of Life on Earth
2. Classification of Living Things
3. Population Dynamics

C ▶ Diversity of Living Things
1. Single-Celled Organisms and Viruses
2. Introduction to Multicellular Organisms
3. Plants
4. Invertebrate Animals
5. Vertebrate Animals

D ▶ Ecology
1. Ecosystems and Biomes
2. Interactions Within Ecosystems
3. Human Impact on Ecosystems

E ▶ Human Biology
1. Systems, Support, and Movement
2. Absorption, Digestion, and Exchange
3. Transport and Protection
4. Control and Reproduction
5. Growth, Development, and Health

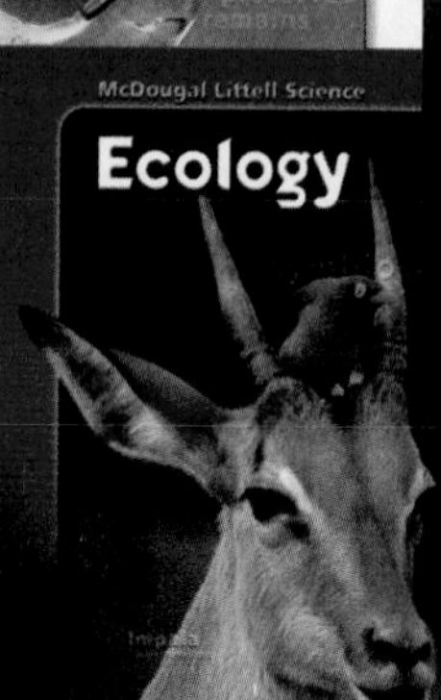

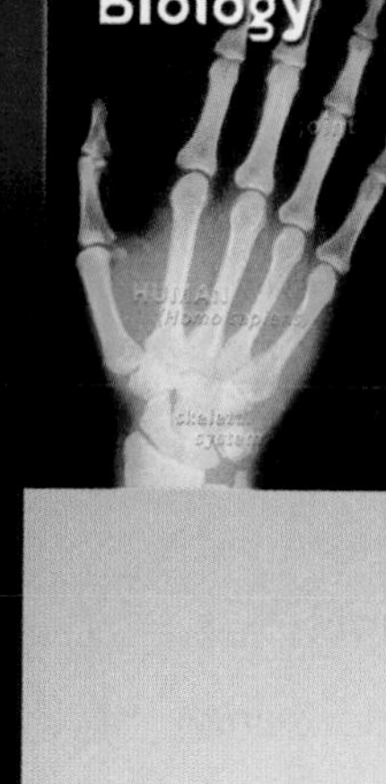

Earth Science Titles

A ▶ Earth's Surface
1. Views of Earth Today
2. Minerals
3. Rocks
4. Weathering and Soil Formation
5. Erosion and Deposition

B ▶ The Changing Earth
1. Plate Tectonics
2. Earthquakes
3. Mountains and Volcanoes
4. Views of Earth's Past
5. Natural Resources

C ▶ Earth's Waters
1. The Water Planet
2. Freshwater Resources
3. Ocean Systems
4. Ocean Environments

D ▶ Earth's Atmosphere
1. Earth's Changing Atmosphere
2. Weather Patterns
3. Weather Fronts and Storms
4. Climate and Climate Change

E ▶ Space Science
1. Exploring Space
2. Earth, Moon, and Sun
3. Our Solar System
4. Stars, Galaxies, and the Universe

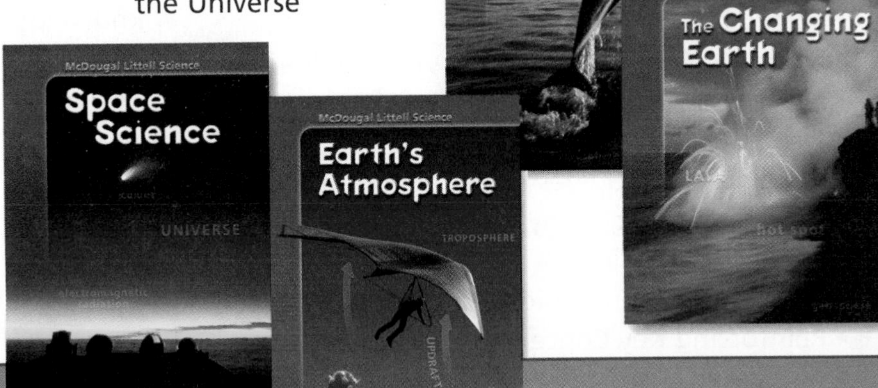

Physical Science Titles

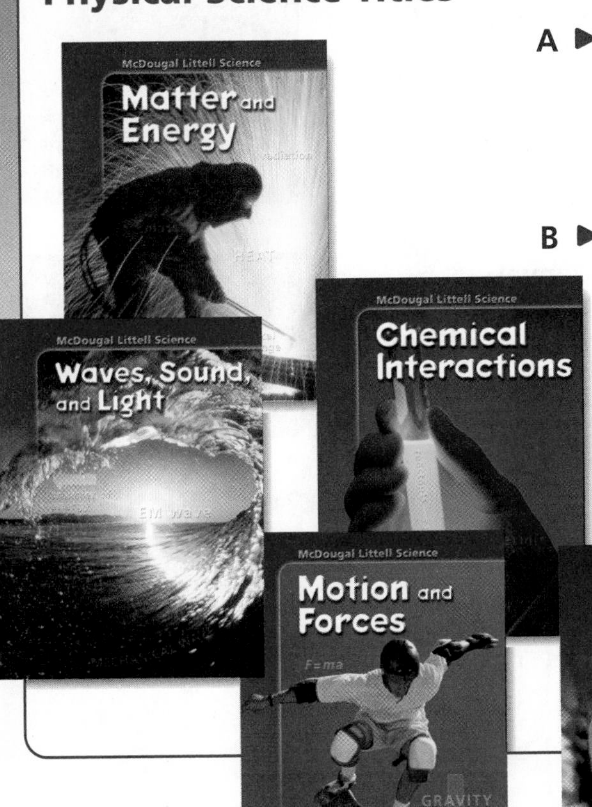

A ▶ Matter and Energy
1. Introduction to Matter
2. Properties of Matter
3. Energy
4. Temperature and Heat

B ▶ Chemical Interactions
1. Atomic Structure and the Periodic Table
2. Chemical Bonds and Compounds
3. Chemical Reactions
4. Solutions
5. Carbon in Life and Materials

C ▶ Motion and Forces
1. Motion
2. Forces
3. Gravity, Friction, and Pressure
4. Work and Energy
5. Machines

D ▶ Waves, Sound, and Light
1. Waves
2. Sound
3. Electromagnetic Waves
4. Light and Optics

E ▶ Electricity and Magnetism
1. Electricity
2. Circuits and Electronics
3. Magnetism

Teaching Resources

A wealth of print and technology resources help you adapt the program to your teaching style and to the specific needs of your students.

Book-Specific Print Resources

Unit Resource Book provides all of the teaching resources for the unit organized by chapter and section.

- Family Letters
- *Scientific American Frontiers* Video Guide
- Unit Projects
- Lesson Plans
- Reading Study Guides (Levels A and B)
- Spanish Reading Study Guides
- Challenge Readings
- Challenge and Extension Activities
- Reinforcing Key Concepts
- Vocabulary Practice
- Math Support and Practice
- Investigation Datasheets
- Chapter Investigations (Levels A, B, and C)
- Additional Investigations (Levels A, B, and C)
- Summarizing the Chapter

McDougal Littell Science

Unit Resource Book

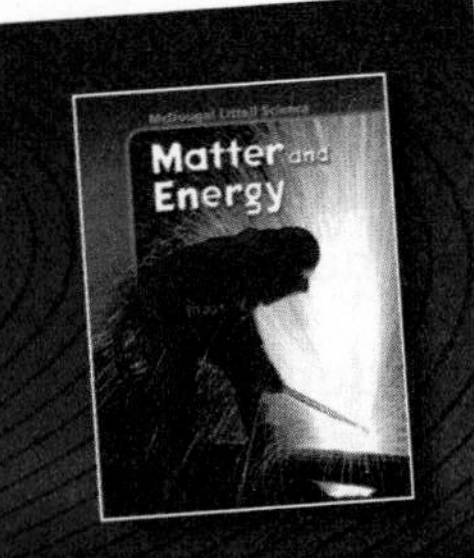

Matter and Energy

- Family Letters (English and Spanish)
- *Scientific American Frontiers* Video Guides
- Unit Projects (with Rubrics)
- Lesson Plans
- Reading Study Guides (Levels A and B and Spanish)
- Challenge Activities and Readings
- Reinforcing Key Concepts
- Vocabulary Practice and Decoding Support
- Math Support and Practice
- Investigation Datasheets
- Chapter Investigations (Levels A, B, and C)
- Additional Investigations (Levels A, B, and C)

Unit Assessment Book contains complete resources for assessing student knowledge and performance.

- Chapter Diagnostic Tests
- Section Quizzes
- Chapter Tests (Levels A, B, and C)
- Alternative Assessments
- Unit Tests (Levels A, B, and C)

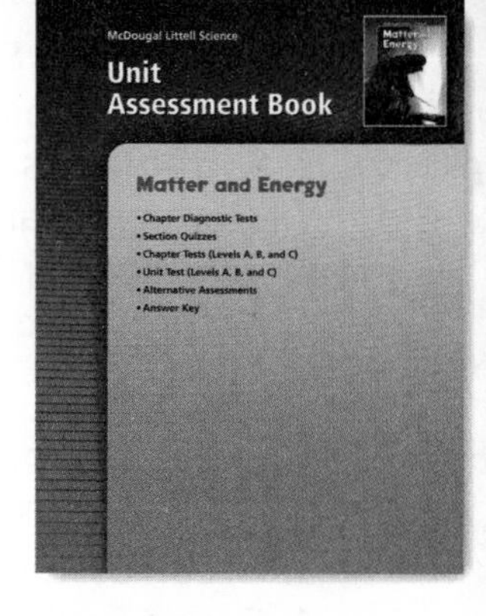

Unit Transparency Book includes instructional visuals for each chapter.

- Three-Minute Warm-Ups
- Note-Taking Models
- Daily Vocabulary Scaffolding
- Chapter Outlines
- Big Idea Flow Charts
- Chapter Teaching Visuals

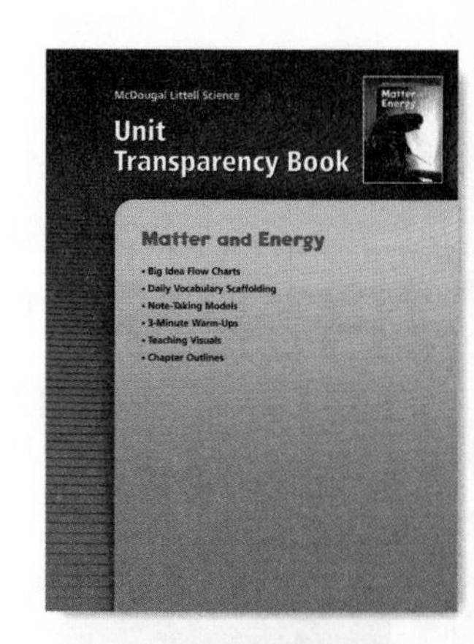

Unit Lab Manual

Unit Note-Taking/Reading Study Guide

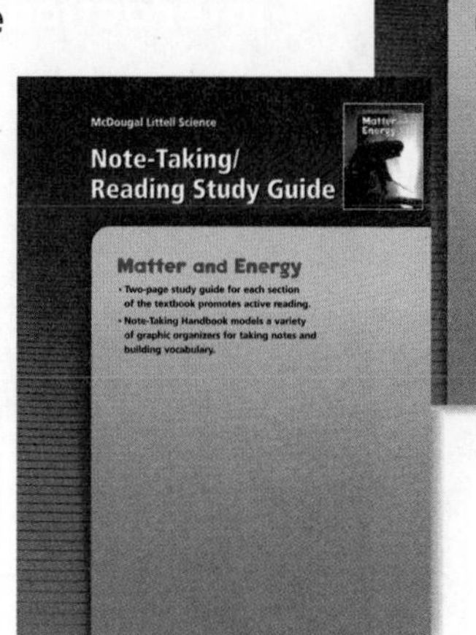

Program-Wide Print Resources

Process and Lab Skills

Problem Solving and Critical Thinking

Standardized Test Practice

Science Toolkit

City Science

Visual Glossary

Multi-Language Glossary

English Learners Package

***Scientific American Frontiers* Video Guide**

How Stuff Works Express
This quarterly magazine offers opportunities to explore current science topics.

Technology Resources

***Scientific American Frontiers* Video Program**
Each specially-tailored segment from this award-winning PBS series correlates to a unit; available on VHS and DVD

Audio CDs Complete chapter texts read in both English and Spanish

Lab Generator CD-ROM
A searchable database of all activities from the program plus additional labs for each unit; edit and print your own version of labs

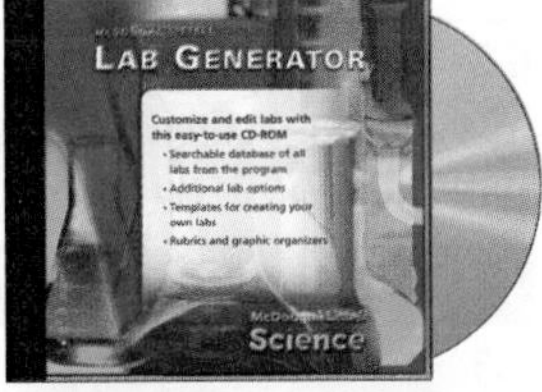

Test Generator CD-ROM

eEdition CD-ROM

EasyPlanner CD-ROM

Content Review CD-ROM

Power Presentations CD-ROM

Online Resources

- ClassZone.com
- Content Review Online
- eEdition Plus Online
- EasyPlanner Plus Online
- eTest Plus Online

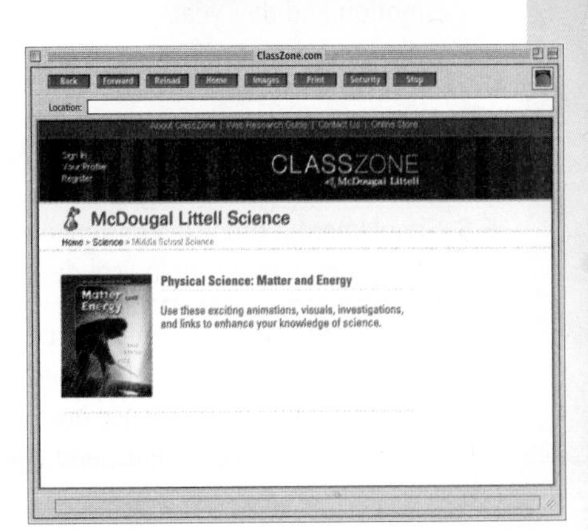

Correlation to National Science Education Standards

This chart provides an overview of how the five Physical Science modules of *McDougal Littell Science* address the National Science Education Standards.

A Matter and Energy
B Chemical Interactions
C Motion and Forces
D Waves, Sound, and Light
E Electricity and Magnetism

A. Science as Inquiry	Book, Chapter, and Section
A.1–A.8 **Abilities necessary to do scientific inquiry** Identify questions for investigation; design and conduct investigations; use evidence; think critically and logically; analyze alternative explanations; communicate; use mathematics.	All books (pp. R2–R44), All Chapter Investigations, All Think Science features
A.9 **Understandings about scientific inquiry** Different kinds of investigations for different questions; investigations guided by current scientific knowledge; importance of mathematics and technology for data gathering and analysis; importance of evidence, logical argument, principles, models, and theories; role of legitimate skepticism; scientific investigations lead to new investigations.	All books (pp. xxii–xxv) A3.1, B2.2, C4.2, D3.2, E3.1

B. Physical Science	Book, Chapter, and Section
B.1 **Properties and changes of properties in matter** Physical properties; substances, elements, and compounds; chemical reactions.	A1.1, A1.2, A1.3, A1.4, A2.1, A2.2, B1, B3.2, B4.1, B4.2, B4.3, B5.1, B5.3, C3.4
B.2 **Motions and forces** Position, speed, direction of motion; balanced and unbalanced forces.	C1.1, C1.2, C1.3, C2.1, C2.3, C3.1, C3.2, C3.3, C3.4, C4.1, E5.1
B.3 **Transfer of energy** Energy transfer; forms of energy; heat and light; electrical circuits; sun as source of Earth's energy.	A3.1, A3.2, A3.3, A4.1, A4.2, A4.3, B3.3, C4.2, D1.1, D3.3, D3.4, D4.1, D4.2, D4.3, E1.1, E1.2, E1.3, E2.1, E2.2

C. Life Science	Book, Chapter, and Section
C.1 **Structure and function in living systems** Systems; structure and function; levels of organization; cells and cell activities; specialization; human body systems; disease.	B1.1 (Connecting Sciences), B5.2, C5.2 (Connecting Sciences)

D. Earth and Space Science	Book, Chapter, and Section
D.1 **Earth's changing atmosphere**	B4.2 (Connecting Sciences)
D.3 **Earth in the solar system** Sun, planets, asteroids, comets; regular and predictable motion and day, year, phases of the moon, and eclipses; gravity and orbits; sun as source of energy for earth; cause of seasons.	A3.2, A3.3, C3.1, D3.3

E. Science and Technology	Book, Chapter, and Section
E.1–E.5 **Abilities of technological design** Identify problems; design a solution or product; implement a proposed design; evaluate completed designs or products; communicate the process of technological design.	A2.3, A3.1, A4.3, B (p. 5), B4.2, C1.2, C2.1, C3.2, C5.3, D2.4, D3.1, D3.3, D4.4, E2.3
E.6 **Understandings about science and technology** Similarities and differences between scientific inquiry and technological design; contributions of people in different cultures; reciprocal nature of science and technology; nonexistence of perfectly designed solutions; constraints, benefits, and unintended consequences of technological designs.	All books (pp. xxvi–xxvii) All books (Frontiers in Science, Timelines in Science) A.1.2, A3.3, B3.4, B3.1, B3.3, C5.3, D4.4, E1.2, E1.3

F.	Science in Personal and Social Perspectives	Book, Chapter, and Section
F.1	**Personal health** Exercise; fitness; hazards and safety; tobacco, alcohol, and other drugs; nutrition; STDs; environmental health	B4.2, B5.2
F.2	**Populations, resources, and environments** Overpopulation and resource depletion; environmental degradation.	A3.1
F.3	**Natural hazards** Earthquakes, landslides, wildfires, volcanic eruptions, floods, storms; hazards from human activity; personal and societal challenges.	B3.2, D3.2, E1.2
F.4	**Risks and benefits** Risk analysis; natural, chemical, biological, social, and personal hazards; decisions based on risks and benefits.	B4.3
F.5	**Science and technology in society** Science's influence on knowledge and world view; societal challenges and scientific research; technological influences on society; contributions from people of different cultures and times; work of scientists and engineers; ethical codes; limitations of science and technology.	All books (Timelines in Science) A1.2, A3.2, A3.3, B3.4, B4.4, C5.3, D2, D3.2, D3.3, D4.4, E1.2, E1.3

G.	History and Nature of Science	Book, Chapter, and Section
G.1	**Science as a human endeavor** Diversity of people w.orking in science, technology, and related fields; abilities required by science	All books (pp. xxii–xxv; Frontiers in Science)
G.2	**Nature of science** Observations, experiments, and models; tentative nature of scientific ideas; differences in interpretation of evidence; evaluation of results of investigations, experiments, observations, theoretical models, and explanations; importance of questioning, response to criticism, and communication.	B1.2, B2.1, B2.3, E3.2
G.3	**History of science** Historical examples of inquiry and relationships between science and society; scientists and engineers as valued contributors to culture; challenges of breaking through accepted ideas.	All books (Frontiers in Science; Timelines in Science) B1.2, B3.2, C2.1, D2.4

Correlations to Benchmarks

This chart provides an overview of how the five Physical Science modules of *McDougal Littell Science* address the National Science Education Standards.

A Matter and Energy
B Chemical Interactions
C Motion and Forces
D Waves, Sound, and Light
E Electricity and Magnetism

1. The Nature of Science	Book, Chapter, and Section
	The Nature of Science (pp. xxii–xxv); E2.3; Think Science Features: A3.1, B2.2, C2.1, C4.2, D3.2, E3.1; Scientific Thinking Handbook (pp. R2–R9); Lab Handbook (pp. R10–R35)

3. The Nature of Technology	Book, Chapter, and Section
	The Nature of Technology (pp. xxvi–xxvii); A3.3, B4.4, D4.4, E1, E2.3, E3.2, E3.3, E3.4; Timelines in Science Features

4. The Physical Setting	Book, Chapter, and Section
4.B THE EARTH	A3.1, A4.3, C3.1
4.D STRUCTURE OF MATTER	
4.D.1 All matter is made of atoms; atoms of any element are alike but different from atoms of other elements; different arrangements of atoms into groups compose all substances.	A1.2, A1.3, B1.1, B2.1, B2.2
4.D.2 Equal volumes of different substances usually have different weights.	A2.1, A2.3
4.D.3 Atoms and molecules are perpetually in motion; increased temperature means greater average energy of motion; states of matter: solids, liquids, gases.	A1.2, A4.1
4.D.4 Temperature and acidity of a solution influence reaction rates. Many substances dissolve in water, which may facilitate reactions between them.	B3.1, B4.2, B4.3
4.D.5 Greek philosopheres' scientific ideas about elements; most elements tend to combine with others, so few elements are found in their pure form.	B1.1
4.D.6 Groups of elements have similar properties; oxidation; some elements, like carbon and hydrogen, don't fit into any category and are essential elements of living matter.	B1.2, B1.3, B3.1, B5.1, B5.2
4.D.7 Conservation of matter: the total weight of a closed system remains the same because the total number of atoms stays the same regardless of how they interact with one another.	B3.2
4.E ENERGY TRANSFORMATIONS	
4.E.1 Energy cannot be created or destroyed, but only changed from one form into another.	A3.2, C4.2
4.E.2 Most of what goes on in the universe involves energy transformations.	A3, A4.2, A4.3
4.E.3 Heat can be transferred through materials by the collisions of atoms or across space by radiation; convection currents transfer heat in fluid materials.	A4.2, A4.3
4.E.4 Energy appears in many different forms, including heat energy, chemical energy, mechanical energy, and gravitational energy.	A3.1, A4.2, A4.3, C4.2
4.F MOTION	
4.F.1 Light from the Sun is made up of many different colors of light; objects that give off or reflect light have a different mix of colors.	D3.3, D3.4
4.F.2 Something can be "seen" when light waves emitted or reflected by it enter the eye.	D4.1, D4.3

4.F.3 An unbalanced force acting on an object changes its speed or direction of motion, or both. If the force acts toward a single center, the object's path may curve into an orbit around the center.	C2.1, C2.2, C3.1
4.F.4 Vibrations in materials set up wavelike disturbances (such as sound) that spread away from the source; waves move at different speeds in different materials.	D1, D2.1, D2.2, D3.1, D3.4
4.F.5 Human eyes respond to only a narrow range of wavelengths of electromagnetic radiation—visible light. Differences of wavelengths within that range are perceived as differences in color.	D3.2, D3.4, D4.3
4.G FORCES OF NATURE	
4.G.1 Objects exerts gravitational forces on one another, but these forces depend on the mass and distance of objects, and may be too small to detect.	C3.1
4.G.2 The Sun's gravitational pull holds Earth and other planets in their orbits; planets' gravitational pull keeps their moons in orbit around them.	C3.1
4.G.3 Electric currents and magnets can exert a force on each other.	E3.1, E3.2, E3.3
5. The Living Environment	**Book, Chapter, and Section**
5.E Flow of Matter and Energy	B5.2
8. The Designed World	**A2.1, A2.3, A3.3, B3.4, B4.4, B5.3, C5.3, E2.3, E3.2**
9. The Mathematical World	**All Math in Science Features, E2.3**
10. Historical Perspectives	**B1, B2, B3.2, C1.1, C2, D4.4**
12. Habits of Mind	**Book, Chapter, and Section**
12.A VALUES AND ATTITUDES	Think Science Features: A3.1, B2.2, C2.1, C4.2, D3.2, E3.1
12.B Computation and Estimation	All Math in Science Features, Lab Handbook (pp. R10–R35)
12.C Manipulation and Observation	All Investigates and Chapter Investigations
12.D Communication Skills	All Chapter Investigations, Lab Handbook (pp. R10–R35)
12.E Critical-Response Skills	Think Science Features: A3.1, B2.2, C2.1, C4.2, D3.2, E3.1; Scientific Thinking Handbook (pp. R2–R9)

Planning the Unit

The Pacing Guide provides suggested pacing for all chapters in the unit as well as the two unit features shown below.

Frontiers in Science

- Features cutting-edge research as an engaging point of entry into the unit
- Connects to an accompanying *Scientific American Frontiers* video and viewing guide
- Introduces three options for unit projects.

The stream of ions glows blue as it is shot out of ...

Ion Engines for Long Vo...

Rocket engines must provide huge amount... spacecraft away from Earth and keep them... can weigh more than the spacecraft thems... tists and engineers are always looking for... give spacecraft and other vehicles the ener...

One method of powering spacecraft us... particles called ions. The atoms of a gas—u... made into ions. An electric field is then use... of the engine at a very high speed—faster than 100,000 kilometers per hour (62,000 mi/h). This stream of rapidly moving ions works like the gases coming out of a jet engine on a plane—propelling the spacecraft in the direction opposite to the ion stream.

An advantage of ion propulsion is that its fuel is much lighter than the chemical fuel used in rockets. Ion propulsion does not provide enough thrust to be used for a rocket launch, but it can be used to move a spacecraft through long distances in outer space. This method of propulsion provides a small force to the spacecraft; however, over time the spacecraft can reach great speeds.

The space probe *Deep Space 1* was the first to use an ion engine to travel between planets. The engine generated enough speed for the probe to follow and photograph comet Borrelly in 2001.

so much fuel is needed to power a spacecraft, the fuel itself can weigh more than the spacecraft. That is why it is necessary for scientists to learn about the various types of energy sources for fueling spacecraft. Have students review the parts of the scientific process on pages xxiv–xxv and ask what part this step falls under. *determining what is known*

Technology Design

Focus students' attention on "Ion Engines for Long Voyages." Discuss how using ions—electrically charged particles—power the spacecraft by creating an ion stream that propels the spacecraft while in orbit. Ask students to name the advantage as well as the disadvantage of using this engine design. *The fuel for ion engines is lighter than chemical fuel used for rockets, but ion propulsion does not provide enough of a force for launching the rocket.*

ADDITIONAL RESOURCES

Technology Resources

Scientific American Frontiers Video: Sunrayce: 12-minute video segment that introduces the unit.

ClassZone.com
CAREER LINK, Physicist, Chemist, Aeronautical Engineer

Guide student viewing and comprehension of the video:
Video Teaching Guide, pp. 1–2; Video Viewing Guide, p. 3; Video Wrap-Up, p. 4

Scientific American Frontiers Video Guide, pp. 39–42

Unit projects procedures and rubrics:
Unit Projects, pp. 5–10

DIFFERENTIATE INSTRUCTION

More Reading Support
A What provides spacecraft with the large amounts of energy they need to move? *rocket engines*
B What is ion propulsion best used for? *long distances*

Below Level Have students write out the steps to power a spacecraft using ions. Steps should include where ions come from, what is created with those ions, and how the ions work to push the spacecraft.

B 2 Unit: Matter and Energy

Frontiers in Science 3 B

TIMELINES in Science

FOCUS

Set Learning Goals

Students will

- Observe how scientists created new theories of temperature and heat by building on earlier observations.
- Learn the characteristics of temperature and heat and how they are measured.
- Write a procedure for an experiment to test a specific method of calculating temperature.

National Science Education Standards

A.9.a–b, A.9.d–g, Understandings About Scientific Inquiry
E.6.a–c Understandings About Science and Technology
F.5.a–e, F.5.g Science and Technology in Society
G.1.a–b Science as a Human Endeavor
G.2.a Nature of Science
G.3.a–c History of Science

TIMELINES in Science

ABOUT TEMPERATURE AND HEAT

Scientific Process

1724 Gabriel Fahrenheit improved on existing technology to measure temperature through his observations and investigations. He found that mercury was more accurate in measuring temperature than the alcohol thermometers of the time. Fahrenheit then developed the mercury thermometer, which we still use today.

Sharing Results

1762 Chemist Joseph Black experimented with the amount of heat a substance can hold. His results helped James Watt, who was working on improvements to the steam engine. Ask students what research Black was conducting that was critical to Watt's work. *boiling and evaporation*

Mathematics Connection

1742 The Celsius scale is the metric scale for measuring temperature. Most countries use the Celsius scale for everyday temperature measurement. Scientists also use this scale for their experiments. In the United States, the Fahrenheit scale is used to measure temperature. To convert a Celsius temperature to a Fahrenheit temperature, multiply the Celsius temperature by 9/5 and then add 32 to the result. Write the formula °F = 9/5(°C) + 32 on the board. Ask students to convert 35°C to Fahrenheit. *95°F*

Application

ALCHEMY While alchemy is no longer considered a true science, for some time it was a major source of chemical knowledge. Because alchemists experimented with turning metals such as lead into gold, they gained wide knowledge about chemical substances. The alchemist's workshop became the forerunner of the modern chemistry laboratory. Alchemists used tools such as funnels, beakers, and balances. Although alchemists failed to produce gold from other materials, their experimentation was successful in other ways. Ask students what impact alchemy had on science and technology. *Alchemy created many chemical processes and tools that are still used by chemists today.*

DIFFERENTIATE INSTRUCTION

Below Level For students who may have difficulty understanding how information is organized on a timeline, point out the dates on the center line. Explain how the dates become more recent when read from left to right. Show them the lines connecting the event boxes to the specific dates on the timeline. Discuss how timelines are a good way to show the order in which events happened.

DIFFERENTIATE INSTRUCTION

Advanced Have students use the temperature conversion formula shown in the Mathematics Connection on this page to convert Fahrenheit to Celsius. Ask students what the formula for the conversion is. *°C = 5/9 (°F − 32)* Then ask students to convert 98.6°F to Celsius. *37°C*

Timelines in Science

- Traces the history of key scientific discoveries
- Highlights interactions between science and technology.

B 96 Unit: Matter and Energy

Timelines in Science 97 B

Matter and Energy Pacing Guide

The following pacing guide shows how the chapters in ***Matter and Energy*** can be adapted to fit your specific course needs.

	TRADITIONAL SCHEDULE (DAYS)	BLOCK SCHEDULE (DAYS)
Frontiers in Science: Fuels of the Future	1	0.5
Chapter 1 Introduction to Matter		
1.1 Matter has mass and volume.	2	1
1.2 Matter is made of atoms.	2	1
1.3 Matter combines to form different substances.	2	1
1.4 Matter exists in different physical states.	3	1.5
Chapter Investigation	1	0.5
Chapter 2 Properties of Matter		
2.1 Matter has observable properties.	2	1
2.2 Changes of state are physical changes.	2	1
2.3 Properties are used to identify substances.	3	1.5
Chapter Investigation	1	0.5
Chapter 3 Energy		
3.1 Energy exists in different forms.	2	1
3.2 Energy can change forms but is never lost.	2	1
3.3 Technology improves the way people use energy.	3	1.5
Chapter Investigation	1	0.5
Timelines in Science: About Temperature and Heat	1	0.5
Chapter 4 Temperature and Heat		
4.1 Temperature depends on particle movement.	2	1
4.2 Energy flows from warmer to cooler objects.	2	1
4.3 The transfer of energy as heat can be controlled.	3	1.5
Chapter Investigation	1	0.5
Total Days for Module	**36**	**18**

Planning the Chapter

Complete planning support precedes each chapter.

Previewing Content

- Section-by-section science background notes
- Common Misconceptions notes

CHAPTER

Introduction to Matter

Physical Science
UNIFYING PRINCIPLES

PRINCIPLE 1
Matter is made of particles too small to see.

PRINCIPLE 2
Matter changes form and moves from place to place.

PRINCIPLE 3
Energy chang form to anoth cannot be cre destroyed.

PRINCIPLE 4

Unit: Matter and Energy
BIG IDEAS

CHAPTER 1 Introduction to Matter
Everything that has mass and takes up space is matter.

CHAPTER 2 Properties of Matter
Matter has properties that can be changed by physical and chemical processes.

CHAPTER 3 Energy
Energy has differen but it is always co

CHAPTER 1
KEY CONCEPTS

SECTION 1.1
Matter has mass and volume.
1. All objects are made of matter.
2. Mass is a measure of the amount of matter.
3. Volume is a measure of the space matter occupies.

SECTION 1.2
Matter is made of atoms.
1. Atoms are extremely small.
2. Atoms and molecules are always in motion.

SECTION
Matter com different su
1. Matter ca mixed.
2. Parts of m the same throughou

The Big Idea Flow Chart is available on p. T1 in the **UNIT TRANSPARENCY BOOK**

Previewing Content

SECTION

Matter has mass and volume. pp. 9–15

1. All objects are made of matter.
Anything that has **mass** and takes up space is **matter.** All the objects, liquids, gases, and living things in the universe are made of matter. Energy is not matter. However, under special circumstances, such as a nuclear reaction, energy can become matter and matter can become energy.

Previewing Content

SECTION 1.3
Matter combines to form different substances. pp. 21–26

1. Matter can be pure or mixed.
Matter that contains only one kind of atom or molecule is pure. Matter often contains two or more substances mixed together. Substances can be composed of elements, compounds, or mixtures.
- An **element** is a substance that contains only one kind of atom. Gold is the element represented in the diagram on the left below.

Element: Gold

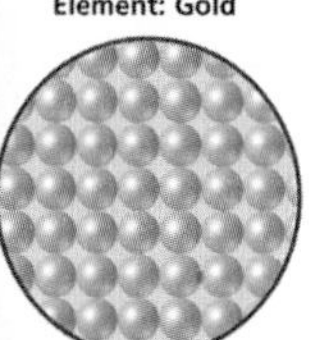
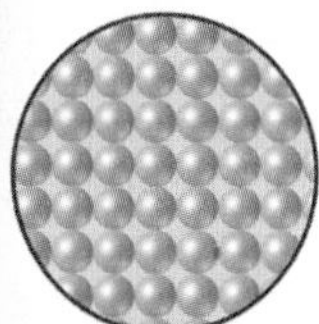

Compound: Dry Ice

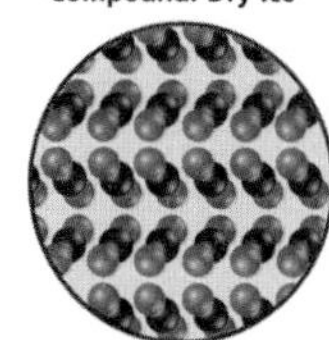
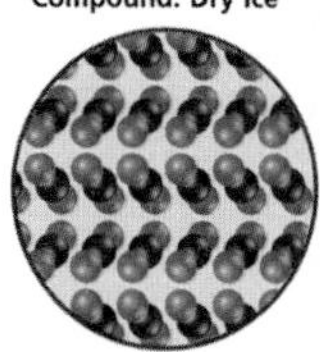

- A **compound** is a substance that consists of two or more different types of atoms bonded together as shown in the diagram on the right above. Water molecules are compounds because they contain two kinds of atom bonded covalently. A molecule of oxygen is not a compound. Some compounds, such as table salt, are bonded ionically.
- A **mixture** is a combination of different substances that retain their individual properties and can be separated by physical means.

2. Parts of mixtures can be the same or different throughout.
Mixtures can be either heterogeneous or homogeneous.
- A heterogeneous mixture has different properties in different parts of the mixture because the substances in different parts of the mixture vary.
- A homogeneous mixture has substances evenly spread out throughout the mixture.

SECTION 1.4
Matter exists in different physical states. pp. 27–33

1. Particle arrangement and motion determine the state of matter.
Solid, liquid, and gas are three common **states of matter.** When a substance changes from one state to another, the arrangement of its molecules changes. The distance between molecules and the attraction they have for one another change.

2. Solid, liquid, and gas are common states of matter.
The state of a substance depends on the space between its particles and the way in which the particles move.
- A **solid** has particles that are close together. The particles are attached to one another and can vibrate in place, but they cannot move from place to place.
- A **liquid** has particles that are attracted to one another and are close together. The particles can slide over one another and move from one place to another.
- A **gas** has particles that are not close to one another and can move about freely.

The diagrams below show the arrangement and motion of particles in different states of matter.

Solid Liquid Gas

3. Solids have a definite volume and shape.
A solid has a fixed volume and shape. The particles in some solids are in regular patterns and form crystals.

4. Liquids have a definite volume but no definite shape.
A liquid has a definite volume because its particles are close enough together that they cannot move about freely, although they slide past each other. A liquid takes the shape of the container that it is in.

5. Gases have no definite volume or shape.
A gas has no definite volume or shape. The volume, pressure, and temperature of a gas are related to one another, and changing one can change the others.

Common Misconceptions

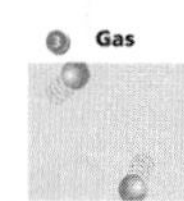
MISCONCEPTION DATABASE
CLASSZONE.COM Background on student misconceptions

ATOMS AND COLOR Students often think that individual atoms and molecules have the same properties as the substance they make up. For example, students might think a gold atom is hard and solid or a gas molecule is transparent.

This misconception is addressed in the Teacher Demo on p. 22.

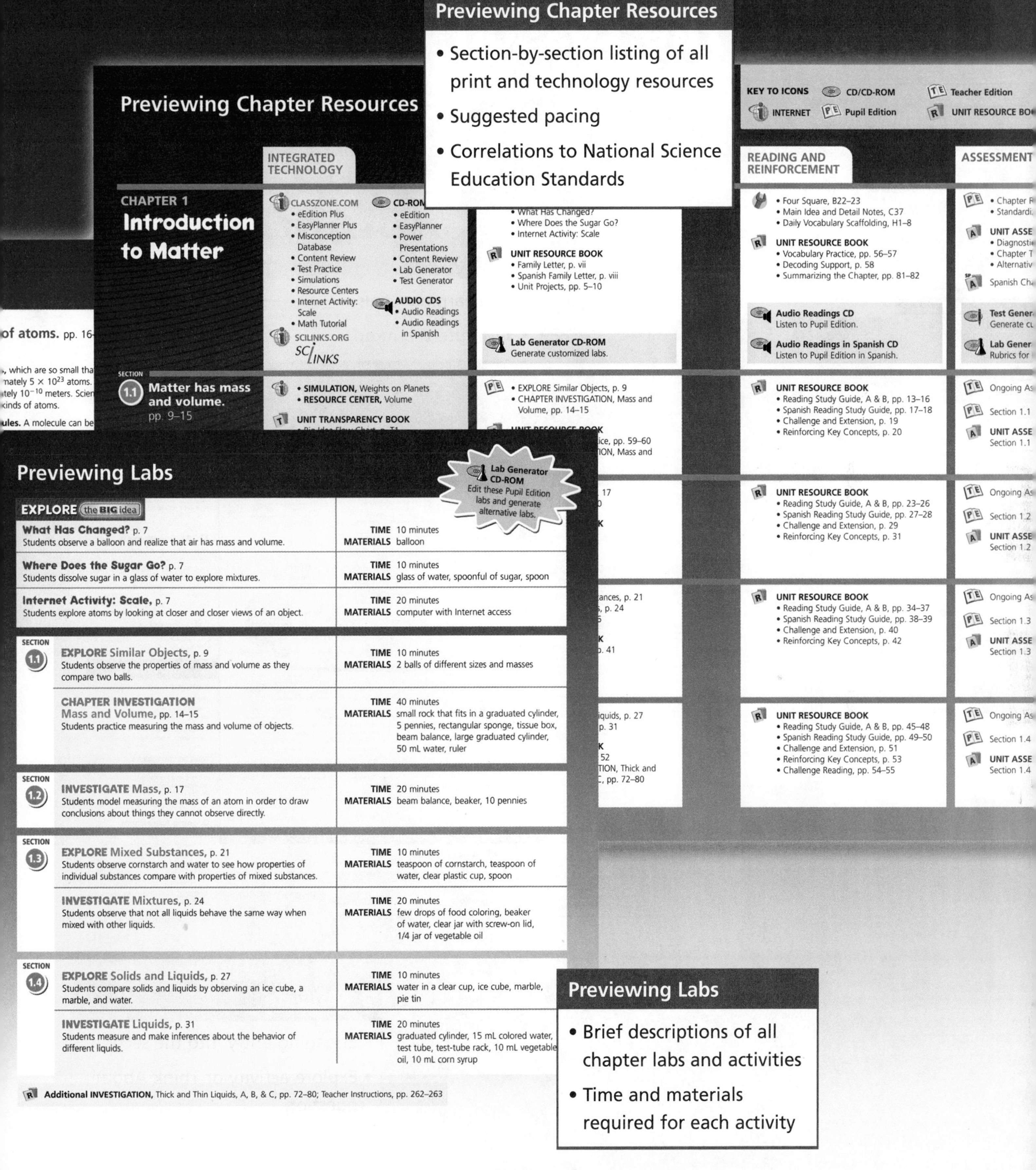

Previewing Chapter Resources
• Section-by-section listing of all print and technology resources
• Suggested pacing
• Correlations to National Science Education Standards
Previewing Labs
• Brief descriptions of all chapter labs and activities
• Time and materials required for each activity

Previewing Labs

Lab Generator CD-ROM: Edit these Pupil Edition labs and generate alternative labs.

EXPLORE the BIG idea	
What Has Changed? p. 7 Students observe a balloon and realize that air has mass and volume.	TIME 10 minutes MATERIALS balloon
Where Does the Sugar Go? p. 7 Students dissolve sugar in a glass of water to explore mixtures.	TIME 10 minutes MATERIALS glass of water, spoonful of sugar, spoon
Internet Activity: Scale, p. 7 Students explore atoms by looking at closer and closer views of an object.	TIME 20 minutes MATERIALS computer with Internet access
SECTION 1.1 **EXPLORE** Similar Objects, p. 9 Students observe the properties of mass and volume as they compare two balls.	TIME 10 minutes MATERIALS 2 balls of different sizes and masses
CHAPTER INVESTIGATION Mass and Volume, pp. 14–15 Students practice measuring the mass and volume of objects.	TIME 40 minutes MATERIALS small rock that fits in a graduated cylinder, 5 pennies, rectangular sponge, tissue box, beam balance, large graduated cylinder, 50 mL water, ruler
SECTION 1.2 **INVESTIGATE** Mass, p. 17 Students model measuring the mass of an atom in order to draw conclusions about things they cannot observe directly.	TIME 20 minutes MATERIALS beam balance, beaker, 10 pennies
SECTION 1.3 **EXPLORE** Mixed Substances, p. 21 Students observe cornstarch and water to see how properties of individual substances compare with properties of mixed substances.	TIME 10 minutes MATERIALS teaspoon of cornstarch, teaspoon of water, clear plastic cup, spoon
INVESTIGATE Mixtures, p. 24 Students observe that not all liquids behave the same way when mixed with other liquids.	TIME 20 minutes MATERIALS few drops of food coloring, beaker of water, clear jar with screw-on lid, 1/4 jar of vegetable oil
SECTION 1.4 **EXPLORE** Solids and Liquids, p. 27 Students compare solids and liquids by observing an ice cube, a marble, and water.	TIME 10 minutes MATERIALS water in a clear cup, ice cube, marble, pie tin
INVESTIGATE Liquids, p. 31 Students measure and make inferences about the behavior of different liquids.	TIME 20 minutes MATERIALS graduated cylinder, 15 mL colored water, test tube, test-tube rack, 10 mL vegetable oil, 10 mL corn syrup

Additional INVESTIGATION, Thick and Thin Liquids, A, B, & C, pp. 72–80; Teacher Instructions, pp. 262–263

Chapter 1: **Introduction to Matter** 5D

Planning the Lesson

Point-of-use support for each lesson provides a wealth of teaching options.

1. Prepare

- Concept and vocabulary review
- Note-taking and vocabulary strategies

2. Focus

- Set Learning Goals
- 3-Minute Warm-up

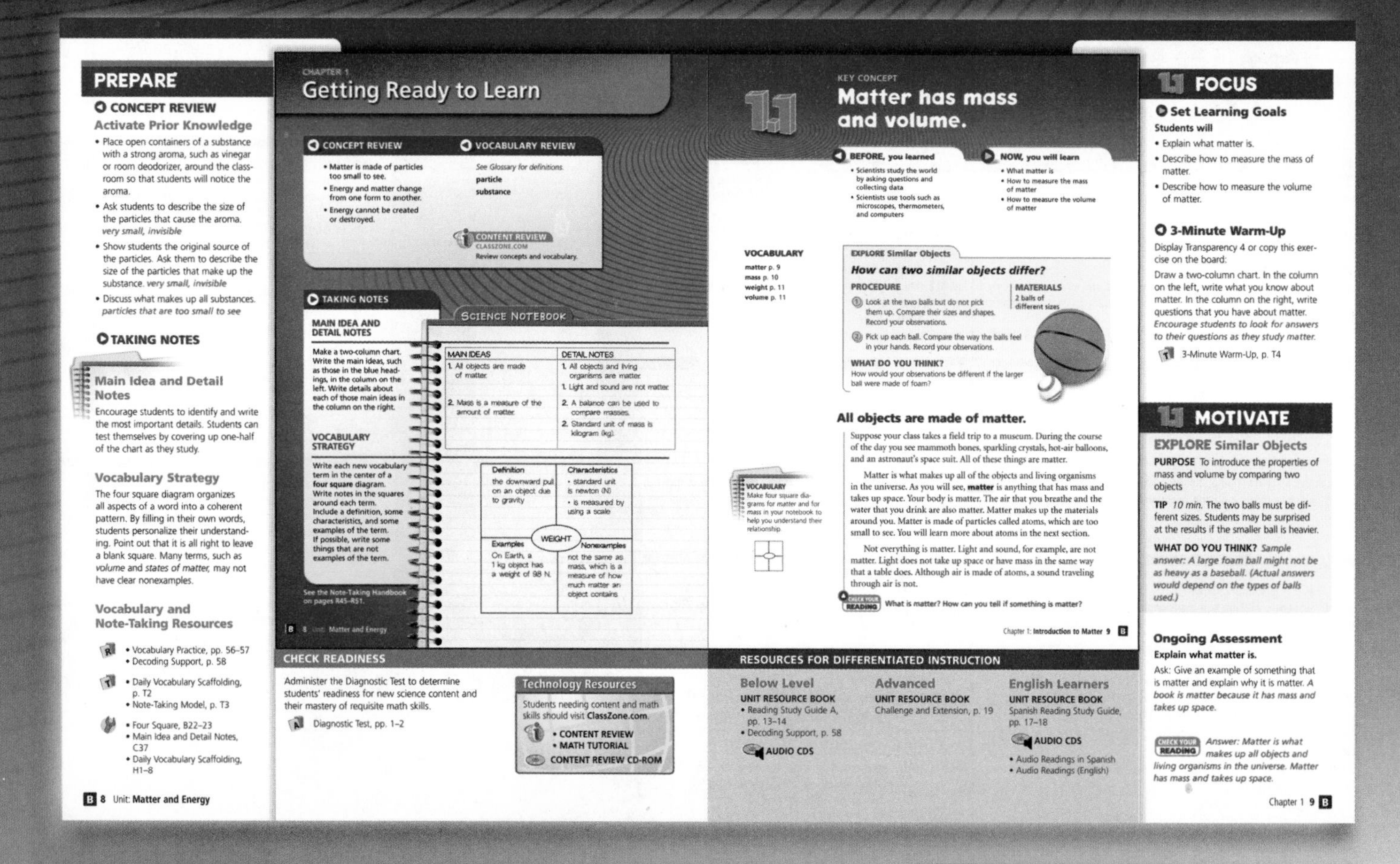

PREPARE

CONCEPT REVIEW

Activate Prior Knowledge

- Place open containers of a substance with a strong aroma, such as vinegar or room deodorizer, around the classroom so that students will notice the aroma.
- Ask students to describe the size of the particles that cause the aroma. *very small, invisible*
- Show students the original source of the particles. Ask them to describe the size of the particles that make up the substance. *very small, invisible*
- Discuss what makes up all substances. *particles that are too small to see*

TAKING NOTES

Main Idea and Detail Notes

Encourage students to identify and write the most important details. Students can test themselves by covering up one-half of the chart as they study.

Vocabulary Strategy

The four square diagram organizes all aspects of a word into a coherent pattern. By filling in their own words, students personalize their understanding. Point out that it is all right to leave a blank square. Many terms, such as *volume* and *states of matter,* may not have clear nonexamples.

Vocabulary and Note-Taking Resources

- Vocabulary Practice, pp. 56–57
- Decoding Support, p. 58
- Daily Vocabulary Scaffolding, p. T2
- Note-Taking Model, p. T3
- Four Square, B22–23
- Main Idea and Detail Notes, C37
- Daily Vocabulary Scaffolding, H1–8

B 8 Unit: **Matter and Energy**

CHAPTER 1

Getting Ready to Learn

CONCEPT REVIEW

- Matter is made of particles too small to see.
- Energy and matter change from one form to another.
- Energy cannot be created or destroyed.

VOCABULARY REVIEW

See Glossary for definitions.

particle

substance

CONTENT REVIEW
CLASSZONE.COM
Review concepts and vocabulary.

TAKING NOTES

MAIN IDEA AND DETAIL NOTES

Make a two-column chart. Write the main ideas, such as those in the blue headings, in the column on the left. Write details about each of those main ideas in the column on the right.

VOCABULARY STRATEGY

Write each new vocabulary term in the center of a **four square** diagram. Write notes in the squares around each term. Include a definition, some characteristics, and some examples of the term. If possible, write some things that are not examples of the term.

See the Note-Taking Handbook on pages R45–R51.

SCIENCE NOTEBOOK

MAIN IDEAS	DETAIL NOTES
1. All objects are made of matter.	1. All objects and living organisms are matter. 1. Light and sound are not matter.
2. Mass is a measure of the amount of matter.	2. A balance can be used to compare masses. 2. Standard unit of mass is kilogram (kg).

Definition	Characteristics
the downward pull on an object due to gravity	• standard unit is newton (N) • is measured by using a scale
Examples	**Nonexamples**
On Earth, a 1 kg object has a weight of 9.8 N.	not the same as mass, which is a measure of how much matter an object contains

WEIGHT

B 8 Unit: Matter and Energy

CHECK READINESS

Administer the Diagnostic Test to determine students' readiness for new science content and their mastery of requisite math skills.

Diagnostic Test, pp. 1–2

Technology Resources

Students needing content and math skills should visit **ClassZone.com.**

- CONTENT REVIEW
- MATH TUTORIAL

CONTENT REVIEW CD-ROM

KEY CONCEPT

1.1 Matter has mass and volume.

BEFORE, you learned

- Scientists study the world by asking questions and collecting data
- Scientists use tools such as microscopes, thermometers, and computers

NOW, you will learn

- What matter is
- How to measure the mass of matter
- How to measure the volume of matter

VOCABULARY

matter p. 9
mass p. 10
weight p. 11
volume p. 11

EXPLORE Similar Objects

How can two similar objects differ?

PROCEDURE

1. Look at the two balls but do not pick them up. Compare their sizes and shapes. Record your observations.
2. Pick up each ball. Compare the way the balls feel in your hands. Record your observations.

MATERIALS

2 balls of different sizes

WHAT DO YOU THINK?

How would your observations be different if the larger ball were made of foam?

All objects are made of matter.

Suppose your class takes a field trip to a museum. During the course of the day you see mammoth bones, sparkling crystals, hot-air balloons, and an astronaut's space suit. All of these things are matter.

Matter is what makes up all of the objects and living organisms in the universe. As you will see, **matter** is anything that has mass and takes up space. Your body is matter. The air that you breathe and the water that you drink are also matter. Matter makes up the materials around you. Matter is made of particles called atoms, which are too small to see. You will learn more about atoms in the next section.

Not everything is matter. Light and sound, for example, are not matter. Light does not take up space or have mass in the same way that a table does. Although air is made of atoms, a sound traveling through air is not.

VOCABULARY Make four square diagrams for *matter* and for *mass* in your notebook to help you understand their relationship.

CHECK YOUR READING What is matter? How can you tell if something is matter?

Chapter 1: **Introduction to Matter** 9 B

RESOURCES FOR DIFFERENTIATED INSTRUCTION

Below Level

UNIT RESOURCE BOOK

- Reading Study Guide A, pp. 13–14
- Decoding Support, p. 58

AUDIO CDS

Advanced

UNIT RESOURCE BOOK

Challenge and Extension, p. 19

English Learners

UNIT RESOURCE BOOK

Spanish Reading Study Guide, pp. 17–18

AUDIO CDS

- Audio Readings in Spanish
- Audio Readings (English)

1.1 FOCUS

Set Learning Goals

Students will

- Explain what matter is.
- Describe how to measure the mass of matter.
- Describe how to measure the volume of matter.

3-Minute Warm-Up

Display Transparency 4 or copy this exercise on the board:

Draw a two-column chart. In the column on the left, write what you know about matter. In the column on the right, write questions that you have about matter. *Encourage students to look for answers to their questions as they study matter.*

3-Minute Warm-Up, p. T4

1.1 MOTIVATE

EXPLORE Similar Objects

PURPOSE To introduce the properties of mass and volume by comparing two objects

TIP *10 min.* The two balls must be different sizes. Students may be surprised at the results if the smaller ball is heavier.

WHAT DO YOU THINK? *Sample answer: A large foam ball might not be as heavy as a baseball. (Actual answers would depend on the types of balls used.)*

Ongoing Assessment

Explain what matter is.

Ask: Give an example of something that is matter and explain why it is matter. *A book is matter because it has mass and takes up space.*

CHECK YOUR READING *Answer: Matter is what makes up all objects and living organisms in the universe. Matter has mass and takes up space.*

Chapter 1 9 B

3. Motivate

- Engaging entry into the section
- Explore activity or Think About question

1.1 INSTRUCT

Develop Critical Thinking

APPLY Tell students that enough mass cubes are not available for the entire class. Have students apply what they know about measuring mass to invent standard units of mass based on common objects. Ask them to find the mass of objects based on the invented standard units.

Teach from Visuals

To help students interpret the photographs and illustrations comparing the size and mass of a bowling ball and a basketball, ask:

- How do the sizes of the bowling ball and the basketball in the photographs compare? *They are about the same.*
- How can you tell from the illustrations which ball has more mass? *More standard cubes are needed to balance the pan with the bowling ball than the pan with the basketball.*

Ongoing Assessment

Describe how to measure the mass of matter.

Ask: How would you find the mass of an object? *Compare its mass with the mass of standard units on a balance.*

CHECK YOUR READING *Answer: Mass is a measure of how much matter an object contains.*

MAIN IDEA AND DETAILS As you read, write the blue headings on the left side of a two-column chart. Add details in the other column.

Mass is a measure of the amount of matter.

Different objects contain different amounts of matter. **Mass** is a measure of how much matter an object contains. A metal teaspoon, for example, contains more matter than a plastic teaspoon. Therefore, a metal teaspoon has a greater mass than a plastic teaspoon. An elephant has more mass than a mouse.

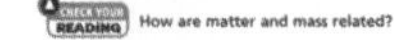

CHECK YOUR READING How are matter and mass related?

Measuring Mass

When you measure mass, you compare the mass of the object with a standard amount, or unit, of mass. The standard unit of mass is the kilogram (kg). A large grapefruit has a mass of about one-half kilogram. Smaller masses are often measured in grams (g). There are 1000 grams in a kilogram. A penny has a mass of between two and three grams.

How can you compare the masses of two objects? One way is to use a pan balance, as shown below. If two objects balance each other on a pan balance, then they contain the same amount of matter. If a basketball balances a metal block, for example, then the basketball and the block have the same mass. Beam balances work in a similar way, but instead of comparing the masses of two objects, you compare the mass of an object with a standard mass on the beam.

A bowling ball and a basketball are about the same size, but a bowling ball has more mass.

Measuring Weight

When you hold an object such as a backpack full of books, you feel it pulling down on your hands. This is because Earth's gravity pulls the backpack toward the ground. Gravity is the force that pulls two masses toward each other. In this example, the two masses are Earth and the backpack. **Weight** is the downward pull on an object due to gravity. If the pull of the backpack is strong, you would say that the backpack weighs a lot.

Weight is measured by using a scale, such as a spring scale like the one shown on the right, that tells how hard an object is pushing or pulling on it. The standard scientific unit for weight is the newton (N). A common unit for weight is the pound (lb).

Gravity is pulling down on both the girl and the backpack. The heavier the backpack is, the stronger the pull of gravity is on it.

Mass and weight are closely related, but they are not the same. Mass describes the amount of matter an object has, and weight describes how strongly gravity is pulling on that matter. On Earth, a one-kilogram object has a weight of 9.8 newtons (2.2 lb). When a person says that one kilogram is equal to 2.2 pounds, he or she is really saying that one kilogram has a weight of 2.2 pounds on Earth. On the Moon, however, gravity is one-sixth as strong as it is on Earth. On the Moon, the one-kilogram object would have a weight of 1.6 newtons (0.36 lb). The amount of matter in the object, or its mass, is the same on Earth as it is on the Moon, but the pull of gravity is different.

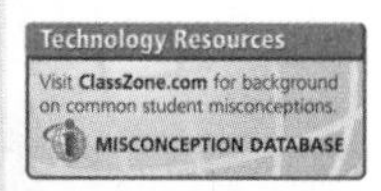

SIMULATION CLASSZONE.COM Compare weights on different planets.

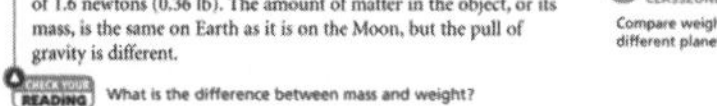

CHECK YOUR READING What is the difference between mass and weight?

Volume is a measure of the space matter occupies.

Matter takes up space. A bricklayer stacks bricks on top of each other to build a wall. No two bricks can occupy the same place because the matter in each brick takes up space.

The amount of space that matter in an object occupies is called the object's **volume.** The bowling ball and the basketball shown on page 10 take up approximately the same amount of space. Therefore, the two balls have about the same volume. Although the basketball is hollow, it is not empty. Air fills up the space inside the basketball. Air and other gases take up space and have volume.

Teach Difficult Concepts

Students often have trouble understanding the difference between mass and weight. Have students find the weight of an object by using a spring scale. Have them weigh the object as they gently pull down or push up on the object. Compare this weight with the object's weight on a planet with different gravity. Ask:

- How is the object's weight affected by gravity? ***It weighs more where gravity is stronger and less where gravity is weaker.***
- How is the object's mass affected by gravity? ***It is not.***

Address Misconceptions

IDENTIFY Ask students to identify all the matter in a blown-up balloon and in a balloon filled with water. If students do not include the air and water inside the balloons, they may hold the misconception that liquids and gases are not matter.

CORRECT Slowly let the air out of a balloon. Let the water out of another balloon. Have students observe the change in the volume of the balloons.

REASSESS Ask students what was taking up space in the balloons. *air and water* Point out that air and water are matter because they take up space and have mass.

Technology Resources

Visit **ClassZone.com** for background on common student misconceptions.

MISCONCEPTION DATABASE

EXPLORE the BIG idea

Revisit "What Has Changed" on p. 7. Have students explain their results.

Ongoing Assessment

CHECK YOUR READING *Answer: Mass is a measure of how much matter an object contains, but weight is the downward pull on an object due to*

DIFFERENTIATE INSTRUCTION

More Reading Support

A To measure mass, what do you compare the object's mass to? *stan-*

English Learners Phrasal verbs such as "makes up" and "takes up" are used throughout this chapter (for example, on p. 9). Make sure students understand not to read "up" and other adverbs or prepositions in phrasal verbs as literal directions. If students are still confused, offer synonyms for phrasal [illegible]or instance, "makes up" means composes, and "takes [illegible]ans occupies. English learners may also lack background [illegible]dge of a field trip (p. 9).

DIFFERENTIATE INSTRUCTION

More Reading Support

B Which is affected by gravity, mass or weight? *weight*

C What do you call the amount of space an

Below Level Have students make a table that compares mass and weight. Ask them to include a definition of each, the standard units used to measure them, and the tools used to measure them.

4. Instruct

- Teaching strategies
- Reading support
- Ongoing assessment
- Addressing misconceptions
- Differentiated instruction activities and tips

Measuring Volume by Displacement

Although a box has a regular shape, a rock does not. There is no simple formula for calculating the volume of something with an irregular shape. Instead, you can make use of the fact that two objects cannot be in the same place at the same time. This method of measuring is called displacement.

1. Add water to a graduated cylinder. Note the volume of the water by reading the water level on the cylinder.
2. Submerge the irregular object in the water. Because the object and the water cannot share the same space, the water is displaced, or moved upward. Note the new volume of the water with the object in it.
3. Subtract the volume of the water before you added the object from the volume of the water and the object together. The result is the volume of the object. The object displaces a volume of water equal to the volume of the object.

You measure the volume of a liquid by measuring how much space it takes up in a container. The volume of a liquid usually is measured in liters (L) or milliliters (mL). One liter is equal to 1000 milliliters. Milliliters and cubic centimeters are equivalent. This can be written as $1 \text{ mL} = 1 \text{ cm}^3$. If you had a box with a volume of one cubic centimeter and you filled it with water, you would have one milliliter of water.

water rises

Measure the volume of water without the rock.

Measure the volume of water with the rock in it.

In the first photograph, the graduated cylinder contains 50 mL of water. Placing a rock in the cylinder causes the water level to rise from 50 mL to 55 mL. The difference is 5 mL; therefore, the volume of the rock is 5 cm^3.

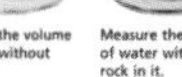

1.1 Review

KEY CONCEPTS

1. Give three examples of matter.
2. What do weight and mass measure?
3. How can you measure the volume of an object that has an irregular shape?

CRITICAL THINKING

4. **Calculate** What is the volume of a box that is 12 cm long, 6 cm wide, and 4 cm high?
5. **Synthesize** What is the relationship between the units of measurement for the volume of a liquid and of a solid object?

CHALLENGE

6. **Infer** Why might a small increase in the dimensions of an object cause a large change in its volume?

ANSWERS

1. Sample answer: air, water, a rock

2. Weight measures the downward pull on an object due to gravity. Mass measures how much matter an object contains.

3. Submerge the object in water and subtract the volume of the water before the object was added from the volume of the water and the object together.

4. 12 cm · 6 cm · 4 cm = 288 cm^3

5. One cubic centimeter of solid is equivalent to one milliliter of liquid.

6. Each of the three dimensions increases, so, for example, doubling the dimensions increases the volume by a multiple of eight.

Ongoing Assessment

Describe how to measure the volume of matter.

Ask: How would you find the volume of a box? of a small rock? *Multiply the box's length by its width and height. Put the small stone in a graduated cylinder of water and determine how much water the rock displaces.*

Teacher Demo

Place a clear glass that is full to its brim with water in a shallow tray. Ask students to predict what will happen if you slowly drop a stone into the glass. Ask them to explain their predictions.

Reinforce the BIG idea

Have students relate the section to the Big Idea.

Reinforcing Key Concepts, p. 20

1.1 ASSESS & RETEACH

Assess

Section 1.1 Quiz, p. 3

Reteach

Write the heading *Matter* on the board. Ask a volunteer to define *matter* and write the definition on the board. Add these two subheads under *Matter: Mass* and *Volume*. Ask a volunteer to define *mass* and *volume* and write the definitions on the board. Ask students to describe how each is measured, which tools are used for the measurements, and which units are used.

Technology Resources

Have students visit **ClassZone.com** for reteaching of Key Concepts.

CONTENT REVIEW

CONTENT REVIEW CD-ROM

5. Assess & Reteach

- Answers to Section Review
- Reteaching activity
- Resources for review and assessment

Lab Materials List

The following charts list the consumables, nonconsumables, and equipment needed for all activities. Quantities are per group of four students. Lab aprons, goggles, water, books, paper, pens, pencils, and calculators are assumed to be available for all activities.

Materials kits are available. For more information, please call McDougal Littell at 1-800-323-5435.

Consumables

Description	Quantity per Group	Explore *page*	Investigate *page*	Chapter Investigation *page*
aluminum foil	7–8 ft		89, 112	84, 122
baking powder	1 tsp	58		
baking soda	1 tsp	58		
bottle, plastic, 1 liter	1		107	
bottle, plastic, pint	2			122
box, tissue	1			14
can, aluminum soda	1			84
cardboard, 12" x 24"	1		75	
clay, modeling	2 sticks	41	107	84, 122
coffee filter, basket	1		61	
corn syrup, dark	10 mL		31	
cornstarch	2 tsp	21	47	
crouton	1			84
cup, clear plastic	12	21, 27, 58	47, 61, 89, 112	
foam packing pellet	1 cup			122
food coloring	1 bottle		24, 31, 107	
ice cube	1	27		
index card	3	86	61	
Isopropyl alcohol	250 mL		107	
Lugol's iodine solution	6 mL		47	
match, wood	4–5			84
paper clip	1			84
pepper	1 tsp		61	
pie plate, aluminum	3	27	61	84
plastic garbage bag, black, 4" x 4" piece	1		89	
plastic garbage bag, white, 4" x 4" piece	1		89	
plastic wrap	3–4 ft			122
rice cake, caramel	1/4			84
rubber band, large	30	103	89	122
salt, table	1 tsp		61	

Description	Quantity per Group	Explore *page*	Investigate *page*	Chapter Investigation *page*
sand	1 1/2 lb	71	61	122
soil, potting	1/4 lb			122
sponge	1			14
spoon, plastic	3	21	47, 61	
stearic acid	30 mL			56
stopper, cork	1			84
straw, clear drinking	1		107	
tape, packing	8"		75	
vegetable oil	1 cup		24, 31	
vitamin C tablet, 100 mg	1		47	
wire, copper, uninsulated	18"			56

Nonconsumables

Description	Quantity per Group	Explore *page*	Investigate *page*	Chapter Investigation *page*
balance, triple beam	1		17, 75, 112	14, 84
baseball	1	9		
basketball	1	9		
beaker, 50 mL	1		17	
beaker, 100 mL	1		112	
beaker, 200 mL	2	116		122
beaker, 500 mL	2	116	24	
bowl, large plastic	1	71	107	
can opener	1			84
cloth, wool, 8" x 8"	1		61	
coffee can, 1 lb	1			122
coin, penny	10		17, 112	14
comb, plastic	1		61	
dowel rod, 1/4" diameter	6"			84
eyedropper	2		47	
funnel	1		61	

Description	Quantity per Group	Explore *page*	Investigate *page*	Chapter Investigation *page*
graduated cylinder, 100 mL	1		31, 47, 61, 112	14, 84, 122
jar, baby food with lid	1		24	
marble, metal	1	27		
meter stick	1		75	
pebble	1	71		
ring stand with ring	1			84
rock, medium	1	71		
rock, small	1			14
ruler, metric	1	86		14
scissors	1		89	
solar calculator with no battery backup	1	86		
stopwatch	1	116	89, 112	122
test tube rack	1		31	56
test tube, large	1		31	56
thermometer	2	116	89, 112	122
thermometer, 12"	1			56, 84
tongs, test tube	1			56
toy car, large	1		75	
washer, metal 1"	20		75	

Safety Equipment

Description		Explore *page*	Investigate *page*	Chapter Investigation *page*
gloves				122

McDougal Littell Science

Matter and Energy

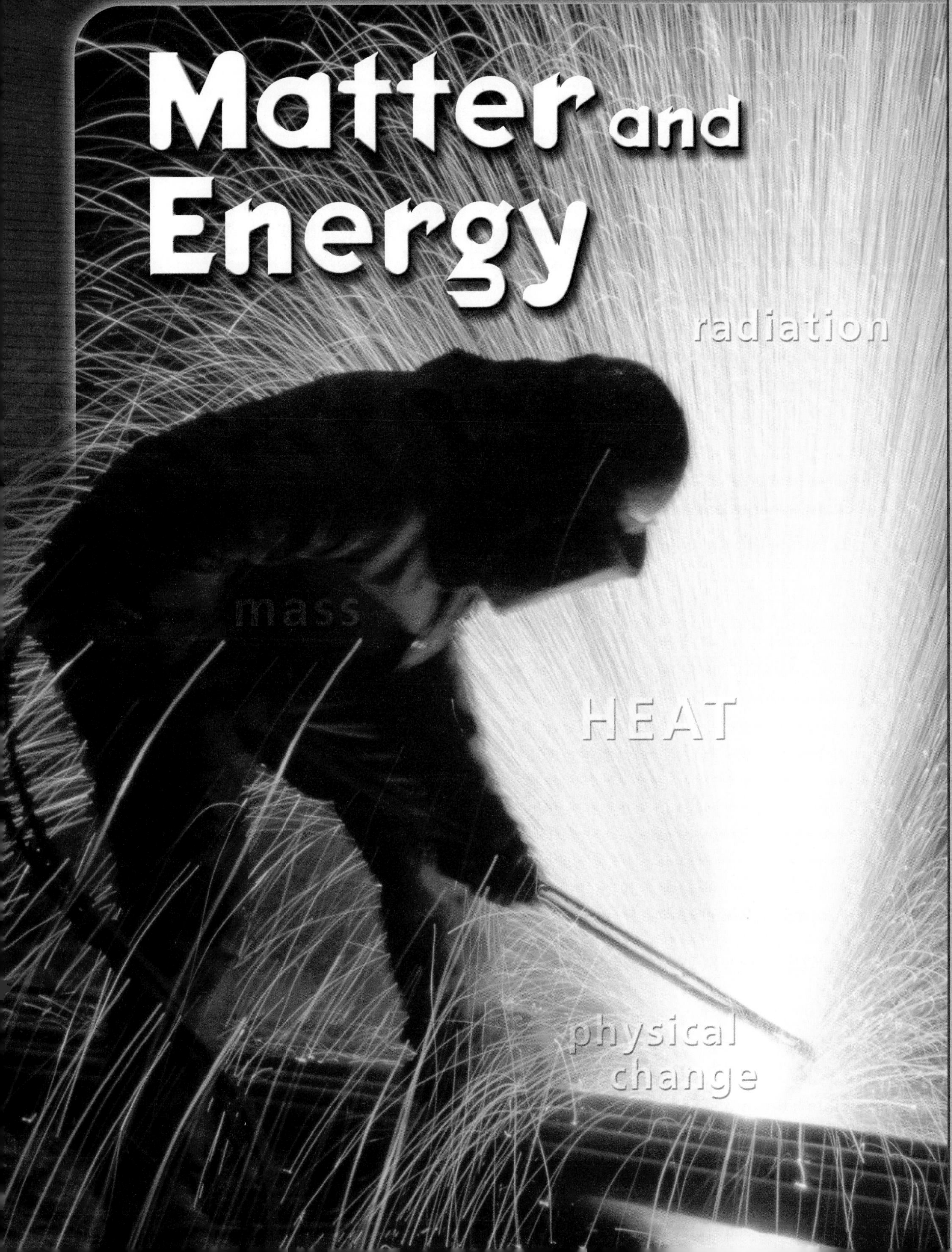

radiation

mass

HEAT

physical change

PHYSICAL SCIENCE

A ▶ Matter and Energy
B ▶ Chemical Interactions
C ▶ Motion and Forces
D ▶ Waves, Sound, and Light
E ▶ Electricity and Magnetism

LIFE SCIENCE

A ▶ Cells and Heredity
B ▶ Life Over Time
C ▶ Diversity of Living Things
D ▶ Ecology
E ▶ Human Biology

EARTH SCIENCE

A ▶ Earth's Surface
B ▶ The Changing Earth
C ▶ Earth's Waters
D ▶ Earth's Atmosphere
E ▶ Space Science

ISBN: 0-618-33444-0 1 2 3 4 5 6 7 8 VJM 08 07 06 05 04

Internet Web Site: http://www.mcdougallittell.com

Science Consultants

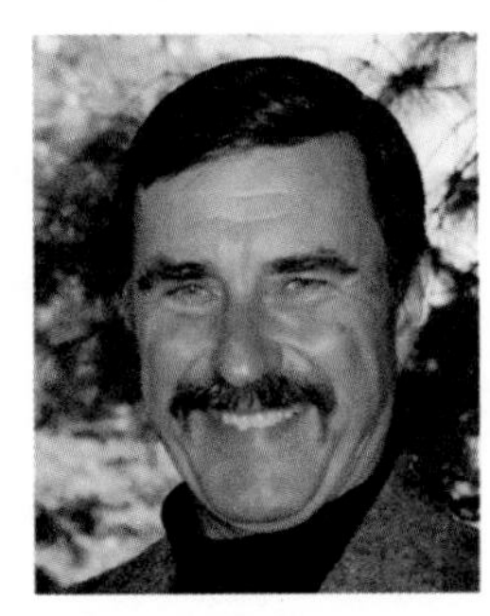

Chief Science Consultant

James Trefil, Ph.D. is the Clarence J. Robinson Professor of Physics at George Mason University. He is the author or co-author of more than 25 books, including *Science Matters* and *The Nature of Science.* Dr. Trefil is a member of the American Association for the Advancement of Science's Committee on the Public Understanding of Science and Technology. He is also a fellow of the World Economic Forum and a frequent contributor to *Smithsonian* magazine.

Rita Ann Calvo, Ph.D. is Senior Lecturer in Molecular Biology and Genetics at Cornell University, where for 12 years she also directed the Cornell Institute for Biology Teachers. Dr. Calvo is the 1999 recipient of the College and University Teaching Award from the National Association of Biology Teachers.

Kenneth Cutler, M.S. is the Education Coordinator for the Julius L. Chambers Biomedical Biotechnology Research Institute at North Carolina Central University. A former middle school and high school science teacher, he received a 1999 Presidential Award for Excellence in Science Teaching.

Instructional Design Consultants

Douglas Carnine, Ph.D. is Professor of Education and Director of the National Center for Improving the Tools of Educators at the University of Oregon. He is the author of seven books and over 100 other scholarly publications, primarily in the areas of instructional design and effective instructional strategies and tools for diverse learners. Dr. Carnine also serves as a member of the National Institute for Literacy Advisory Board.

Linda Carnine, Ph.D. consults with school districts on curriculum development and effective instruction for students struggling academically. A former teacher and school administrator, Dr. Carnine also co-authored a popular remedial reading program.

Donald Steely, Ph.D. serves as principal investigator at the Oregon Center for Applied Science (ORCAS) on federal grants for science and language arts programs. His background also includes teaching and authoring of print and multimedia programs in science, mathematics, history, and spelling.

Sam Miller, Ph.D. is a middle school science teacher and the Teacher Development Liaison for the Eugene, Oregon, Public Schools. He is the author of curricula for teaching science, mathematics, computer skills, and language arts.

Vicky Vachon, Ph.D. consults with school districts throughout the United States and Canada on improving overall academic achievement with a focus on literacy. She is also co-author of a widely used program for remedial readers.

iii

Content Reviewers

John Beaver, Ph.D.
Ecology
Professor, Director of Science Education Center
College of Education and Human Services
Western Illinois University
Macomb, IL

Donald J. DeCoste, Ph.D.
Matter and Energy, Chemical Interactions
Chemistry Instructor
University of Illinois
Urbana-Champaign, IL

Dorothy Ann Fallows, Ph.D., MSc
Diversity of Living Things, Microbiology
Partners in Health
Boston, MA

Michael Foote, Ph.D.
The Changing Earth, Life Over Time
Associate Professor
Department of the Geophysical Sciences
The University of Chicago
Chicago, IL

Lucy Fortson, Ph.D.
Space Science
Director of Astronomy
Adler Planetarium and Astronomy Museum
Chicago, IL

Elizabeth Godrick, Ph.D.
Human Biology
Professor, CAS Biology
Boston University
Boston, MA

Isabelle Sacramento Grilo, M.S.
The Changing Earth
Lecturer, Department of the Geological Sciences
Montana State University
Bozeman, MT

David Harbster, MSc
Diversity of Living Things
Professor of Biology
Paradise Valley Community College
Phoenix, AZ

Richard D. Norris, Ph.D.
Earth's Waters
Professor of Paleobiology
Scripps Institution of Oceanography
University of California, San Diego
La Jolla, CA

Donald B. Peck, M.S.
Motion and Forces; Waves, Sound, and Light; Electricity and Magnetism
Director of the Center for Science Education (retired)
Fairleigh Dickinson University
Madison, NJ

Javier Penalosa, Ph.D.
Diversity of Living Things, Plants
Associate Professor, Biology Department
Buffalo State College
Buffalo, NY

Raymond T. Pierrehumbert, Ph.D.
Earth's Atmosphere
Professor in Geophysical Sciences (Atmospheric Science)
The University of Chicago
Chicago, IL

Brian J. Skinner, Ph.D.
Earth's Surface
Eugene Higgins Professor of Geology and Geophysics
Yale University
New Haven, CT

Nancy E. Spaulding, M.S.
Earth's Surface, The Changing Earth, Earth's Waters
Earth Science Teacher (retired)
Elmira Free Academy
Elmira, NY

Steven S. Zumdahl, Ph.D.
Matter and Energy, Chemical Interactions
Professor Emeritus of Chemistry
University of Illinois
Urbana-Champaign, IL

Susan L. Zumdahl, M.S.
Matter and Energy, Chemical Interactions
Chemistry Education Specialist
University of Illinois
Urbana-Champaign, IL

Safety Consultant

Juliana Texley, Ph.D.
Former K–12 Science Teacher and School Superintendent
Boca Raton, FL

English Language Advisor

Judy Lewis, M.A.
Director, State and Federal Programs for reading proficiency and high risk populations
Rancho Cordova, CA

iv

Teacher Panel Members

Carol Arbour
Tallmadge Middle School,
Tallmadge, OH

Patty Belcher
Goodrich Middle School,
Akron, OH

Gwen Broestl
Luis Munoz Marin Middle School,
Cleveland, OH

Al Brofman
Tehipite Middle School,
Fresno, CA

John Cockrell
Clinton Middle School,
Columbus, OH

Jenifer Cox
Sylvan Middle School,
Citrus Heights, CA

Linda Culpepper
Martin Middle School,
Charlotte, NC

Kathleen Ann DeMatteo
Margate Middle School,
Margate, FL

Melvin Figueroa
New River Middle School,
Ft. Lauderdale, FL

Doretha Grier
Kannapolis Middle School,
Kannapolis, NC

Robert Hood
Alexander Hamilton Middle School,
Cleveland, OH

Scott Hudson
Coverdale Elementary School,
Cincinnati, OH

Loretta Langdon
Princeton Middle School,
Princeton, NC

Carlyn Little
Glades Middle School,
Miami, FL

Ann Marie Lynn
Amelia Earhart Middle School,
Riverside, CA

James Minogue
Lowe's Grove Middle School,
Durham, NC

Joann Myers
Buchanan Middle School,
Tampa, FL

Barbara Newell
Charles Evans Hughes Middle School,
Long Beach, CA

Anita Parker
Kannapolis Middle School,
Kannapolis, NC

Greg Pirolo
Golden Valley Middle School,
San Bernardino, CA

Laura Pottmyer
Apex Middle School,
Apex, NC

Lynn Prichard
Booker T. Washington Middle Magnet
School, Tampa, FL

Jacque Quick
Walter Williams High School,
Burlington, NC

Robert Glenn Reynolds
Hillman Middle School,
Youngstown, OH

Theresa Short
Abbott Middle School,
Fayetteville, NC

Rita Slivka
Alexander Hamilton Middle School,
Cleveland, OH

Marie Sofsak
B F Stanton Middle School,
Alliance, OH

Nancy Stubbs
Sweetwater Union Unified School District,
Chula Vista, CA

Sharon Stull
Quail Hollow Middle School,
Charlotte, NC

Donna Taylor
Okeeheelee Middle School,
West Palm Beach, FL

Sandi Thompson
Harding Middle School,
Lakewood, OH

Lori Walker
Audubon Middle School & Magnet Center,
Los Angeles, CA

Teacher Lab Evaluators

Jill Brimm-Byrne
Albany Park Academy,
Chicago, IL

Gwen Broestl
Luis Munoz Marin Middle School,
Cleveland, OH

Al Brofman
Tehipite Middle School,
Fresno, CA

Michael A. Burstein
The Rashi School,
Newton, MA

Trudi Coutts
Madison Middle School,
Naperville, IL

Jenifer Cox
Sylvan Middle School,
Citrus Heights, CA

Larry Cwik
Madison Middle School,
Naperville, IL

Jennifer Donatelli
Kennedy Junior High School,
Lisle, IL

Paige Fullhart
Highland Middle School,
Libertyville, IL

Sue Hood
Glen Crest Middle School,
Glen Ellyn, IL

Ann Min
Beardsley Middle School,
Crystal Lake, IL

Aileen Mueller
Kennedy Junior High School,
Lisle, IL

Nancy Nega
Churchville Middle School,
Elmhurst, IL

Oscar Newman
Sumner Math and Science Academy,
Chicago, IL

Marina Penalver
Moore Middle School,
Portland, ME

Lynn Prichard
Booker T. Washington Middle Magnet
School, Tampa, FL

Jacque Quick
Walter Williams High School,
Burlington, NC

Seth Robey
Gwendolyn Brooks Middle School,
Oak Park, IL

Kevin Steele
Grissom Middle School,
Tinley Park, IL

v

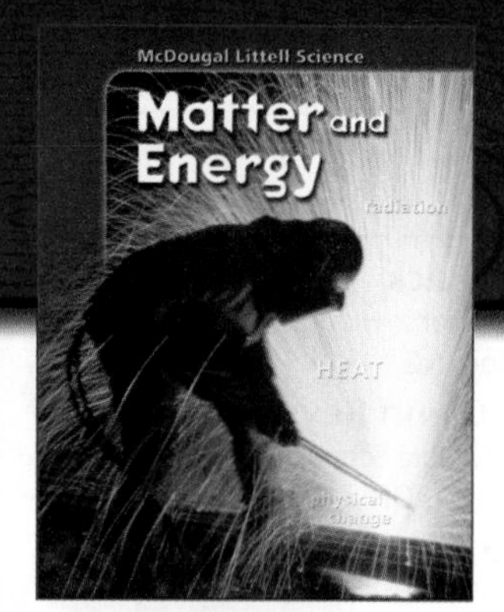

eEdition

Matter and Energy

Unit Features

SCIENTIFIC AMERICAN

What properties could help you identify this sculpture as sugar?
page 38

vi

What different forms of energy are shown in this photograph? page 68

Features

Visual Highlights

Internet Resources @ ClassZone.com

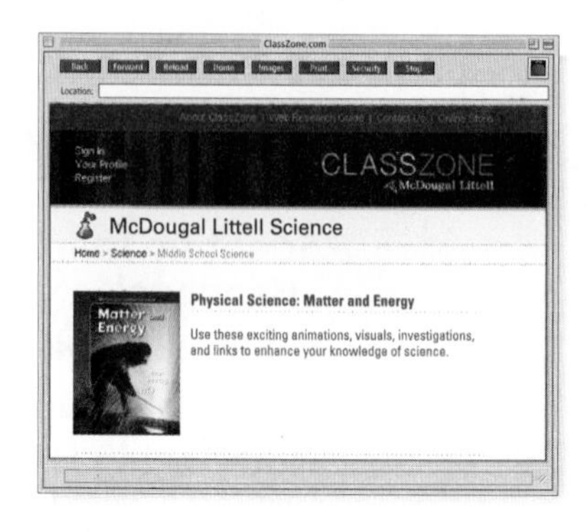

INVESTIGATIONS AND ACTIVITIES

EXPLORE THE BIG IDEA

Chapter Opening Inquiry

CHAPTER INVESTIGATION

Full-Period Labs

EXPLORE

Introductory Inquiry Activities

INVESTIGATE

Skill Labs

Standards and Benchmarks

Each chapter in **Matter and Energy** covers some of the learning goals that are described in the *National Science Education Standards* (NSES) and the Project 2061 *Benchmarks for Science Literacy*. Selected content and skill standards are shown below in shortened form. The following National Science Education Standards are covered on pages xii–xxvii, in Frontiers in Science, and in Timelines in Science, as well as in chapter features and laboratory investigations: Understandings About Scientific Inquiry (A.9), Understandings About Science and Technology (E.6), Science and Technology in Society (F.5), Science as a Human Endeavor (G.1), Nature of Science (G.2), and History of Science (G.3).

Content Standards

1 Introduction to Matter

National Science Education Standards

B.1.c | There are more than 100 known elements that combine to produce compounds.

Project 2061 Benchmarks

4.D.1 | All matter is made up of atoms.

4.D.3 | Atoms and molecules are always in motion.
- In solids, they vibrate.
- In liquids, they slide past one another.
- In gases, they move freely.

2 Properties of Matter

National Science Education Standards

B.1.a |
- A substance has characteristic properties.
- A mixture often can be separated into the original substances using these properties.

Project 2061 Benchmarks

4.D.2 | Equal volumes of different substances usually have different weights.

8.B.1 | The choice of materials for a job depends on their properties.

3 Energy

National Science Education Standards

B.3.a | Energy is a property of substances that is often associated with
- heat
- light
- electricity
- mechanical motion
- sound
- atomic nuclei
- chemical compounds

Energy is transferred in many ways.

Project 2061 Benchmarks

4.E.1 | Energy cannot be created or destroyed, but it can be changed from one form to another.

4.E.2 | Most of what goes on in the universe involves energy transformations.

4.E.4 | Energy has many different forms, including
- heat
- chemical
- mechanical
- gravitational

8.C.1	• As energy changes from one form to another, some energy is always converted to heat. • Some systems transform energy with less heat loss than others.

4 Temperature and Heat

	National Science Education Standards
B.3.b	Heat flows from warmer objects to cooler ones, until both reach the same temperature.
	Project 2061 Benchmarks
4.E.2	Energy in the form of heat is almost always one of the products of an energy transformation.
4.E.3	Heat can be transferred through materials by the collisions of atoms or across space by radiation.
4.E.4	• Energy appears in different forms. • Heat energy is the disorderly motion of molecules.

Process and Skill Standards

	National Science Education Standards
A.1	Identify questions that can be answered through investigation.
A.2	Design and conduct a scientific investigation.
A.3	Use appropriate tools and techniques to gather and interpret data.
A.4	Use evidence to describe, predict, explain, and model.
A.5	Use critical thinking to find relationships between results and interpretations.
A.6	Consider alternative explanations and predictions.
A.7	Communicate procedures, results, and conclusions.
A.8	Use mathematics in scientific investigations.
E.1	Identify a problem to be solved.
E.2	Design a solution or product.
E.3	Implement the proposed solution.
E.4	Evaluate the solution or design.

	Project 2061 Benchmarks
1.C.1	Contributions to science and technology have been made by different people, in different cultures, at different times.
2.B.1	Mathematics contributes to science and technology.
3.C.4	Technology has influenced the course of history.
9.A.3	How decimals should be written depends on how precise the measurements are.
9.A.5	The expression *a*/*b* can mean different things: *a* divided by *b* or *a* compared to *b*.
11.C.4	Use equations to summarize observed changes.
12.B.1	Find what percentage one number is of another.
12.B.7	Determine the appropriate unit for an answer. Convert units.
12.B.8	Round a calculation to the correct number of significant figures.
12.C.1	Compare amounts proportionally.
12.C.3	Using appropriate units, use and read instruments that measure length, volume, weight, time, rate, and temperature.
12.D.1	Use tables and graphs to organize information and identify relationships.
12.D.2	Read, interpret, and describe tables and graphs.
12.D.4	Understand information that includes different types of charts and graphs, including circle charts, bar graphs, line graphs, data tables, diagrams, and symbols.

Introducing Physical Science

Scientists are curious. Since ancient times, they have been asking and answering questions about the world around them. Scientists are also very suspicious of the answers they get. They carefully collect evidence and test their answers many times before accepting an idea as correct.

In this book you will see how scientific knowledge keeps growing and changing as scientists ask new questions and rethink what was known before. The following sections will help get you started.

UNIFYING PRINCIPLES of Physical Science

What Is Physical Science?

In the simplest terms, physical science is the study of what things are made of and how they change. It combines the studies of both physics and chemistry. Physics is the science of matter, energy, and forces. It includes the study of topics such as motion, light, and electricity and magnetism. Chemistry is the study of the structure and properties of matter, and it especially focuses on how substances change into different substances.

The text and pictures in this book will help you learn key concepts and important facts about physical science. A variety of activities will help you investigate these concepts. As you learn, it helps to have a big picture of physical science as a framework for this new information. The four unifying principles listed below will give you this big picture. Read the next few pages to get an overview of each of these principles and a sense of why they are so important.

- **Matter is made of particles too small to see.**
- **Matter changes form and moves from place to place.**
- **Energy changes from one form to another, but it cannot be created or destroyed.**
- **Physical forces affect the movement of all matter on Earth and throughout the universe.**

the BIG idea

Each chapter begins with a big idea. Keep in mind that each big idea relates to one or more of the unifying principles.

UNIFYING PRINCIPLE

Matter is made of particles too small to see.

This simple statement is the basis for explaining an amazing variety of things about the world. For example, it explains why substances can exist as solids, liquids, and gases, and why wood burns but iron does not. Like the tiles that make up this mosaic picture, the particles that make up all substances combine to make patterns and structures that can be seen. Unlike these tiles, the individual particles themselves are far too small to see.

What It Means

To understand this principle better, let's take a closer look at the two key words: *matter* and *particles.*

Matter

Objects you can see and touch are all around you. The materials that these objects are made of are called **matter.** All living things—even you—are also matter. Even though you can't see it, the air around you is matter too. Scientists often say that matter is anything that has mass and takes up space. **Mass** is a measure of the amount of matter in an object. We use the word **volume** to refer to the amount of space an object or a substance takes up.

Particles

The tiny particles that make up all matter are called **atoms.** Just how tiny are atoms? They are far too small to see, even through a powerful microscope. In fact, an atom is more than a million times smaller than the period at the end of this sentence.

There are more than 100 basic kinds of matter called **elements.** For example, iron, gold, and oxygen are three common elements. Each element has its own unique kind of atom. The atoms of any element are all alike but different from the atoms of any other element.

Many familiar materials are made of particles called molecules. In a **molecule,** two or more atoms stick together to form a larger particle. For example, a water molecule is made of two atoms of hydrogen and one atom of oxygen.

Why It's Important

Understanding atoms and molecules makes it possible to explain and predict the behavior of matter. Among other things, this knowledge allows scientists to

- explain why different materials have different characteristics
- predict how a material will change when heated or cooled
- figure out how to combine atoms and molecules to make new and useful materials

UNIFYING PRINCIPLE

Matter changes form and moves from place to place.

You see matter change form every day. You see the ice in your glass of juice disappear without a trace. You see a black metal gate slowly develop a flaky, orange coating. Matter is constantly changing and moving.

What It Means

Remember that matter is made of tiny particles called atoms. Atoms are constantly moving and combining with one another. All changes in matter are the result of atoms moving and combining in different ways.

Matter Changes and Moves

You can look at water to see how matter changes and moves. A block of ice is hard like a rock. Leave the ice out in sunlight, however, and it changes into a puddle of water. That puddle of water can eventually change into water vapor and disappear into the air. The water vapor in the air can become raindrops, which may fall on rocks, causing them to weather and wear away. The water that flows in rivers and streams picks up tiny bits of rock and carries them from one shore to another. Understanding how the world works requires an understanding of how matter changes and moves.

Matter Is Conserved

No matter was lost in any of the changes described above. The ice turned to water because its molecules began to move more quickly as they got warmer. The bits of rock carried away by the flowing river were not gone forever. They simply ended up farther down the river. The puddles of rainwater didn't really disappear; their molecules slowly mixed with molecules in the air.

Under ordinary conditions, when matter changes form, no matter is created or destroyed. The water created by melting ice has the same mass as the ice did. If you could measure the water vapor that mixes with the air, you would find it had the same mass as the water in the puddle did.

Why It's Important

Understanding how mass is conserved when matter changes form has helped scientists to

- describe changes they see in the world
- predict what will happen when two substances are mixed
- explain where matter goes when it seems to disappear

xvii

Energy changes from one form to another, but it cannot be created or destroyed.

When you use energy to warm your food or to turn on a flashlight, you may think that you "use up" the energy. Even though the camp-stove fuel is gone and the flashlight battery no longer functions, the energy they provided has not disappeared. It has been changed into a form you can no longer use. Understanding how energy changes forms is the basis for understanding how heat, light, and motion are produced.

What It Means

Changes that you see around you depend on energy. **Energy,** in fact, means the ability to cause change. The electrical energy from an outlet changes into light and heat in a light bulb. Plants change the light energy from the Sun into chemical energy, which animals use to power their muscles.

Energy Changes Forms

Using energy means changing energy. You probably have seen electric energy changing into light, heat, sound, and mechanical energy in household appliances. Fuels like wood, coal, and oil contain chemical energy that produces heat when burned. Electric power plants make electrical energy from a variety of energy sources, including falling water, nuclear energy, and fossil fuels.

Energy Is Conserved

Energy can be converted into forms that can be used for specific purposes. During the conversion, some of the original energy is converted into unwanted forms. For instance, when a power plant converts the energy of falling water into electrical energy, some of the energy is lost to friction and sound.

Similarly, when electrical energy is used to run an appliance, some of the energy is converted into forms that are not useful. Only a small percentage of the energy used in a light bulb, for instance, produces light; most of the energy becomes heat. Nonetheless, the total amount of energy remains the same through all these conversions.

The fact that energy does not disappear is a law of physical science. The **law of conservation of energy** states that energy cannot be created or destroyed. It can only change form.

Why It's Important

Understanding that energy changes form but does not disappear has helped scientists to

- predict how energy will change form
- manage energy conversions in useful ways
- build and improve machines

UNIFYING PRINCIPLE

Physical forces affect the movement of all matter on Earth and throughout the universe.

What makes the world go around? The answer is simple: forces. Forces allow you to walk across the room, and forces keep the stars together in galaxies. Consider the forces acting on the rafts below. The rushing water is pushing the rafts forward. The force from the people paddling helps to steer the rafts.

What It Means

A **force** is a push or a pull. Every time you push or pull an object, you're applying a force to that object, whether or not the object moves. There are several forces—several pushes and pulls—acting on you right now. All these forces are necessary for you to do the things you do, even sitting and reading.

- You are already familiar with the force of gravity. **Gravity** is the force of attraction between two objects. Right now gravity is at work pulling you to Earth and Earth to you. The Moon stays in orbit around Earth because gravity holds it close.
- A contact force occurs when one object pushes or pulls another object by touching it. If you kick a soccer ball, for instance, you apply a contact force to the ball. You apply a contact force to a shopping cart that you push down a grocery aisle or a sled that you pull up a hill.
- **Friction** is the force that resists motion between two surfaces pressed together. If you've ever tried to walk on an icy sidewalk, you know how important friction can be. If you lightly rub your finger across a smooth page in a book and then across a piece of sandpaper, you can feel how the different surfaces produce different frictional forces. Which is easier to do?
- There are other forces at work in the world too. For example, a compass needle responds to the magnetic force exerted by Earth's magnetic field, and objects made of certain metals are attracted by magnets. In addition to magnetic forces, there are electrical forces operating between particles and between objects. For example, you can demonstrate electrical forces by rubbing an inflated balloon on your hair. The balloon will then stick to your head or to a wall without additional means of support.

Why It's Important

Although some of these forces are more obvious than others, physical forces at work in the world are necessary for you to do the things you do. Understanding forces allows scientists to

- predict how objects will move
- design machines that perform complex tasks
- predict where planets and stars will be in the sky from one night to the next

The Nature of Science

You may think of science as a body of knowledge or a collection of facts. More important, however, science is an active process that involves certain ways of looking at the world.

Scientific Habits of Mind

Scientists are curious. They are always asking questions. Scientists have asked questions such as, "What is the smallest form of matter?" and "How do the smallest particles behave?" These and other important questions are being investigated by scientists around the world.

Scientists are observant. They are always looking closely at the world around them. Scientists once thought the smallest parts of atoms were protons, neutrons, and electrons. Later, protons and neutrons were found to be made of even smaller particles called quarks.

Scientists are creative. They draw on what they know to form possible explanations for a pattern, an event, or an interesting phenomenon that they have observed. Then scientists create a plan for testing their ideas.

Scientists are skeptical. Scientists don't accept an explanation or answer unless it is based on evidence and logical reasoning. They continually question their own conclusions and the conclusions suggested by other scientists. Scientists trust only evidence that is confirmed by other people or methods.

Scientists cannot always make observations with their own eyes. They have developed technology, such as this particle detector, to help them gather information about the smallest particles of matter.

Scientists ask questions about the physical world and seek answers through carefully controlled procedures. Here a researcher works with supercooled magnets.

Science Processes at Work

You can think of science as a continuous cycle of asking and seeking answers to questions about the world. Although there are many processes that scientists use, scientists typically do each of the following:

- Ask a question
- Determine what is known
- Investigate
- Interpret results
- Share results

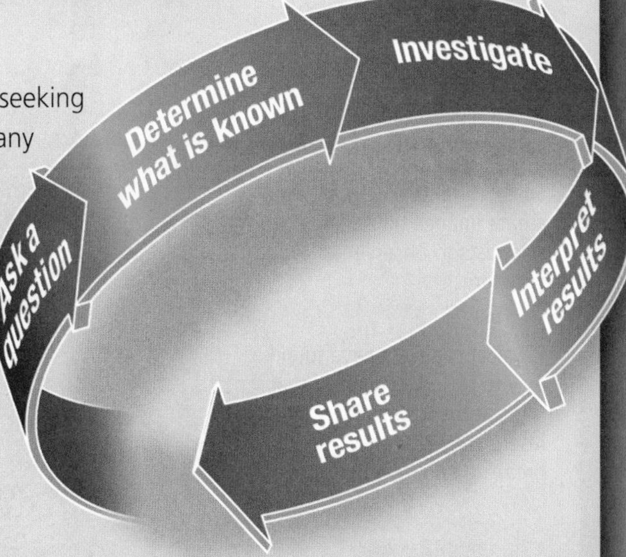

Ask a Question

It may surprise you that asking questions is an important skill. A scientific process may start when a scientist asks a question. Perhaps scientists observe an event or a process that they don't understand, or perhaps answering one question leads to another.

Determine What Is Known

When beginning an inquiry, scientists find out what is already known about a question. They study results from other scientific investigations, read journals, and talk with other scientists. A scientist working on subatomic particles is most likely a member of a large team using sophisticated equipment. Before beginning original research, the team analyzes results from previous studies.

Investigate

Investigating is the process of collecting evidence. Two important ways of investigating are observing and experimenting.

Observing is the act of noting and recording an event, a characteristic, or anything else detected with an instrument or with the senses. A researcher may study the properties of a substance by handling it, finding its mass, warming or cooling it, stretching it, and so on. For information about the behavior of subatomic particles, however, a researcher may rely on technology such as scanning tunneling microscopes, which produce images of structures that cannot be seen with the eye.

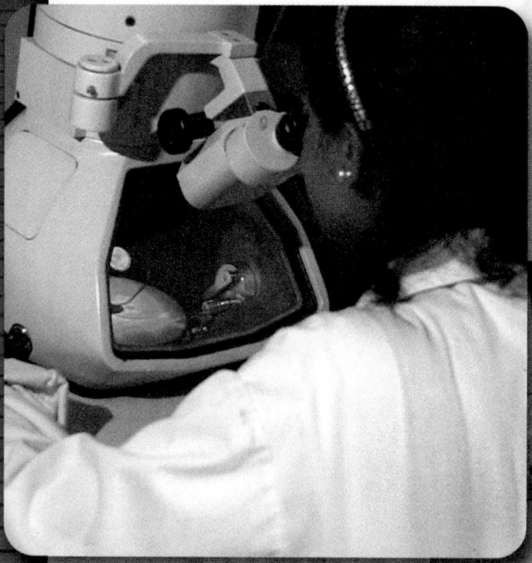

An **experiment** is an organized procedure to study something under controlled conditions. In order to study the effect of wing shape on the motion of a glider, for instance, a researcher would need to conduct controlled studies in which gliders made of the same materials and with the same masses differed only in the shape of their wings.

Scanning tunneling microscopes create images that allow scientists to observe molecular structure.

Physical chemists have found a way to observe chemical reactions at the atomic level. Using lasers, they can watch bonds breaking and new bonds forming.

Forming hypotheses and making predictions are two of the skills involved in scientific investigations. A **hypothesis** is a tentative explanation for an observation, a phenomenon, or a scientific problem that can be tested by further investigation. For example, in the mid-1800s astronomers noticed that the planet Uranus departed slightly from its expected orbit. One astronomer hypothesized that the irregularities in the planet's orbit were due to the gravitational effect of another planet—one that had not yet been detected. A **prediction** is an expectation of what will be observed or what will happen. A prediction can be used to test a hypothesis. The astronomers predicted that they would discover a new planet in the position calculated, and their prediction was confirmed with the discovery of the planet Neptune.

Interpret Results

As scientists investigate, they analyze their evidence, or data, and begin to draw conclusions. **Analyzing data** involves looking at the evidence gathered through observations or experiments and trying to identify any patterns that might exist in the data. Scientists often need to make additional observations or perform more experiments before they are sure of their conclusions. Many times scientists make new predictions or revise their hypotheses.

Often scientists use computers to help them analyze data. Computers reveal patterns that might otherwise be missed.

Scientists use computers to create models of objects or processes they are studying. This model shows carbon atoms forming a sphere.

Share Results

An important part of scientific investigation is sharing results of experiments. Scientists read and publish in journals and attend conferences to communicate with other scientists around the world. Sharing data and procedures gives them a way to test one another's results. They also share results with the public through newspapers, television, and other media.

The Nature of Technology

When you think of technology, you may think of cars, computers, and cell phones, as well as refrigerators, radios, and bicycles. Technology is not only the machines and devices that make modern lives easier, however. It is also a process in which new methods and devices are created. Technology makes use of scientific knowledge to design solutions to real-world problems.

Science and Technology

Science and technology go hand in hand. Each depends upon the other. Even designing a device as simple as a toaster requires knowledge of how heat flows and which materials are the best conductors of heat. Just as technology based on scientific knowledge makes our lives easier, some technology is used to advance scientific inquiry itself. For example, researchers use a number of specialized instruments to help them collect data. Microscopes, telescopes, spectrographs, and computers are just a few of the tools that help scientists learn more about the world. The more information these tools provide, the more devices can be developed to aid scientific research and to improve modern lives.

The Process of Technological Design

The process of technology involves many choices. For example, how does an automobile engineer design a better car? Is a better car faster? safer? cheaper? Before designing any new machine, the engineer must decide exactly what he or she wants the machine to do as well as what may be given up for the machine to do it. A faster car may get people to their destinations more quickly, but it may cost more and be less safe. As you study the technological process, think about all the choices that were made to build the technologies you use.

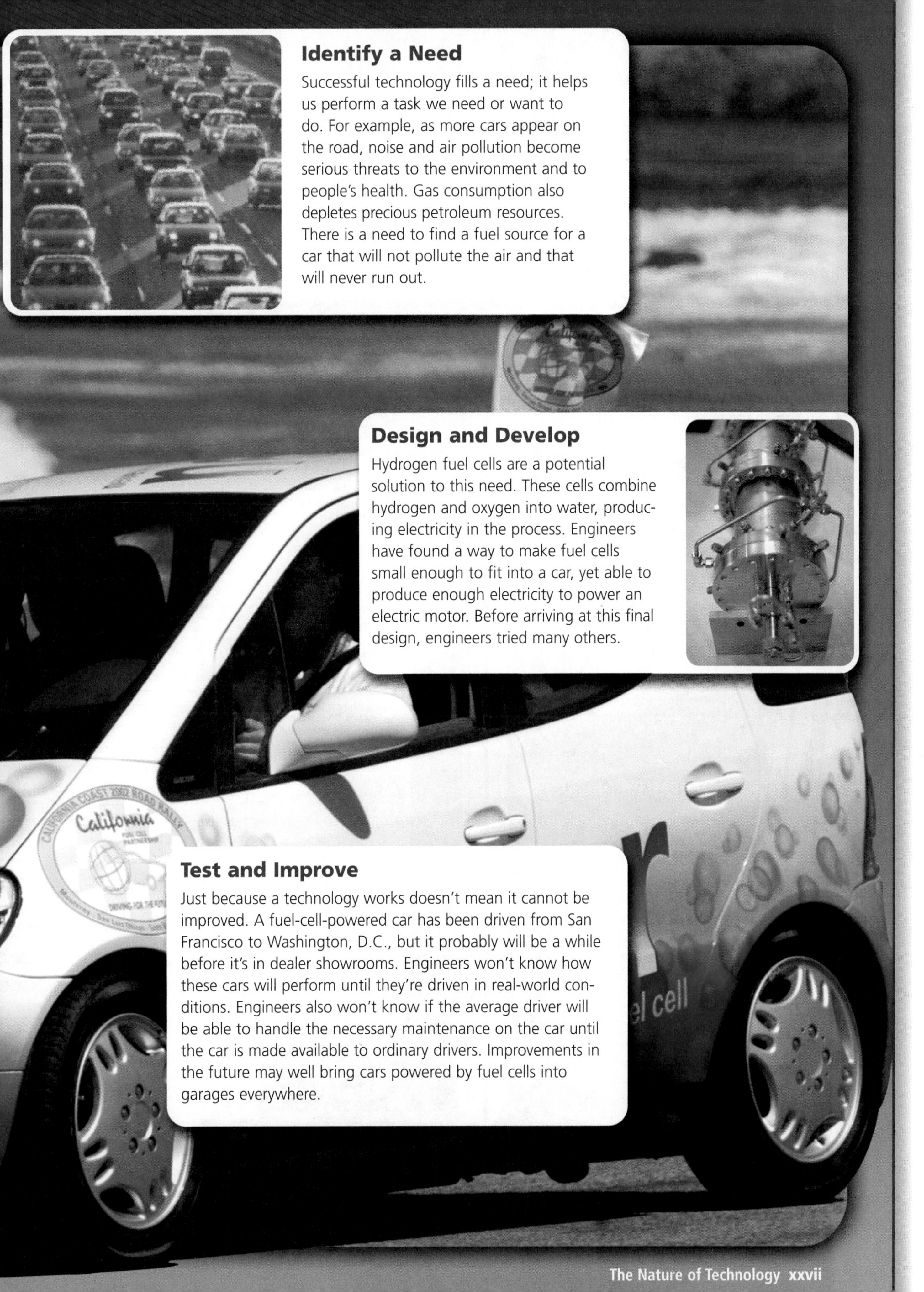

Identify a Need

Successful technology fills a need; it helps us perform a task we need or want to do. For example, as more cars appear on the road, noise and air pollution become serious threats to the environment and to people's health. Gas consumption also depletes precious petroleum resources. There is a need to find a fuel source for a car that will not pollute the air and that will never run out.

Design and Develop

Hydrogen fuel cells are a potential solution to this need. These cells combine hydrogen and oxygen into water, producing electricity in the process. Engineers have found a way to make fuel cells small enough to fit into a car, yet able to produce enough electricity to power an electric motor. Before arriving at this final design, engineers tried many others.

Test and Improve

Just because a technology works doesn't mean it cannot be improved. A fuel-cell-powered car has been driven from San Francisco to Washington, D.C., but it probably will be a while before it's in dealer showrooms. Engineers won't know how these cars will perform until they're driven in real-world conditions. Engineers also won't know if the average driver will be able to handle the necessary maintenance on the car until the car is made available to ordinary drivers. Improvements in the future may well bring cars powered by fuel cells into garages everywhere.

Using McDougal Littell Science

Reading Text and Visuals

This book is organized to help you learn. Use these boxed pointers as a path to help you learn and remember the **Big Ideas** and **Key Concepts.**

Take notes.

Use the strategies on the **Getting Ready to Learn** page.

Read the Big Idea.

As you read **Key Concepts** for the chapter, relate them to **the Big Idea.**

CHAPTER 2 Proper… Matte…

the BIG idea

Matter has properties that can be changed by physical and chemical processes.

Key Concepts

SECTION 2.1 Matter has observable properties. Learn how to recognize physical and chemical properties.

SECTION 2.2 Changes of state are physical changes. Learn how energy is related to changes of state.

SECTION 2.3 Properties are used to identify substances. Learn how the properties of substances can be used to identify them and to separate mixtures.

Internet Preview
CLASSZONE.COM
Chapter 2 online resources: Content Review, Simulation, three Resource Centers, Math Tutorial, Test Practice

A 38 Unit: Matter and Energy

CHAPTER 2
Getting Ready to Learn

CONCEPT REVIEW
- Everything is made of matter.
- Matter has mass and volume.
- Atoms combine to form molecules.

VOCABULARY REVIEW
mass p. 10
volume p. 11
molecule p. 18
states of matter p. 27

CONTENT REVIEW
CLASSZONE.COM
Review concepts and vocabulary.

TAKING NOTES

MAIN IDEA WEB

Write each new blue heading in a box. Then write notes in boxes around the center box that give important terms and details about that heading.

VOCABULARY STRATEGY

Think about a vocabulary term as a **magnet word** diagram. Write related terms and ideas in boxes around it.

See the Note-Taking Handbook on pages R45–R51.

SCIENCE NOTEBOOK

color, shape, size, texture, volume, mass

melting poin…

Physical properties describe a substa…

density: a measure of the amount of matter in a given volume

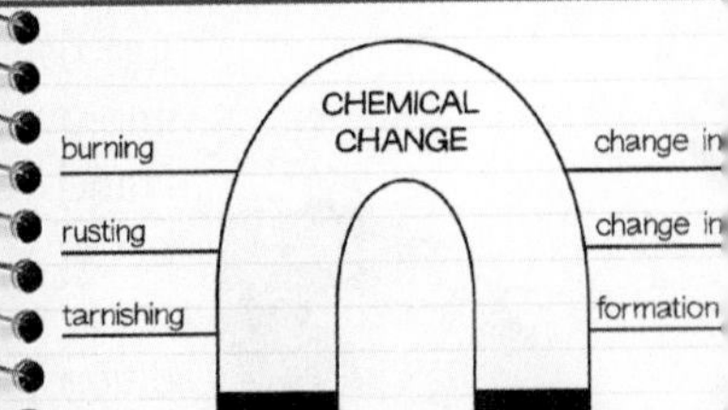

A 40 Unit: Matter and Energy

xxviii

Read each heading.

See how it fits into the outline of the chapter.

KEY CONCEPT

2.1 Matter has observable properties.

BEFORE, you learned

- Matter has mass and volume
- Matter is made of atoms
- Matter exists in different states

NOW, you will learn

- About physical and chemical properties
- About physical changes
- About chemical changes

VOCABULARY

EXPLORE Physical Properties

How can a substance be changed?

PROCEDURE

1. Observe the clay. Note its physical characteristics, such as color, shape, texture, and size.
2. Change the shape of the clay. Note which characteristics changed and which ones stayed the same.

MATERIAL

rectangular piece of clay

WHAT DO YOU THINK?

- How did reshaping the clay change its physical characteristics?
- How were the mass and the volume of the clay affected?

Physical properties describe a substance.

What words would you use to describe a table? a chair? the sandwich you ate for lunch? You would probably say something about the shape, color, and size of each item. Next you might consider whether it is hard or soft, smooth or rough to the touch. Normally, when describing an object, you identify the characteristics of the object that you can observe without changing the identity of the object.

The characteristics of a substance that can be observed without changing the identity of the substance are called **physical properties.** In science, observation can include measuring and handling a substance. All of your senses can be used to detect physical properties. Color, shape, size, texture, volume, and mass are a few of the physical properties you probably have encountered.

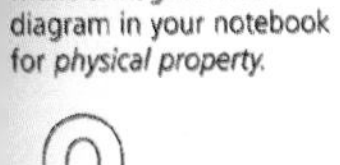

VOCABULARY
Make a magnet word diagram in your notebook for *physical property.*

CHECK YOUR READING Describe some of the physical properties of your desk.

Chapter 2: **Properties of Matter 41** A

Remember what you know.

Think about concepts you learned earlier and preview what you'll learn now.

Try the activities.

They will introduce you to science concepts.

Learn the vocabulary.

Take notes on each term.

Answer the questions.

Check Your Reading questions will help you remember what you read.

Reading Text and Visuals

Read one paragraph at a time.

Look for a topic sentence that explains the main idea of the paragraph. Figure out how the details relate to that idea. One paragraph might have several important ideas; you may have to reread to understand.

Answer the questions.

Check Your Reading questions will help you remember what you read.

Study the visuals.

- Read the title.
- Read all labels and captions.
- Figure out what the picture is showing. Notice colors, arrows, and lines.
- Answer the question. **Reading Visuals** questions will help you understand the picture.

Physical Properties

How do you know which characteristics are physical properties? Just ask yourself whether observing the property involves changing the substance to a different substance. For example, you can stretch a rubber band. Does stretching the rubber band change what it is made of? No. The rubber band is still a rubber band before and after it is stretched. It may look a little different, but it is still a rubber band.

Mass and volume are two physical properties. Measuring these properties does not change the identity of a substance. For example, a lump of clay might have a mass of 200 grams (g) and a volume of 100 cubic centimeters (cm^3). If you were to break the clay in half, you would have two 100 g pieces of clay, each with a volume of 50 cm^3. You can bend and shape the clay too. Even if you were to mold a realistic model of a car out of the clay, it still would be a piece of clay. Although you have changed some of the properties of the object, such as its shape and volume, you have not changed the fact that the substance you are observing is clay.

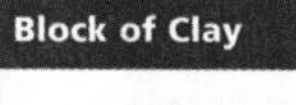

Because all formulas for volume involve the multiplication of three measurements, volume has a unit that is cubed (such as cm^3).

CHECK YOUR READING Which physical properties listed above are found by taking measurements? Which are not?

Physical Properties

Physical properties of clay—such as volume, mass, color, texture, and shape—can be observed without changing the fact that the substance is clay.

Block of Clay

Shaped Clay

READING VISUALS COMPARE AND CONTRAST Which physical properties do the two pieces of clay have in common? Which are different?

42 Unit: **Matter and Energy**

Doing Labs

To understand science, you have to see it in action. Doing labs helps you understand how things really work.

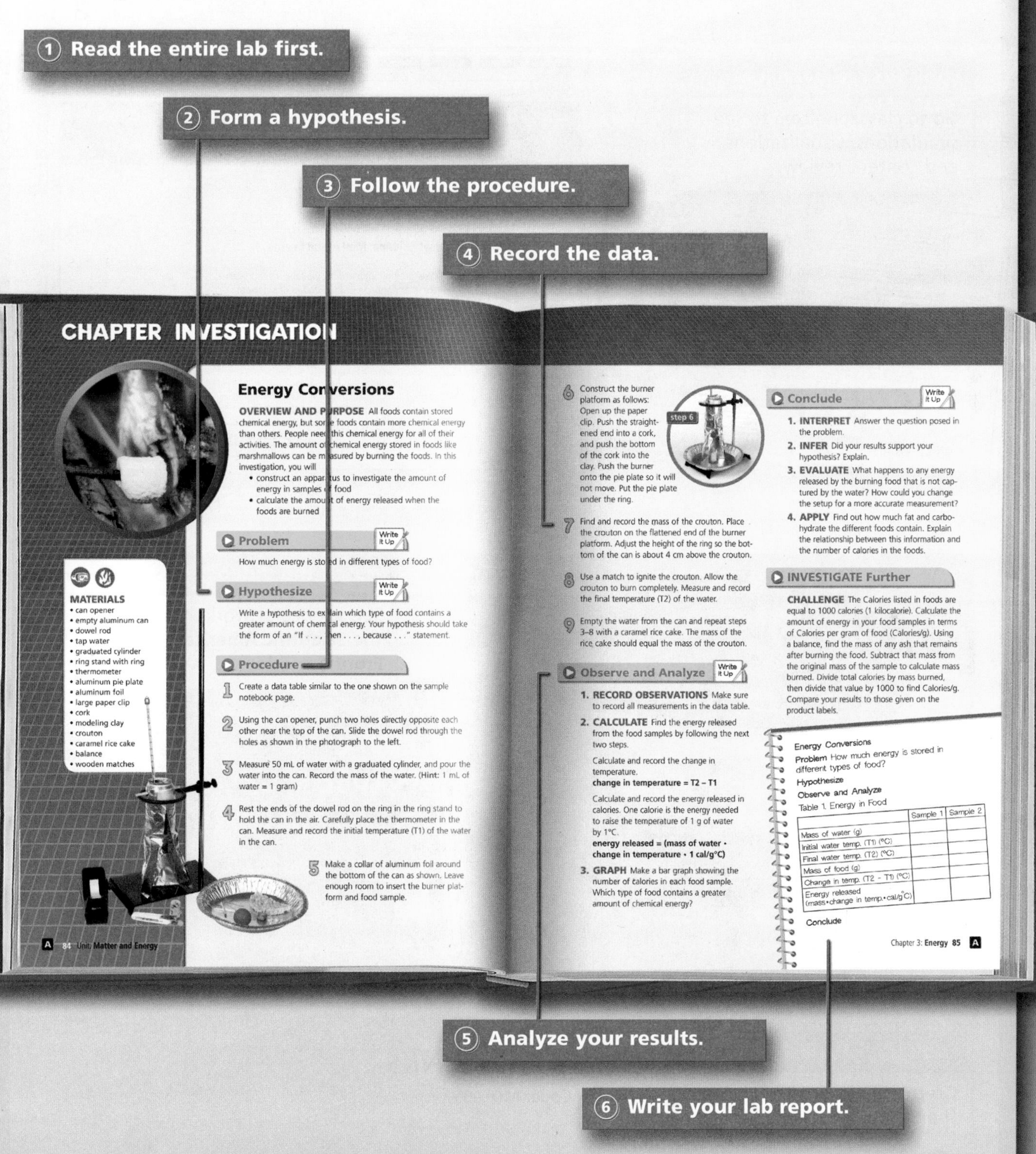

Using Technology

The Internet is a great source of information about up-to-date science. The ClassZone Web site and SciLinks have exciting sites for you to explore. Video clips and simulations can make science come alive.

Look for red banners.

Go to **classzone.com** to see simulations, visualizations, and content review.

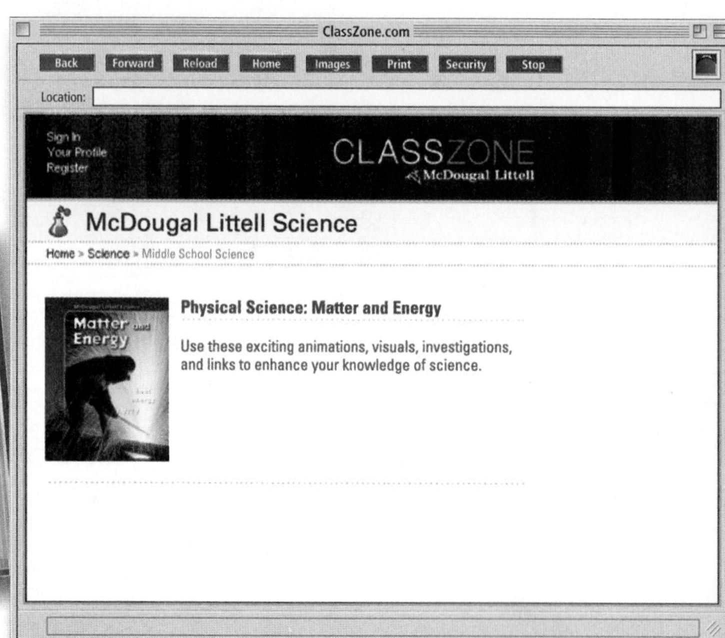

Watch the videos.

See science at work in the **Scientific American Frontiers video.**

Look up SciLinks.

Go to **scilinks.org** to explore the topic.

Forces **Code: MDL005**

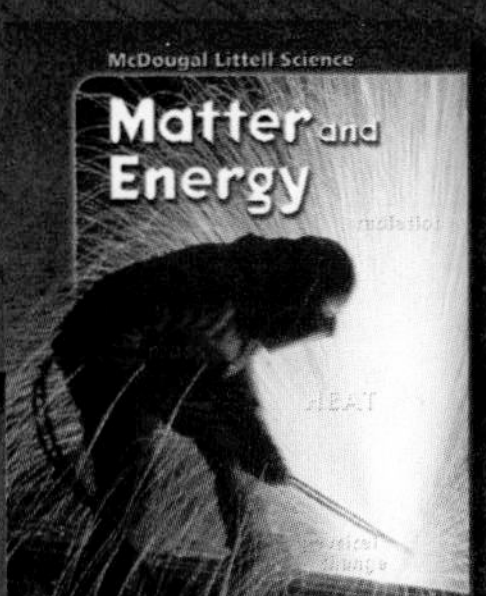

Matter and Energy Contents Overview

Unit Features

FRONTIERS in Science

VIDEO SUMMARY

SCIENTIFIC AMERICAN FRONTIERS

"Sunrayce," a segment of the *Scientific American Frontiers* series that aired on PBS stations, follows a solar-powered car race from Texas to Minnesota. Contestants are teams of student engineers and scientists from 34 colleges and universities. Each team has designed and built a car with an electric engine that is fueled by direct solar power and/or batteries that can be charged by solar energy. The cars are covered with solar panels to collect sunlight.

Teams have just seven days to complete the 1102-mile race. This can be especially challenging when sunlight is the only source of power and rain clouds can spell disaster. The video follows the team from California State University, Los Angeles, as they prepare for the race and strategize to use the maximum amount of sunlight.

National Science Education Standards

A.9.a–d Understandings About Scientific Inquiry

E.6.a–f Understandings About Science and Technology

F.5.a–e Science and Technology in Society

G.1.a–b Science as a Human Endeavor

G.2.a Nature of Science

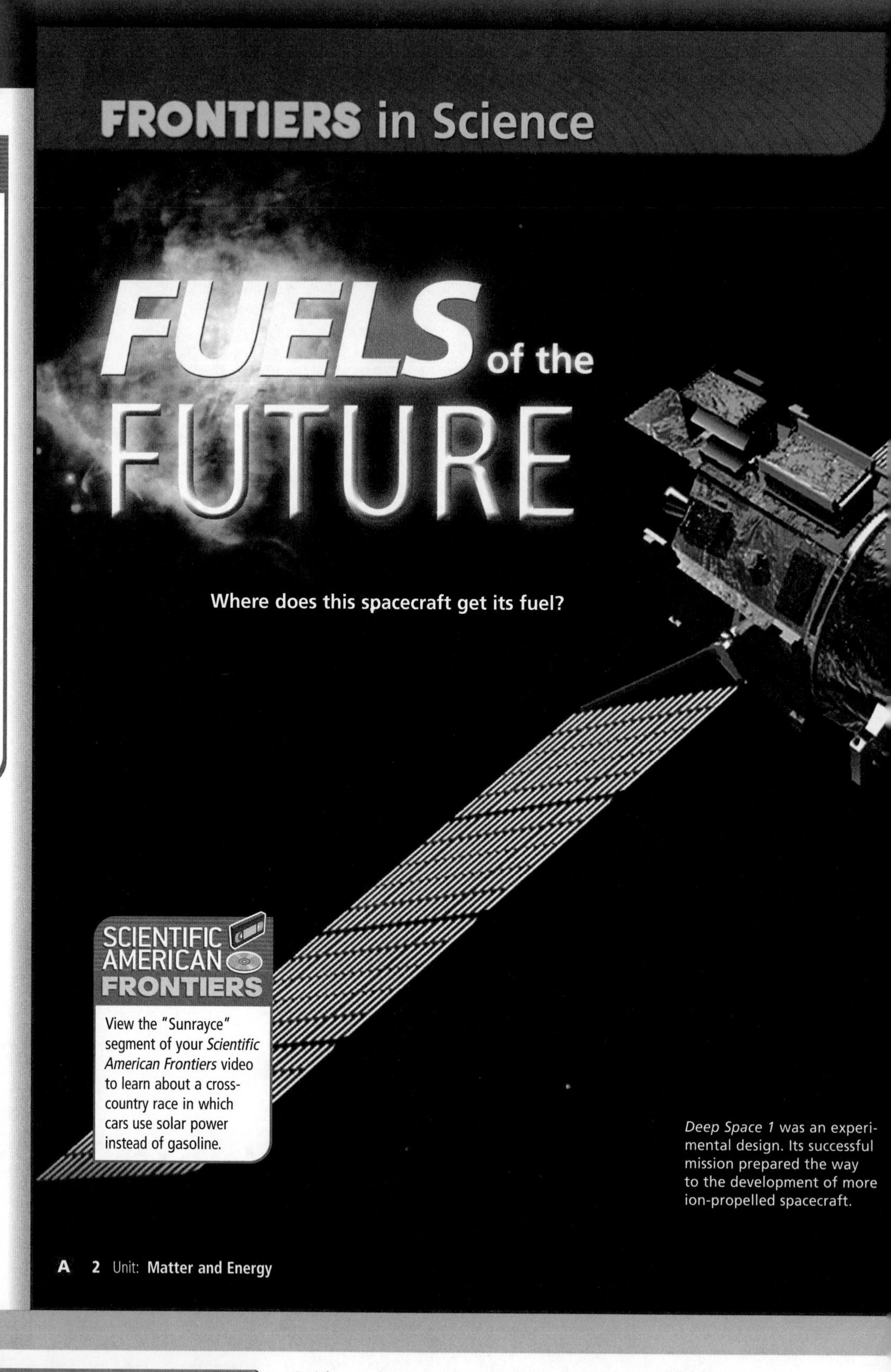

FRONTIERS in Science
FUELS of the FUTURE
Where does this spacecraft get its fuel?
SCIENTIFIC AMERICAN FRONTIERS
View the "Sunrayce" segment of your *Scientific American Frontiers* video to learn about a cross-country race in which cars use solar power instead of gasoline.
A 2 Unit: Matter and Energy

Deep Space 1 was an experimental design. Its successful mission prepared the way to the development of more ion-propelled spacecraft.

ADDITIONAL RESOURCES

Technology Resources

Scientific American Frontiers Video: *Sunrayce:* 12-minute video segment that introduces the unit.

ClassZone.com
CAREER LINK, Physicist, Chemist, Aeronautical Engineer

Guide student viewing and comprehension of the video:

 Video Teaching Guide, pp. 1–2; Video Viewing Guide, p. 3; Video Wrap-Up, p. 4

Scientific American Frontiers Video Guide, pp. 39–42

Unit projects procedures and rubrics:

 Unit Projects, pp. 5–10

The stream of ions glows blue as it is shot out of an ion propulsion engine.

Ion Engines for Long Voyages

Rocket engines must provide huge amounts of energy to move spacecraft away from Earth and keep them orbit. The fuel required can weigh more than the spacecraft themselves. That is why scientists and engineers are always looking for more efficient ways to give spacecraft and other vehicles the energy to move.

One method of powering spacecraft uses electrically charged particles called ions. The atoms of a gas—usually xenon—are first made into ions. An electric field is then used to pull these ions out of the engine at a very high speed—faster than 100,000 kilometers per hour (62,000 mi/h). This stream of rapidly moving ions works like the gases coming out of a jet engine on a plane—propelling the spacecraft in the direction opposite to the ion stream.

An advantage of ion propulsion is that its fuel is much lighter than the chemical fuel used in rockets. Ion propulsion does not provide enough thrust to be used for a rocket launch, but it can be used to move a spacecraft through long distances in outer space. This method of propulsion provides a small force to the spacecraft; however, over time the spacecraft can reach great speeds.

The space probe *Deep Space 1* was the first to use an ion engine to travel between planets. The engine generated enough speed for the probe to follow and photograph comet Borrelly in 2001.

DIFFERENTIATE INSTRUCTION

More Reading Support

A What provides spacecraft with the large amounts of energy they need to move? *rocket engines*

B What is ion propulsion best used for? *long distances*

Below Level Have students write out the steps to power a spacecraft using ions. Steps should include where ions come from, what is created with those ions, and how the ions work to push the spacecraft.

FOCUS

Set Learning Goals

Students will

- Examine how spacecraft are powered.
- Explore alternative types of fuels.
- Determine how solar energy works.

Remind students that frontiers are undeveloped fields for discovery or research, and that the "Sunrayce" video shows engineers and scientists testing designs to find answers or solve problems. Have students look at the photographs of the spacecraft and engine and predict what problem the engineers and scientists are trying to solve.

INSTRUCT

Scientific Process

Emphasize to students that in order to move spacecraft out into orbit, large amounts of energy are required. Because so much fuel is needed to power a spacecraft, the fuel itself can weigh more than the spacecraft. That is why it is necessary for scientists to learn about the various types of energy sources for fueling spacecraft. Have students review the parts of the scientific process on pages xxiv–xxv and ask what part this step falls under. *determining what is known*

Technology Design

Focus students' attention on "Ion Engines for Long Voyages." Discuss how using ions—electrically charged particles—power the spacecraft by creating an ion stream that propels the spacecraft while in orbit. Ask students to name the advantage as well as the disadvantage of using this engine design. *The fuel for ion engines is lighter than chemical fuel used for rockets, but ion propulsion does not provide enough of a force for launching the rocket.*

Teach from Visuals

Have students look at the photograph of the solar sails. Ask:

- What is one problem with using solar energy for powering spacecraft? *Once the spacecraft moves far from the Sun, it does not collect as much sunlight and loses energy.*
- For what purpose might scientists use solar sails? *Sample answer: Solar sails are used to reflect sunlight to provide the solar energy to move spacecraft farther through space to the outer planets.*

Technology Design

Remind students about the problem to be solved: providing enough power for spacecraft to move far from Earth and the Sun. Ask: How did the scientists improve the recovery of solar energy? *using sails as mirrors to reflect sunlight* How did they further supplement the energy source? *by beaming energy from Earth to the solar sails through microwaves and laser lights*

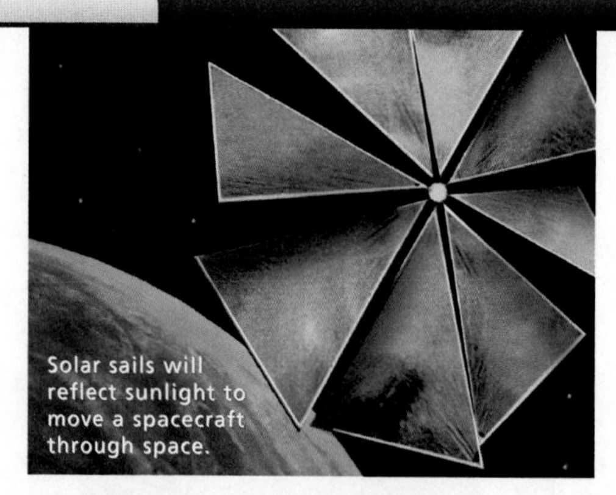

Solar sails will reflect sunlight to move a spacecraft through space.

Running on Sunlight

C

Solar energy is used for travel in outer space, where there is plenty of sunlight and very little friction to slow down a spacecraft. However, once a spacecraft travels far away from the Sun—as far as the outer planets Jupiter and Saturn—the amount of energy reaching it is far less than the energy it was getting near Earth. The sunlight can be helpful only if solar cells on the vehicle can collect enough of it. One solution is to reflect sunlight. Scientists are developing solar sails, which will act like enormous mirrors. The pressure of reflected sunlight on the sails can be used to move a large ship through space—even far from the Sun.

Beaming Energy from Earth

Another way to power a spacecraft is to send energy to it all the way from Earth. This idea is called beamed energy propulsion. A beam delivers energy to solar sails on the spacecraft. The energy can be in the form of microwaves—the same energy that heats food in a microwave oven or delivers calls on a cell phone. Or it can be in the form of laser light, a very concentrated beam of visible light. This method has already been used successfully to power very small vehicles, 10 centimeters (4 in.) long. Experiments are under way with larger spacecraft.

Combined Technologies

Some recent space flights have combined common and experimental technologies. For example, the *Cassini* space probe has two regular rocket engines for propulsion. Other energy comes from three generators powered by radioactive decay. This combination of engines allowed *Cassini* to be the largest and most complicated spacecraft ever launched. Its goal is to explore Saturn.

View the "Sunrayce" segment of your *Scientific American Frontiers* video to see what is involved in solar-car racing.

IN THIS SCENE FROM THE VIDEO Students from California State University, Los Angeles, work on their solar car.

CATCHING THE SUN'S RAYS Since 1990 teams of college students have built and raced solar-powered cars. The races are held every two years to promote awareness of solar energy and to inspire young people to work in science and engineering.

Solar cells on the cars' bodies convert sunlight into electricity. The goal is to make lightweight cars that convert sunlight efficiently. Today's solar cars can reach speeds of up to 75 miles per hour, but the average racing speed is 25 miles per hour. On cloudy or rainy days, the teams conserve power by traveling more slowly—or risk running down their batteries.

In 2003 the American Solar Challenge took place on historic Route 66 from Chicago to Claremont, California. At 3700 kilometers (2300 mi), the ten-day event was the longest solar-car race in the world.

DIFFERENTIATE INSTRUCTION

More Reading Support

C What type of energy uses sunlight for fuel? *solar energy*

D What are two sources of energy beams? *microwaves and laser light*

English Learners English learners may be unfamiliar with some of the vocabulary used to describe how spacecraft and other vehicles are fueled. The examples below use different words to indicate the same idea.

". . . collect enough sunlight to drive a large ship . . ."

"Another way to power a spacecraft . . ."

If read literally, "to drive" and "to power" might be confusing. Help students learn the multiple meanings of words such as these.

Alternative Fuels on Earth

Scientists and inventors have long been looking for practical alternative fuels to power vehicles on Earth as well as in outer space. Most vehicle engines on Earth use gasoline or other fossil fuels. These fuels are based on resources, such as petroleum, that are found in underground deposits. Those deposits will not be replaced for millions of years. Solar energy, by contrast, is endlessly renewable, so it seems to be a good alternative to nonrenewable fossil fuels.

Solar-powered cars rely on solar cells, which convert the energy of sunlight directly into electrical energy that can be stored in batteries. One outstanding solar car was built by Dutch students and entered in the 2001 World Solar Challenge.

The students' car, called the *Nuna,* used several technologies that had been developed for space travel. Its body was reinforced with Kevlar, a space-age material that is also used in satellites, space suits, and bulletproof vests. During the race, the *Nuna* covered 3010 kilometers of desert in Australia, breaking solar-car speed records, and won the race.

Does the development of solar cars like the *Nuna* mean that most people will be driving solar cars soon? Unfortunately, such cars run only when the Sun is shining unless they rely on batteries—and it takes hundreds of pounds of batteries to store the amount of energy in a gallon of gasoline. As with spacecraft, the goal is to design a vehicle in which the fuel doesn't outweigh the vehicle itself.

UNANSWERED Questions

Even as scientists and inventors solve problems in solar technology, new questions arise.

- Can solar technology be made affordable?
- Is solar technology practical for large-scale public transportation?
- Are there any hidden costs to the use of alternative fuels?

UNIT PROJECTS

As you study this unit, work alone or with a group on one of these projects.

Build a Solar Oven

Design and build a solar oven that can boil a quarter cup of water.

- Plan and sketch a design for a solar oven that can reach 100°C.
- Collect materials and assemble your oven. Then conduct trials and improve your design.

Multimedia Presentation

Create an informative program on solar race cars and the way they work.

- Collect information about solar race cars. Research how they are powered.
- Examine why solar cars have specific shapes. Learn how the solar panels and batteries work together.
- Give a multimedia presentation describing what you learned.

Design an Experiment

Design an experiment that compares how well two of the following alternative energy sources move an object: solar energy, wind power, biomass (fuel from plant material,) waste-material fuel, hydrogen fuel cells, heat exchangers.

- Research the energy sources, and pick two types to compare.
- List materials for your experiment. Create a data table and write up your procedure.
- Describe your experiment for the class.

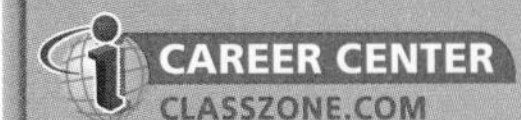

Learn more about careers in electrical engineering.

DIFFERENTIATE INSTRUCTION

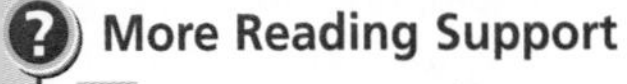

More Reading Support

E What is the difference between fossil fuels and solar energy? *Solar energy is renewable while fossil fuels take millions of years to be replaced.*

Differentiate Unit Projects Projects are appropriate for varying abilities. Allow students to choose the ones that interest them the most. Encourage them to vary the products they produce throughout the year.

Below Level Encourage students to try "Build a Solar Oven."

Advanced Challenge students to complete "Design an Experiment."

UNANSWERED Questions

Have students read the questions and think of some of their own. Remind them that scientists usually end up with more questions—that inquiry is the driving force of science.

- With the class, generate on the board a list of new questions.
- Students can add to the list after they watch the Scientific American Frontiers Video.
- Students can use the list as a springboard for choosing their Unit Projects.

UNIT PROJECTS

Encourage students to pick the project that most appeals to them. Point out that each is long-term and will take several weeks to complete. You might group or pair students to work on projects and in some cases guide student choice. Some of the projects have student choice built into them.

Each project has two worksheet pages, including a rubric. Use the pages to guide students through criteria, process, and schedule.

Unit Projects, pp. 5–10

REVISIT concepts introduced in this article:

Chapter 1
- Matter has mass and volume, pp. 9–13
- Matter is made up of atoms, pp. 16–19
- Matter combines, pp. 21–25

Chapter 2
- Observable properties, pp. 41–48
- Changes of state, pp. 50–55
- Identifying substances, pp. 58–62

Chapter 3
- Energy exists in different forms, pp. 71–76
- Energy is never lost, pp. 78–83
- Technology improves energy use, pp. 86–90

Chapter 4
- Temperature depends on particle movement, pp. 103–108
- Energy flows from warmer to cooler objects, pp. 110–114
- The transfer of energy as heat, pp. 116–121

CHAPTER

Introduction to Matter

Physical Science UNIFYING PRINCIPLES

PRINCIPLE 1
Matter is made of particles too small to see.

PRINCIPLE 2
Matter changes form and moves from place to place.

PRINCIPLE 3
Energy changes from one form to another, but it cannot be created or destroyed.

PRINCIPLE 4
Physical forces affect the movement of all matter on Earth and throughout the universe.

Unit: Matter and Energy BIG IDEAS

CHAPTER 1
Introduction to Matter
Everything that has mass and takes up space is matter.

CHAPTER 2
Properties of Matter
Matter has properties that can be changed by physical and chemical processes.

CHAPTER 3
Energy
Energy has different forms, but it is always conserved.

CHAPTER 4
Temperature and Heat
Heat is a flow of energy due to temperature differences.

CHAPTER 1 KEY CONCEPTS

SECTION 1.1

Matter has mass and volume.

1. All objects are made of matter.
2. Mass is a measure of the amount of matter.
3. Volume is a measure of the space matter occupies.

SECTION 1.2

Matter is made of atoms.

1. Atoms are extremely small.
2. Atoms and molecules are always in motion.

SECTION 1.3

Matter combines to form different substances.

1. Matter can be pure or mixed.
2. Parts of mixtures can be the same or different throughout.

SECTION 1.4

Matter exists in different physical states.

1. Particle arrangement and motion determine the state of matter.
2. Solid, liquid, and gas are common states of matter.

The Big Idea Flow Chart is available on p. T1 in the **UNIT TRANSPARENCY BOOK.**

Previewing Content

SECTION

Matter has mass and volume. pp. 9–15

1. All objects are made of matter.
Anything that has **mass** and takes up space is **matter.** All the objects, liquids, gases, and living things in the universe are made of matter. Energy is not matter. However, under special circumstances, such as a nuclear reaction, energy can become matter and matter can become energy.

2. Mass is a measure of the amount of matter.
Mass is measured by comparing the mass of an object with the mass of known units or standard units of mass. Units of mass are the kilogram and gram. **Weight** is the downward pull of gravity on an object. An object's mass is invariable, but an object's weight varies, depending on the amount of gravity. Weight is measured in newtons.

3. Volume is a measure of the space matter occupies.
Units of **volume** are the cubic centimeter or milliliter. One milliliter is equal to one cubic centimeter. The amount of space that matter takes up can be measured in two ways.

- Use a formula to determine the volume of regular objects. For example, the volume of an object shaped like a square or a rectangular box is length times width times height.
  $V = lwh$
- Use displacement to measure the volume of solids with an irregular shape. Submerge the object in a known amount of water. The increase in the volume of the water is the volume of the object. This technique will not work if the solid can dissolve.

SECTION

Matter is made of atoms. pp. 16–20

1. Atoms are extremely small.
All matter is made up of **atoms,** which are so small that a teaspoon full of water has approximately 5×10^{23} atoms. An atom has a radius of approximately 10^{-10} meters. Scientists have identified more than 100 kinds of atoms.

Atoms combine to form **molecules.** A molecule can be made from two or more of the same kind of atom or from two or more different kinds of atoms. For example, water molecules are made up of hydrogen and oxygen, but ozone molecules are made up of only oxygen. The diagram below shows the makeup of a water molecule and an ozone molecule.

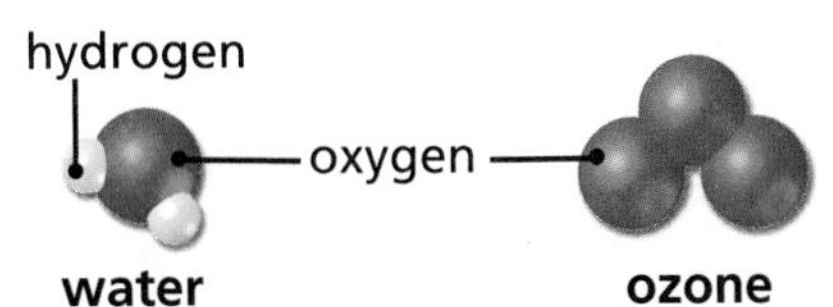

2. Atoms and molecules are always in motion.
Atoms and molecules are always moving. You can see evidence of moving air molecules as they collide with dust or other particles in the air. You can see liquids move when you add a drop of food coloring to water. Even the atoms and molecules in a solid constantly vibrate.

Common Misconceptions

LIQUIDS AND GASES ARE MATTER Students may think that liquids and gases are not matter. Liquids and gases are both matter because they have mass and take up space.

This misconception is addressed on p. 11.

MISCONCEPTION DATABASE
CLASSZONE.COM Background on student misconceptions

SUBSTANCES ARE MADE OF ATOMS AND MOLECULES Students may think that atoms or molecules are merely parts of substances and that there is something other than empty space between the molecules. Rather, substances are composed entirely of atoms and molecules. Molecules of a gas (air) can fill spaces between other molecules. Other space is just empty space.

This misconception is addressed on p. 18.

Previewing Content

SECTION

Matter combines to form different substances. pp. 21–26

1. Matter can be pure or mixed.
Matter that contains only one kind of atom or molecule is pure. Matter often contains two or more substances mixed together. Substances can be composed of elements, compounds, or mixtures.

- An **element** is a substance that contains only one kind of atom. Gold is the element represented in the diagram on the left below.

Element: Gold

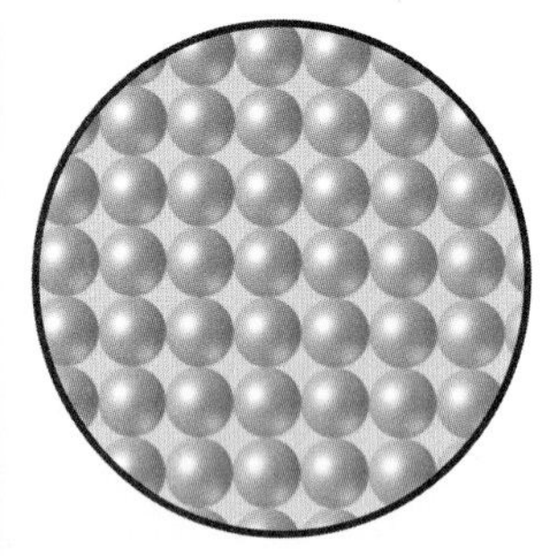

Compound: Dry Ice

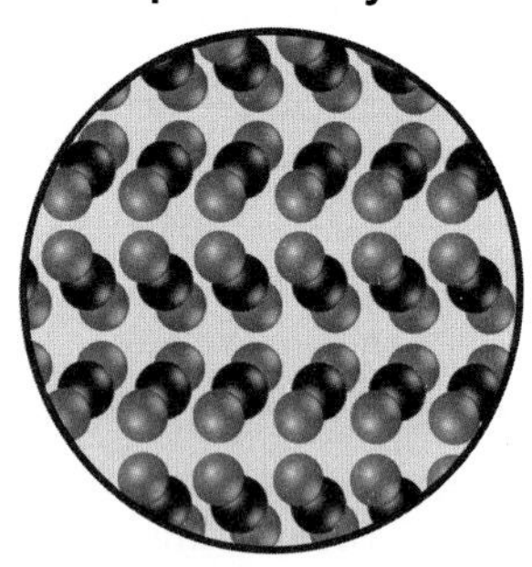

- A **compound** is a substance that consists of two or more different types of atoms bonded together as shown in the diagram on the right above. Water molecules are compounds because they contain two kinds of atom bonded covalently. A molecule of oxygen is not a compound. Some compounds, such as table salt, are bonded ionically.
- A **mixture** is a combination of different substances that retain their individual properties and can be separated by physical means.

2. Parts of mixtures can be the same or different throughout.
Mixtures can be either heterogeneous or homogeneous.

- A heterogeneous mixture has different properties in different parts of the mixture because the substances in different parts of the mixture vary.
- A homogeneous mixture has substances evenly spread out throughout the mixture.

SECTION

Matter exists in different physical states. pp. 27–33

1. Particle arrangement and motion determine the state of matter.
Solid, liquid, and gas are three common **states of matter.** When a substance changes from one state to another, the arrangement of its molecules changes. The distance between molecules and the attraction they have for one another change.

2. Solid, liquid, and gas are common states of matter.
The state of a substance depends on the space between its particles and the way in which the particles move.

- A **solid** has particles that are close together. The particles are attached to one another and can vibrate in place, but they cannot move from place to place.
- A **liquid** has particles that are attracted to one another and are close together. The particles can slide over one another and move from one place to another.
- A **gas** has particles that are not close to one another and can move about freely.

The diagrams below show the arrangement and motion of particles in different states of matter.

1 **Solid**

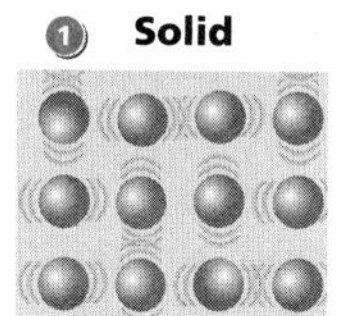

2 **Liquid**

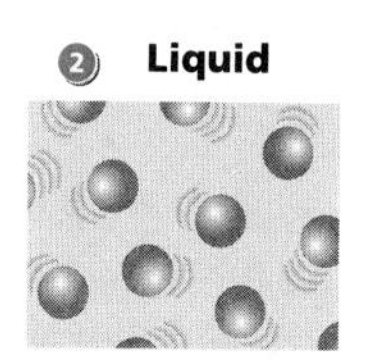

3 **Gas**

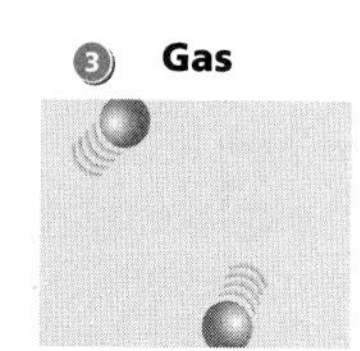

3. Solids have a definite volume and shape.
A solid has a fixed volume and shape. The particles in some solids are in regular patterns and form crystals.

4. Liquids have a definite volume but no definite shape.
A liquid has a definite volume because its particles are close enough together that they cannot move about freely, although they slide past each other. A liquid takes the shape of the container that it is in.

5. Gases have no definite volume or shape.
A gas has no definite volume or shape. The volume, pressure, and temperature of a gas are related to one another, and changing one can change the others.

Common Misconceptions

MISCONCEPTION DATABASE
CLASSZONE.COM Background on student misconceptions

ATOMS AND COLOR Students often think that individual atoms and molecules have the same properties as the substance they make up. For example, students might think a gold atom is hard and solid or a gas molecule is transparent.

This misconception is addressed in the Teacher Demo on p. 22.

Previewing Labs

EXPLORE the BIG idea	
What Has Changed? p. 7 Students observe a balloon and realize that air has mass and volume.	**TIME** 10 minutes **MATERIALS** balloon
Where Does the Sugar Go? p. 7 Students dissolve sugar in a glass of water to explore mixtures.	**TIME** 10 minutes **MATERIALS** glass of water, spoonful of sugar, spoon
Internet Activity: Scale, p. 7 Students explore atoms by looking at closer and closer views of an object.	**TIME** 20 minutes **MATERIALS** computer with Internet access

SECTION 1.1	**EXPLORE Similar Objects,** p. 9 Students observe the properties of mass and volume as they compare two balls.	**TIME** 10 minutes **MATERIALS** 2 balls of different sizes and masses
	CHAPTER INVESTIGATION Mass and Volume, pp. 14–15 Students practice measuring the mass and volume of objects.	**TIME** 40 minutes **MATERIALS** small rock that fits in a graduated cylinder, 5 pennies, rectangular sponge, tissue box, beam balance, large graduated cylinder, 50 mL water, ruler
SECTION 1.2	**INVESTIGATE Mass,** p. 17 Students model measuring the mass of an atom in order to draw conclusions about things they cannot observe directly.	**TIME** 20 minutes **MATERIALS** beam balance, beaker, 10 pennies
SECTION 1.3	**EXPLORE Mixed Substances,** p. 21 Students observe cornstarch and water to see how properties of individual substances compare with properties of mixed substances.	**TIME** 10 minutes **MATERIALS** teaspoon of cornstarch, teaspoon of water, clear plastic cup, spoon
	INVESTIGATE Mixtures, p. 24 Students observe that not all liquids behave the same way when mixed with other liquids.	**TIME** 20 minutes **MATERIALS** few drops of food coloring, beaker of water, clear jar with screw-on lid, 1/4 jar of vegetable oil
SECTION 1.4	**EXPLORE Solids and Liquids,** p. 27 Students compare solids and liquids by observing an ice cube, a marble, and water.	**TIME** 10 minutes **MATERIALS** water in a clear cup, ice cube, marble, pie tin
	INVESTIGATE Liquids, p. 31 Students measure and make inferences about the behavior of different liquids.	**TIME** 20 minutes **MATERIALS** graduated cylinder, 15 mL colored water, test tube, test-tube rack, 10 mL vegetable oil, 10 mL corn syrup

R **Additional INVESTIGATION,** Thick and Thin Liquids, A, B, & C, pp. 72–80; Teacher Instructions, pp. 262–263

Previewing Chapter Resources

CHAPTER 1
Introduction to Matter

INTEGRATED TECHNOLOGY

CLASSZONE.COM
- eEdition Plus
- EasyPlanner Plus
- Misconception Database
- Content Review
- Test Practice
- Simulations
- Resource Centers
- Internet Activity: Scale
- Math Tutorial

SCILINKS.ORG

CD-ROMS
- eEdition
- EasyPlanner
- Power Presentations
- Content Review
- Lab Generator
- Test Generator

AUDIO CDS
- Audio Readings
- Audio Readings in Spanish

LABS AND ACTIVITIES

EXPLORE the Big Idea, p. 7
- What Has Changed?
- Where Does the Sugar Go?
- Internet Activity: Scale

UNIT RESOURCE BOOK
- Family Letter, p. vii
- Spanish Family Letter, p. viii
- Unit Projects, pp. 5–10

Lab Generator CD-ROM
Generate customized labs.

SECTION

1.1 Matter has mass and volume.
pp. 9–15

Time: 3 periods (1.5 blocks)

Lesson Plan, pp. 11–12

Integrated Technology

- **SIMULATION,** Weights on Planets
- **RESOURCE CENTER,** Volume

UNIT TRANSPARENCY BOOK
- Big Idea Flow Chart, p. T1
- Daily Vocabulary Scaffolding, p. T2
- Note-Taking Model, p. T3
- 3-Minute Warm-Up, p. T4

Labs and Activities

- EXPLORE Similar Objects, p. 9
- CHAPTER INVESTIGATION, Mass and Volume, pp. 14–15

UNIT RESOURCE BOOK
- Math Support & Practice, pp. 59–60
- CHAPTER INVESTIGATION, Mass and Volume, pp. 63–71

SECTION

1.2 Matter is made of atoms.
pp. 16–20

Time: 2 periods (1 block)

Lesson Plan, pp. 21–22

Integrated Technology

- **RESOURCE CENTER,** Scanning Tunneling Microscope Images

UNIT TRANSPARENCY BOOK
- Daily Vocabulary Scaffolding, p. T2
- 3-Minute Warm-Up, p. T4

Labs and Activities

- INVESTIGATE Mass, p. 17
- Extreme Science, p. 20

UNIT RESOURCE BOOK
Datasheet, Mass, p. 30

SECTION

1.3 Matter combines to form different substances.
pp. 21–26

Time: 2 periods (1 block)

R Lesson Plan, pp. 32–33

Integrated Technology

- **RESOURCE CENTER,** Mixtures
- **MATH TUTORIAL**

UNIT TRANSPARENCY BOOK
- Daily Vocabulary Scaffolding, p. T2
- 3-Minute Warm-Up, p. T5

Labs and Activities

- EXPLORE Mixed Substances, p. 21
- INVESTIGATE Mixtures, p. 24
- Math in Science, p. 26

UNIT RESOURCE BOOK
- Datasheet, Mixtures, p. 41
- Math Support, p. 61
- Math Practice, p. 62

SECTION

1.4 Matter exists in different physical states.
pp. 27–33

Time: 3 periods (1.5 blocks)

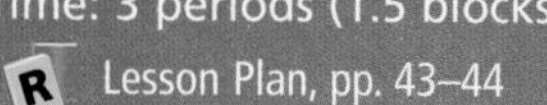
Lesson Plan, pp. 43–44

Integrated Technology

SIMULATION, Gas Behavior

UNIT TRANSPARENCY BOOK
- Big Idea Flow Chart, p. T1
- Daily Vocabulary Scaffolding, p. T2
- 3-Minute Warm-Up, p. T5
- "States of Matter" Visual, p. T6
- Chapter Outline, pp. T7–T8

Labs and Activities

- EXPLORE Solids and Liquids, p. 27
- INVESTIGATE Liquids, p. 31

UNIT RESOURCE BOOK
- Datasheet, Liquids, p. 52
- Additional INVESTIGATION, Thick and Thin Liquids, A, B, & C, pp. 72–80

 KEY TO ICONS

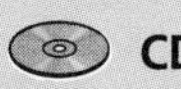 CD/CD-ROM

INTERNET

 Pupil Edition

 Teacher Edition

 UNIT RESOURCE BOOK

 UNIT TRANSPARENCY BOOK

 UNIT ASSESSMENT BOOK

 SPANISH ASSESSMENT BOOK

 SCIENCE TOOLKIT

READING AND REINFORCEMENT	ASSESSMENT	STANDARDS
 • Four Square, B22–23 • Main Idea and Detail Notes, C37 • Daily Vocabulary Scaffolding, H1–8  **UNIT RESOURCE BOOK** • Vocabulary Practice, pp. 56–57 • Decoding Support, p. 58 • Summarizing the Chapter, pp. 81–82 **Audio Readings CD** Listen to Pupil Edition.  **Audio Readings in Spanish CD** Listen to Pupil Edition in Spanish.	 • Chapter Review, pp. 35–36 • Standardized Test Practice, p. 37  **UNIT ASSESSMENT BOOK** • Diagnostic Test, pp. 1–2 • Chapter Test, A, B, & C, pp. 7–18 • Alternative Assessment, pp. 19–20  Spanish Chapter Test, pp. 213–216 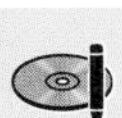  **Test Generator CD-ROM** Generate customized tests.  **Lab Generator CD-ROM** Rubrics for Labs	National Standards A.2–8, A.9.a–c, A.9.e–f, B.1.c, E.6.c, F.5.c See p. 6 for the standards.
UNIT RESOURCE BOOK • Reading Study Guide, A & B, pp. 13–16 • Spanish Reading Study Guide, pp. 17–18 • Challenge and Extension, p. 19 • Reinforcing Key Concepts, p. 20	 Ongoing Assessment, pp. 9–13  Section 1.1 Review, p. 13  **UNIT ASSESSMENT BOOK** Section 1.1 Quiz, p. 3	National Standards A.2–8, A.9.a–c, A.9.e–f
 **UNIT RESOURCE BOOK** • Reading Study Guide, A & B, pp. 23–26 • Spanish Reading Study Guide, pp. 27–28 • Challenge and Extension, p. 29 • Reinforcing Key Concepts, p. 31	 Ongoing Assessment, pp. 16–19  Section 1.2 Review, p. 19  **UNIT ASSESSMENT BOOK** Section 1.2 Quiz, p. 4	National Standards A.2–7, A.9.a–b, A.9.e–f, E.6.c, F.5.c
 **UNIT RESOURCE BOOK** • Reading Study Guide, A & B, pp. 34–37 • Spanish Reading Study Guide, pp. 38–39 • Challenge and Extension, p. 40 • Reinforcing Key Concepts, p. 42	 Ongoing Assessment, pp. 22–24  Section 1.3 Review, p. 25  **UNIT ASSESSMENT BOOK** Section 1.3 Quiz, p. 5	National Standards A.2–8, A.9.a–c, A.9.e–f
 **UNIT RESOURCE BOOK** • Reading Study Guide, A & B, pp. 45–48 • Spanish Reading Study Guide, pp. 49–50 • Challenge and Extension, p. 51 • Reinforcing Key Concepts, p. 53 • Challenge Reading, pp. 54–55	 Ongoing Assessment, pp. 28, 30–32  Section 1.4 Review, p. 33  **UNIT ASSESSMENT BOOK** Section 1.4 Quiz, p. 6	National Standards A.2–7, A.9.a–b, A.9.e–f

Previewing Resources for Differentiated Instruction

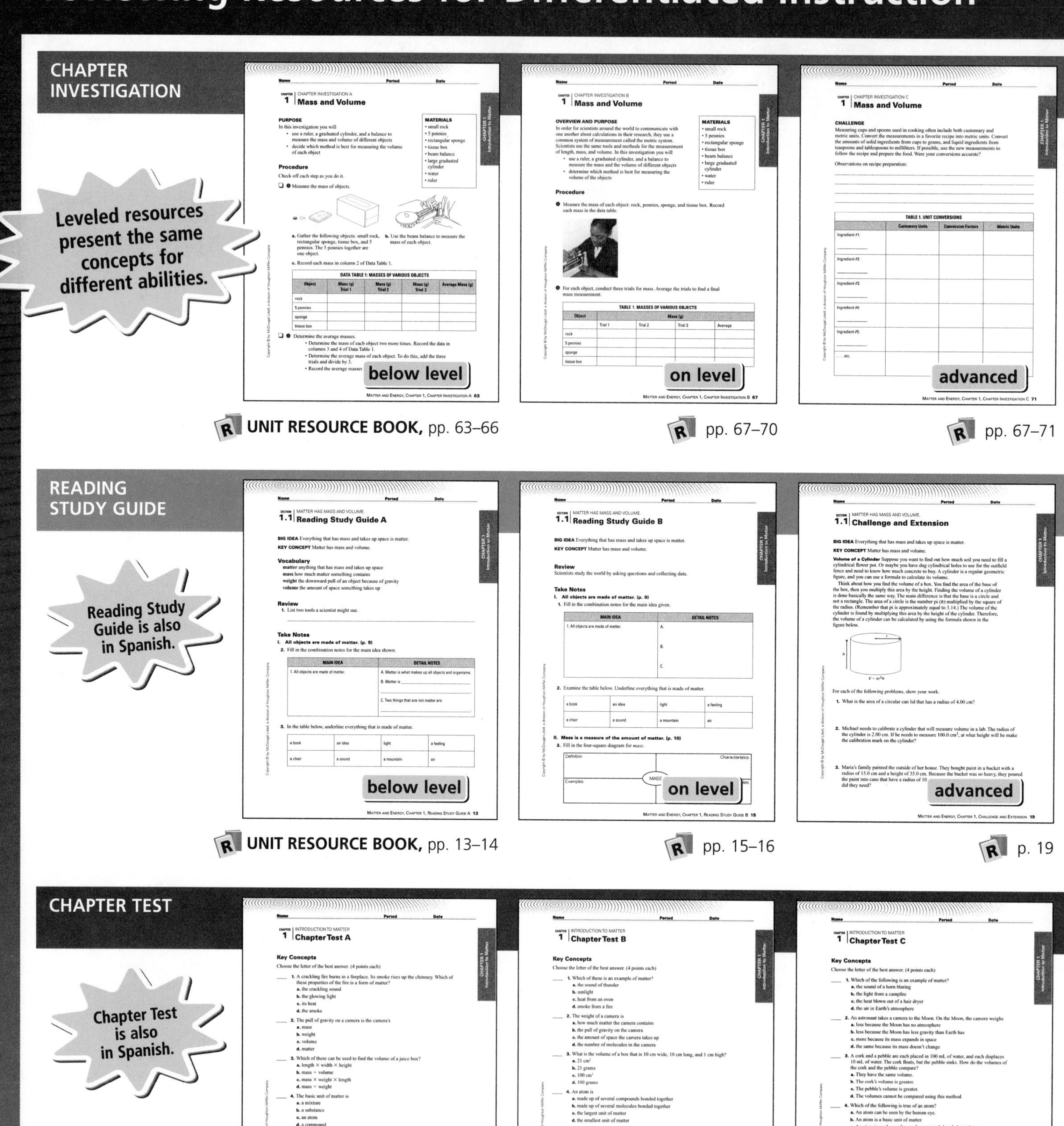

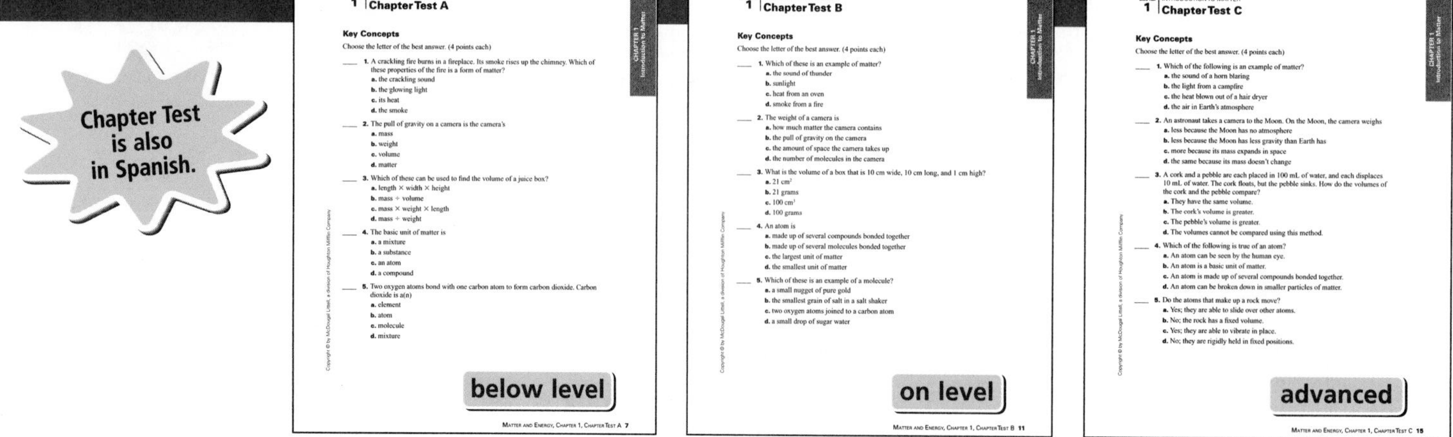

TECHNOLOGY

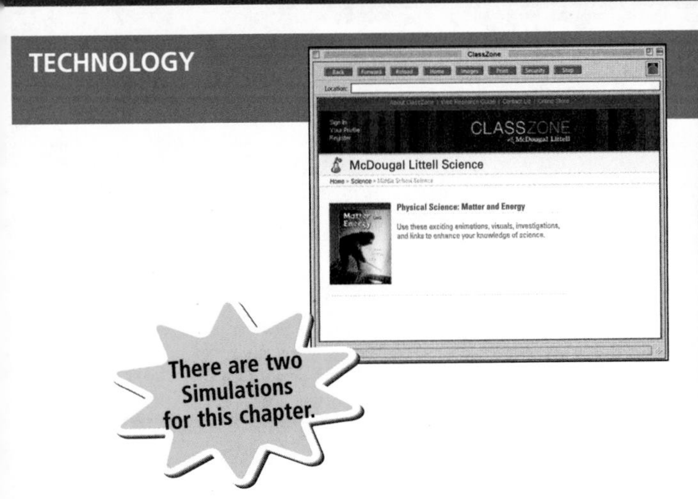

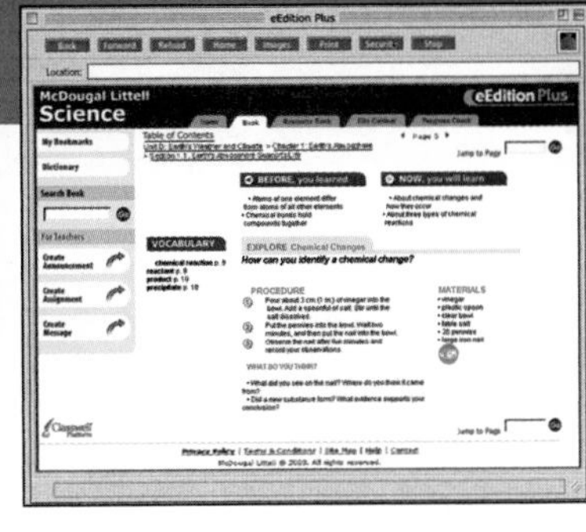

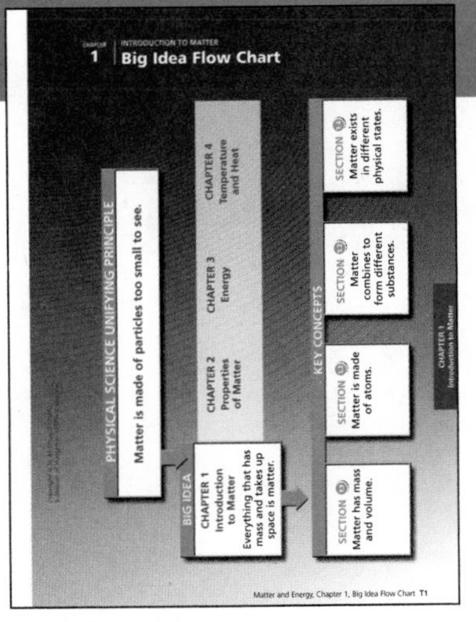

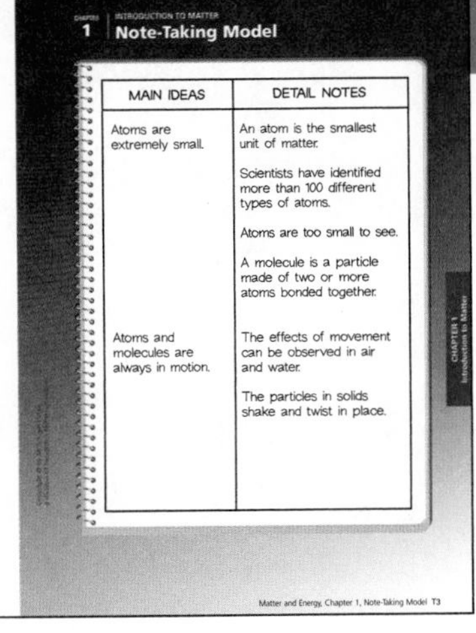

1 Note-Taking Model
MAIN IDEAS
DETAIL NOTES
Atoms are extremely small.
An atom is the smallest unit of matter.
Scientists have identified more than 100 different types of atoms.
Atoms are too small to see.
A molecule is a particle made of two or more atoms bonded together.
Atoms and molecules are always in motion.
The effects of movement can be observed in air and water.
The particles in solids shake and twist in place.
Matter and Energy, Chapter 1, Note-Taking Model T3

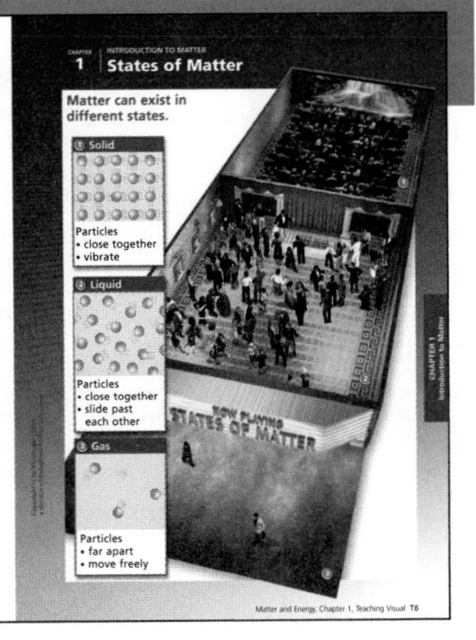

T **UNIT TRANSPARENCY BOOK,** p. T1

T p. T3

T p. T6

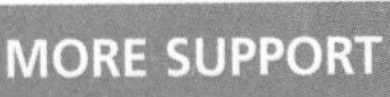

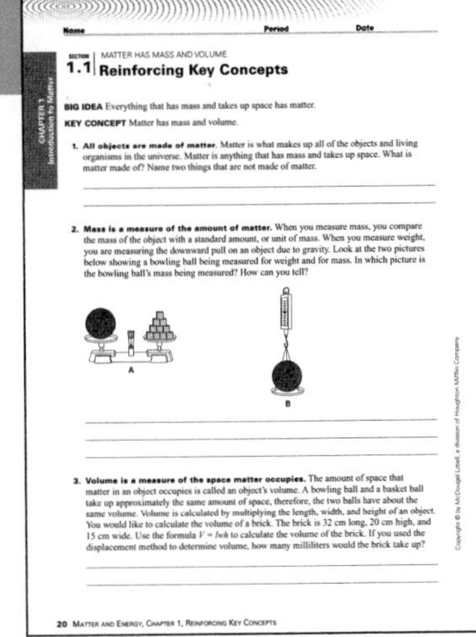
Name Period Date

1.1 MATTER HAS MASS AND VOLUME
Reinforcing Key Concepts

BIG IDEA Everything that has mass and takes up space has matter.

KEY CONCEPT Matter has mass and volume.

1. **All objects are made of matter.** Matter is what makes up all of the objects and living organisms in the universe. Matter is anything that has mass and takes up space. What is matter made of? Name two things that are not made of matter.

2. **Mass is a measure of the amount of matter.** When you measure mass, you compare the mass of the object with a standard amount, or unit of mass. When you measure weight, you are measuring the downward pull on an object due to gravity. Look at the two pictures below showing a bowling ball being measured for weight and for mass. In which picture is the bowling ball's mass being measured? How can you tell?

3. **Volume is a measure of the space matter occupies.** The amount of space that matter in an object occupies is called an object's volume. A bowling ball and a basket ball take up approximately the same amount of space, therefore, the two balls have about the same volume. Volume is calculated by multiplying the length, width, and height of an object. You would like to calculate the volume of a brick. The brick is 32 cm long, 20 cm high, and 15 cm wide. Use the formula $V = lwh$ to calculate the volume of the brick. If you used the displacement method to determine volume, how many milliliters would the brick take up?

20 Matter and Energy, Chapter 1, Reinforcing Key Concepts

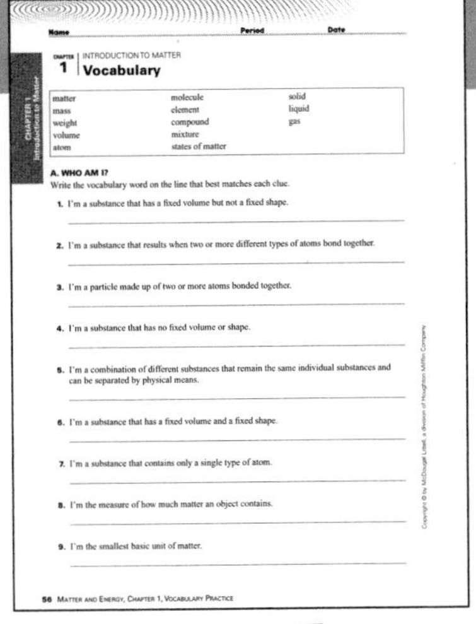
Name Period Date

1 INTRODUCTION TO MATTER
Vocabulary

matter	molecule	solid
mass	element	liquid
weight	compound	gas
volume	mixture	
atom	states of matter	

A. WHO AM I?
Write the vocabulary word on the line that best matches each clue.

1. I'm a substance that has a fixed volume but not a fixed shape.
2. I'm a substance that results when two or more different types of atoms bond together.
3. I'm a particle made up of two or more atoms bonded together.
4. I'm a substance that has no fixed volume or shape.
5. I'm a combination of different substances that remain the same individual substances and can be separated by physical means.
6. I'm a substance that has a fixed volume and a fixed shape.
7. I'm a substance that contains only a single type of atom.
8. I'm the measure of how much matter an object contains.
9. I'm the smallest basic unit of matter.

56 Matter and Energy, Chapter 1, Vocabulary Practice

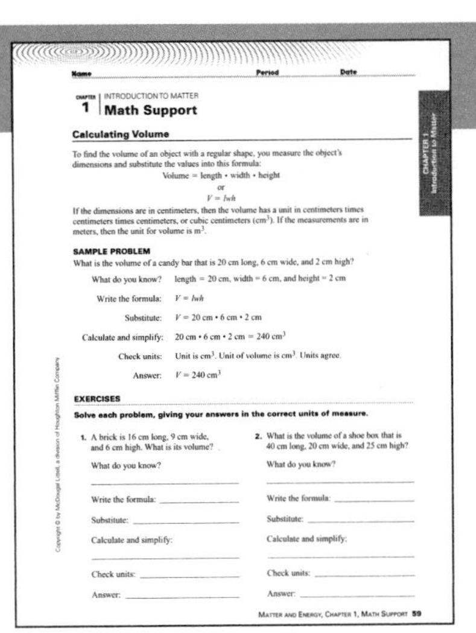
Name Period Date

1 INTRODUCTION TO MATTER
Math Support

Calculating Volume

To find the volume of an object with a regular shape, you measure the object's dimensions and substitute the values into this formula:

Volume = length • width • height
or
$V = lwh$

If the dimensions are in centimeters, then the volume has a unit in centimeters times centimeters times centimeters, or cubic centimeters (cm^3). If the measurements are in meters, then the unit for volume is m^3.

SAMPLE PROBLEM
What is the volume of a candy bar that is 20 cm long, 6 cm wide, and 2 cm high?

What do you know? length = 20 cm, width = 6 cm, and height = 2 cm
Write the formula: $V = lwh$
Substitute: $V = 20\text{ cm} \cdot 6\text{ cm} \cdot 2\text{ cm}$
Calculate and simplify: $20\text{ cm} \cdot 6\text{ cm} \cdot 2\text{ cm} = 240\text{ cm}^3$
Check units: Unit is cm^3. Unit of volume is cm^3. Units agree.
Answer: $V = 240\text{ cm}^3$

EXERCISES
Solve each problem, giving your answers in the correct units of measure.

1. A brick is 16 cm long, 9 cm wide, and 6 cm high. What is its volume?
What do you know?
Write the formula:
Substitute:
Calculate and simplify:
Check units:
Answer:

2. What is the volume of a shoe box that is 40 cm long, 20 cm wide, and 25 cm high?
What do you know?
Write the formula:
Substitute:
Calculate and simplify:
Check units:
Answer:

Matter and Energy, Chapter 1, Math Support 59

R **UNIT RESOURCE BOOK,** p. 20

R p. 59

CHAPTER 1 Introduction to Matter

INTRODUCE

the BIG idea

Have students look at the photograph of the beach and discuss how the question in the box links to the Big Idea:

- Which substances in the photograph have mass and take up space?
- How do the substances differ?
- What isn't matter?

National Science Education Standards

Content

B.1.c Chemical elements do not break down during normal laboratory reactions involving such treatments as heating, exposure to electric current, or reaction with acids. There are more than 100 known elements that combine in a multitude of ways to produce compounds, which account for the living and nonliving substances that we encounter.

Process

A.2–8 Design and conduct an investigation; use tools to gather and interpret data; use evidence to describe, predict, explain, model; think critically to make relationships between evidence and explanation; recognize different explanations and predictions; communicate scientific procedures and explanations; use mathematics.

A.9.a–c, A.9.e–f Understand scientific inquiry by using different investigations, methods, mathematics, and explanations based on logic, evidence, and skepticism.

E.6.c Science drives technology and technology drives science

F.5.c Technology influences society through its products and processes.

CHAPTER

Introduction to Matter

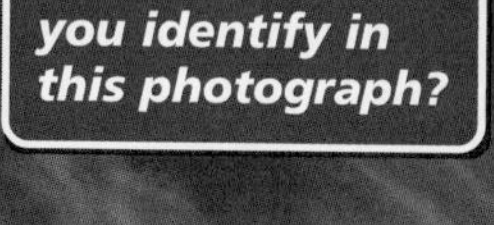

the BIG idea

Everything that has mass and takes up space is matter.

What matter can you identify in this photograph?

Key Concepts

SECTION  1.1 **Matter has mass and volume.** Learn what mass and volume are and how to measure them.

SECTION  1.2 **Matter is made of atoms.** Learn about the movement of atoms and molecules.

SECTION  1.3 **Matter combines to form different substances.** Learn how atoms form compounds and mixtures.

SECTION  1.4 **Matter exists in different physical states.** Learn how different states of matter behave.

Internet Preview

CLASSZONE.COM

Chapter 1 online resources: Content Review, two Simulations, four Resource Centers, Math Tutorial, Test Practice

A 6 Unit: Matter and Energy

INTERNET PREVIEW

CLASSZONE.COM For student use with the following pages:

Review and Practice
- Content Review, pp. 8, 34
- Math Tutorial: Circle Graphs, p. 26
- Test Practice, p. 37

Activities and Resources
- Internet Activity: p. 7
- Simulations: Weights on Different Planets, p. 11; Gas Behavior, p. 33
- Resource Centers: Volume, p. 12; STM Images, p. 20; Mixtures, p. 24

Solids, Liquids, and Gases
Code: MDL061

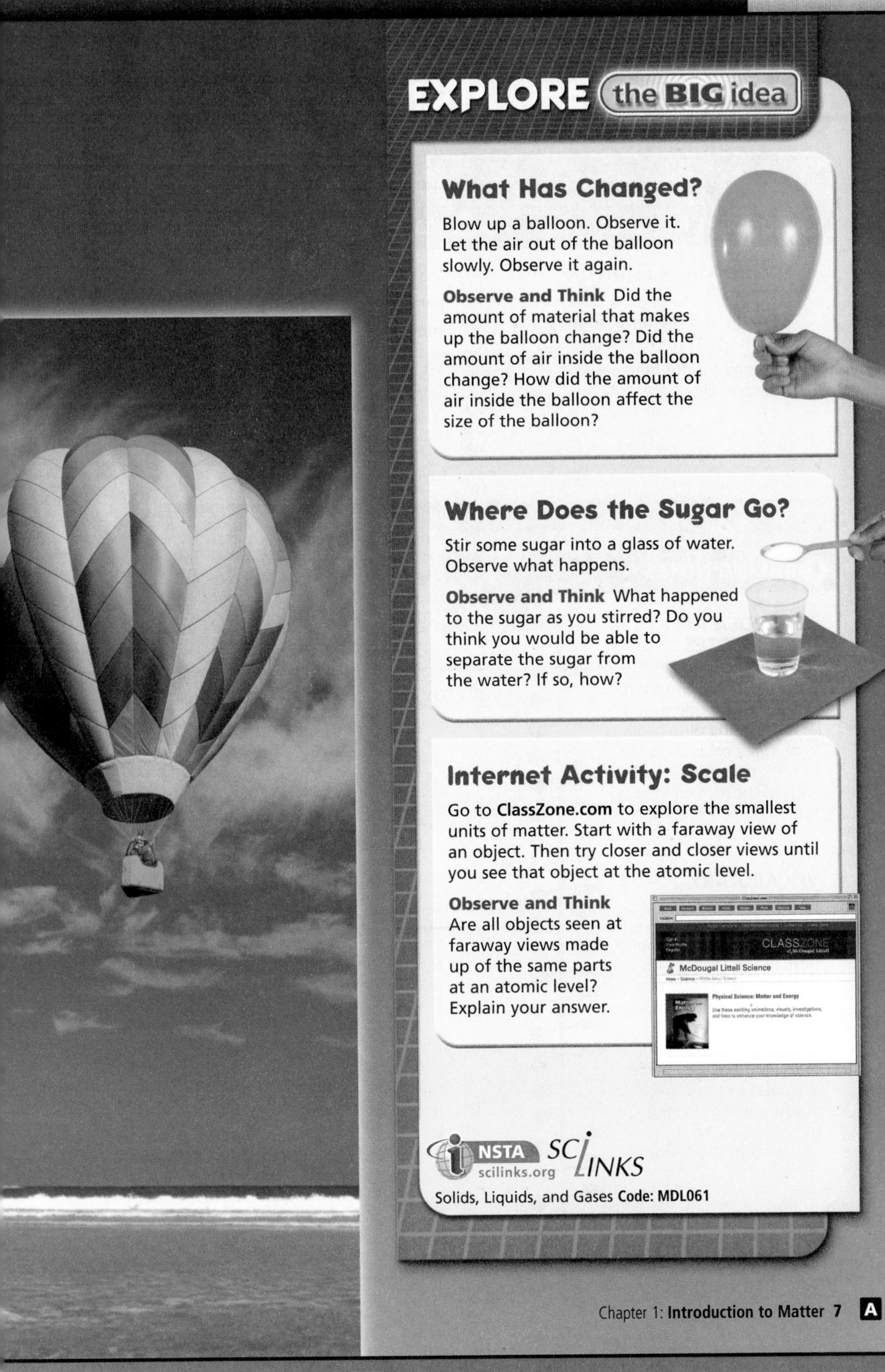

EXPLORE the BIG idea

What Has Changed?

Blow up a balloon. Observe it. Let the air out of the balloon slowly. Observe it again.

Observe and Think Did the amount of material that makes up the balloon change? Did the amount of air inside the balloon change? How did the amount of air inside the balloon affect the size of the balloon?

Where Does the Sugar Go?

Stir some sugar into a glass of water. Observe what happens.

Observe and Think What happened to the sugar as you stirred? Do you think you would be able to separate the sugar from the water? If so, how?

Internet Activity: Scale

Go to **ClassZone.com** to explore the smallest units of matter. Start with a faraway view of an object. Then try closer and closer views until you see that object at the atomic level.

Observe and Think Are all objects seen at faraway views made up of the same parts at an atomic level? Explain your answer.

NSTA scilinks.org SCILINKS

Solids, Liquids, and Gases **Code: MDL061**

TEACHING WITH TECHNOLOGY

Video Camera Students can use a video camera to record real-world examples that model the movement of particles in the different states of matter on p. 28.

Scanning Tunneling Microscope In addition to the ClassZone site, you can search the Internet for other images from SEM (scanning electron microscopes) or STM technology while discussing p. 20.

EXPLORE the BIG idea

These inquiry-based activities are appropriate for use at home or as a supplement to classroom instruction.

What Has Changed?

PURPOSE To introduce the idea that air takes up space and is matter. Students observe a balloon as they blow it up and then slowly let the air out.

TIP *10 min.* For health reasons, do not allow students to share the task of blowing up a balloon. Caution students not to overfill the balloons. Have extra balloons available.

Answer: The amount of matter that made up the balloon did not change. The amount of air inside the balloon changed. The less air in the balloon, the smaller the balloon gets.

REVISIT after p. 11.

Where Does the Sugar Go?

PURPOSE To introduce the concept of two substances forming a mixture. Students dissolve sugar in a glass of water.

TIP *10 min.* If the sugar does not dissolve, have students add more water and stir the mixture for a longer time.

Answer: The sugar dissolved in the water. You could boil the water or let it evaporate. The sugar will be left behind.

REVISIT after p. 25.

Internet Activity: Scale

PURPOSE To introduce students to atoms, the smallest units of matter.

TIP *20 min.* Have students observe more than one object.

Answer: Yes; matter is made up of atoms.

REVISIT after p. 18.

PREPARE

CONCEPT REVIEW

Activate Prior Knowledge

- Place open containers of a substance with a strong aroma, such as vinegar or room deodorizer, around the classroom so that students will notice the aroma.
- Ask students to describe the size of the particles that cause the aroma. *very small, invisible*
- Show students the original source of the particles. Ask them to describe the size of the particles that make up the substance. *very small, invisible*
- Discuss what makes up all substances. *particles that are too small to see*

TAKING NOTES

Main Idea and Detail Notes

Encourage students to identify and write the most important details. Students can test themselves by covering up one-half of the chart as they study.

Vocabulary Strategy

The four square diagram organizes all aspects of a word into a coherent pattern. By filling in their own words, students personalize their understanding. Point out that it is all right to leave a blank square. Many terms, such as *volume* and *states of matter,* may not have clear nonexamples.

Vocabulary and Note-Taking Resources

- Vocabulary Practice, pp. 56–57
- Decoding Support, p. 58

- Daily Vocabulary Scaffolding, p. T2
- Note-Taking Model, p. T3

- Four Square, B22–23
- Main Idea and Detail Notes, C37
- Daily Vocabulary Scaffolding, H1–8

CHAPTER 1

Getting Ready to Learn

CONCEPT REVIEW

- Matter is made of particles too small to see.
- Energy and matter change from one form to another.
- Energy cannot be created or destroyed.

VOCABULARY REVIEW

See Glossary for definitions.

particle

substance

Review concepts and vocabulary.

TAKING NOTES

MAIN IDEA AND DETAIL NOTES

Make a two-column chart. Write the main ideas, such as those in the blue headings, in the column on the left. Write details about each of those main ideas in the column on the right.

VOCABULARY STRATEGY

Write each new vocabulary term in the center of a **four square** diagram. Write notes in the squares around each term. Include a definition, some characteristics, and some examples of the term. If possible, write some things that are not examples of the term.

See the Note-Taking Handbook on pages R45–R51.

SCIENCE NOTEBOOK

MAIN IDEAS	DETAIL NOTES
1. All objects are made of matter.	1. All objects and living organisms are matter. 1. Light and sound are not matter.
2. Mass is a measure of the amount of matter.	2. A balance can be used to compare masses. 2. Standard unit of mass is kilogram (kg).

WEIGHT

Definition	Characteristics
the downward pull on an object due to gravity	• standard unit is newton (N) • is measured by using a scale
Examples	**Nonexamples**
On Earth, a 1 kg object has a weight of 9.8 N.	not the same as mass, which is a measure of how much matter an object contains

CHECK READINESS

Administer the Diagnostic Test to determine students' readiness for new science content and their mastery of requisite math skills.

 Diagnostic Test, pp. 1–2

Technology Resources

Students needing content and math skills should visit **ClassZone.com**.

- **CONTENT REVIEW**
- **MATH TUTORIAL**

 **CONTENT REVIEW CD-ROM**

KEY CONCEPT

Matter has mass and volume.

BEFORE, you learned

- Scientists study the world by asking questions and collecting data
- Scientists use tools such as microscopes, thermometers, and computers

NOW, you will learn

- What matter is
- How to measure the mass of matter
- How to measure the volume of matter

VOCABULARY

matter p. 9
mass p. 10
weight p. 11
volume p. 11

EXPLORE Similar Objects

How can two similar objects differ?

PROCEDURE

1. Look at the two balls but do not pick them up. Compare their sizes and shapes. Record your observations.
2. Pick up each ball. Compare the way the balls feel in your hands. Record your observations.

MATERIALS

2 balls of different sizes

WHAT DO YOU THINK?

How would your observations be different if the larger ball were made of foam?

All objects are made of matter.

Suppose your class takes a field trip to a museum. During the course of the day you see mammoth bones, sparkling crystals, hot-air balloons, and an astronaut's space suit. All of these things are matter.

Matter is what makes up all of the objects and living organisms in the universe. As you will see, **matter** is anything that has mass and takes up space. Your body is matter. The air that you breathe and the water that you drink are also matter. Matter makes up the materials around you. Matter is made of particles called atoms, which are too small to see. You will learn more about atoms in the next section.

Not everything is matter. Light and sound, for example, are not matter. Light does not take up space or have mass in the same way that a table does. Although air is made of atoms, a sound traveling through air is not.

VOCABULARY
Make four square diagrams for *matter* and for *mass* in your notebook to help you understand their relationship.

CHECK YOUR READING What is matter? How can you tell if something is matter?

RESOURCES FOR DIFFERENTIATED INSTRUCTION

Below Level

UNIT RESOURCE BOOK

- Reading Study Guide A, pp. 13–14
- Decoding Support, p. 58

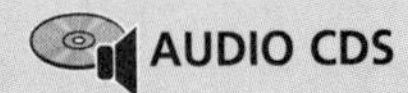

Advanced

UNIT RESOURCE BOOK

Challenge and Extension, p. 19

English Learners

UNIT RESOURCE BOOK

Spanish Reading Study Guide, pp. 17–18

- Audio Readings in Spanish
- Audio Readings (English)

1.1 FOCUS

Set Learning Goals

Students will

- Explain what matter is.
- Describe how to measure the mass of matter.
- Describe how to measure the volume of matter.

3-Minute Warm-Up

Display Transparency 4 or copy this exercise on the board:

Draw a two-column chart. In the column on the left, write what you know about matter. In the column on the right, write questions that you have about matter. *Encourage students to look for answers to their questions as they study matter.*

 3-Minute Warm-Up, p. T4

1.1 MOTIVATE

EXPLORE Similar Objects

PURPOSE To introduce the properties of mass and volume by comparing two objects

TIP *10 min.* The two balls must be different sizes. Students may be surprised at the results if the smaller ball is heavier.

WHAT DO YOU THINK? *Sample answer: A large foam ball might not be as heavy as a baseball. (Actual answers would depend on the types of balls used.)*

Ongoing Assessment

Explain what matter is.

Ask: Give an example of something that is matter and explain why it is matter. *A book is matter because it has mass and takes up space.*

Answer: Matter is what makes up all objects and living organisms in the universe. Matter has mass and takes up space.

1.1 INSTRUCT

Develop Critical Thinking

APPLY Tell students that enough mass cubes are not available for the entire class. Have students apply what they know about measuring mass to invent standard units of mass based on common objects. Ask them to find the mass of objects based on the invented standard units.

Teach from Visuals

To help students interpret the photographs and illustrations comparing the size and mass of a bowling ball and a basketball, ask:

- How do the sizes of the bowling ball and the basketball in the photographs compare? *They are about the same.*
- How can you tell from the illustrations which ball has more mass? *More standard cubes are needed to balance the pan with the bowling ball than the pan with the basketball.*

Ongoing Assessment

Describe how to measure the mass of matter.

Ask: How would you find the mass of an object? *Compare its mass with the mass of standard units on a balance.*

CHECK YOUR READING *Answer: Mass is a measure of how much matter an object contains.*

MAIN IDEA AND DETAILS As you read, write the blue headings on the left side of a two-column chart. Add details in the other column.

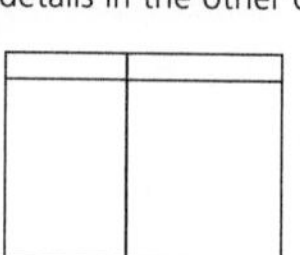

Mass is a measure of the amount of matter.

Different objects contain different amounts of matter. **Mass** is a measure of how much matter an object contains. A metal teaspoon, for example, contains more matter than a plastic teaspoon. Therefore, a metal teaspoon has a greater mass than a plastic teaspoon. An elephant has more mass than a mouse.

CHECK YOUR READING How are matter and mass related?

Measuring Mass

A

When you measure mass, you compare the mass of the object with a standard amount, or unit, of mass. The standard unit of mass is the kilogram (kg). A large grapefruit has a mass of about one-half kilogram. Smaller masses are often measured in grams (g). There are 1000 grams in a kilogram. A penny has a mass of between two and three grams.

How can you compare the masses of two objects? One way is to use a pan balance, as shown below. If two objects balance each other on a pan balance, then they contain the same amount of matter. If a basketball balances a metal block, for example, then the basketball and the block have the same mass. Beam balances work in a similar way, but instead of comparing the masses of two objects, you compare the mass of an object with a standard mass on the beam.

A bowling ball and a basketball are about the same size, but a bowling ball has more mass.

DIFFERENTIATE INSTRUCTION

More Reading Support

A To measure mass, what do you compare the object's mass to? *standard units of mass*

English Learners Phrasal verbs such as "makes up" and "takes up" are used throughout this chapter (for example, on p. 9). Make sure students understand not to read "up" and other adverbs or prepositions in phrasal verbs as literal directions. If students are still confused, offer synonyms for phrasal verbs. For instance, "makes up" means composes, and "takes up" means occupies. English learners may also lack background knowledge of a field trip (p. 9).

Measuring Weight

When you hold an object such as a backpack full of books, you feel it pulling down on your hands. This is because Earth's gravity pulls the backpack toward the ground. Gravity is the force that pulls two masses toward each other. In this example, the two masses are Earth and the backpack. **Weight** is the downward pull on an object due to gravity. If the pull of the backpack is strong, you would say that the backpack weighs a lot.

Gravity is pulling down on both the girl and the backpack. The heavier the backpack is, the stronger the pull of gravity is on it.

Weight is measured by using a scale, such as a spring scale like the one shown on the right, that tells how hard an object is pushing or pulling on it. The standard scientific unit for weight is the newton (N). A common unit for weight is the pound (lb).

Mass and weight are closely related, but they are not the same. Mass describes the amount of matter an object has, and weight describes how strongly gravity is pulling on that matter. On Earth, a one-kilogram object has a weight of 9.8 newtons (2.2 lb). When a person says that one kilogram is equal to 2.2 pounds, he or she is really saying that one kilogram has a weight of 2.2 pounds on Earth. On the Moon, however, gravity is one-sixth as strong as it is on Earth. On the Moon, the one-kilogram object would have a weight of 1.6 newtons (0.36 lb). The amount of matter in the object, or its mass, is the same on Earth as it is on the Moon, but the pull of gravity is different.

SIMULATION CLASSZONE.COM

Compare weights on different planets.

CHECK YOUR READING What is the difference between mass and weight?

Volume is a measure of the space matter occupies.

Matter takes up space. A bricklayer stacks bricks on top of each other to build a wall. No two bricks can occupy the same place because the matter in each brick takes up space.

The amount of space that matter in an object occupies is called the object's **volume.** The bowling ball and the basketball shown on page 10 take up approximately the same amount of space. Therefore, the two balls have about the same volume. Although the basketball is hollow, it is not empty. Air fills up the space inside the basketball. Air and other gases take up space and have volume.

DIFFERENTIATE INSTRUCTION

More Reading Support

B Which is affected by gravity, mass or weight? *weight*

C What do you call the amount of space an object takes up? *volume*

Below Level Have students make a table that compares mass and weight. Ask them to include a definition of each, the standard units used to measure them, and the tools used to measure them.

Teach Difficult Concepts

Students often have trouble understanding the difference between mass and weight. Have students find the weight of an object by using a spring scale. Have them weigh the object as they gently pull down or push up on the object. Compare this weight with the object's weight on a planet with different gravity. Ask:

- How is the object's weight affected by gravity? *It weighs more where gravity is stronger and less where gravity is weaker.*
- How is the object's mass affected by gravity? *It is not.*

Address Misconceptions

IDENTIFY Ask students to identify all the matter in a blown-up balloon and in a balloon filled with water. If students do not include the air and water inside the balloons, they may hold the misconception that liquids and gases are not matter.

CORRECT Slowly let the air out of a balloon. Let the water out of another balloon. Have students observe the change in the volume of the balloons.

REASSESS Ask students what was taking up space in the balloons. *air and water* Point out that air and water are matter because they take up space and have mass.

Technology Resources

Visit **ClassZone.com** for background on common student misconceptions.

MISCONCEPTION DATABASE

EXPLORE the BIG idea

Revisit "What Has Changed" on p. 7. Have students explain their results.

Ongoing Assessment

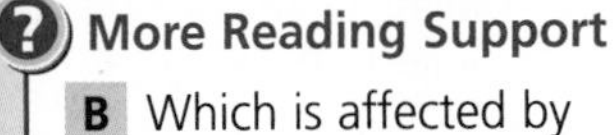

CHECK YOUR READING *Answer: Mass is a measure of how much matter an object contains, but weight is the downward pull on an object due to gravity.*

Real World Example

The amount of some products is determined by volume. For example, bags of mulch used for gardens and yards are sold by cubic feet. A bag often tells consumers how large of an area the bag will cover, although amounts can vary depending on the depth of the mulch. Soil and cement are also sold by volume.

Ongoing Assessment

Answers:

1. 10 cm · 5 cm · 7 cm = 350 cm³

2. 1 m · 0.5 m · 0.5 m = 0.25 m³

- Math Support, p. 59
- Math Practice, p. 60

Find out more about volume.

Determining Volume by Formula

There are different ways to find the volume of an object. For objects that have well-defined shapes, such as a brick or a ball, you can take a few measurements of the object and calculate the volume by substituting these values into a formula.

A rectangular box, for example, has a length, a width, and a height that can be measured. To find the volume of the box, multiply the three values.

$$\text{Volume} = \text{length} \cdot \text{width} \cdot \text{height}$$
$$V = lwh$$

If you measure the length, the width, and the height of the box in centimeters (cm), the volume has a unit of centimeters times centimeters times centimeters, or centimeters cubed (cm^3). If the measurements are meters, the unit of volume is meters cubed (m^3). All measurements must be in the same unit to calculate volume.

Other regular solids, such as spheres and cylinders, also have formulas for calculating volumes. All formulas for volume require multiplying three dimensions. Units for volume are often expressed in terms of a length unit cubed, that is, a length to the third power.

Calculating Volume

Sample Problem

What is the volume of a pizza box that is 8 cm high, 38 cm wide, and 38 cm long?

What do you know?	length = 38 cm, width = 38 cm, height = 8 cm
What do you want to find out?	Volume
Write the formula:	$V = lwh$
Substitute into the formula:	$V = 38\text{ cm} \cdot 38\text{ cm} \cdot 8\text{ cm}$
Calculate and simplify:	$11{,}552\text{ cm} \cdot \text{cm} \cdot \text{cm} = 11{,}552\text{ cm}^3$
Check that your units agree:	Unit is cm^3. Unit of volume is cm^3. Units agree.
Answer:	$11{,}552\text{ cm}^3$

Practice the Math

1. A bar of gold is 10 cm long, 5 cm wide, and 7 cm high. What is its volume?
2. What is the volume of a large block of wood that is 1 m long, 0.5 m high, and 50 cm wide?

A 12 Unit: Matter and Energy

DIFFERENTIATE INSTRUCTION

More Reading Support

D A formula can be used to find the volume of which type of objects? *regular objects with well-defined shapes*

Advanced Have students find the formulas for calculating the volume of a sphere and a cylinder. Encourage them to use the formulas to find the volume of objects in the classroom.

Challenge and Extension, p. 19

Measuring Volume by Displacement

Although a box has a regular shape, a rock does not. There is no simple formula for calculating the volume of something with an irregular shape. Instead, you can make use of the fact that two objects cannot be in the same place at the same time. This method of measuring is called displacement.

1. Add water to a graduated cylinder. Note the volume of the water by reading the water level on the cylinder.
2. Submerge the irregular object in the water. Because the object and the water cannot share the same space, the water is displaced, or moved upward. Note the new volume of the water with the object in it.
3. Subtract the volume of the water before you added the object from the volume of the water and the object together. The result is the volume of the object. The object displaces a volume of water equal to the volume of the object.

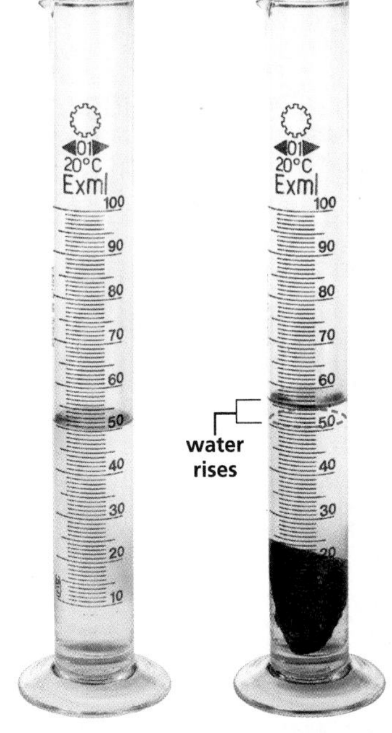

Measure the volume of water without the rock.

Measure the volume of water with the rock in it.

You measure the volume of a liquid by measuring how much space it takes up in a container. The volume of a liquid usually is measured in liters (L) or milliliters (mL). One liter is equal to 1000 milliliters. Milliliters and cubic centimeters are equivalent. This can be written as $1 \text{ mL} = 1 \text{ cm}^3$. If you had a box with a volume of one cubic centimeter and you filled it with water, you would have one milliliter of water.

In the first photograph, the graduated cylinder contains 50 mL of water. Placing a rock in the cylinder causes the water level to rise from 50 mL to 55 mL. The difference is 5 mL; therefore, the volume of the rock is 5 cm^3.

1.1 Review

KEY CONCEPTS

1. Give three examples of matter.
2. What do weight and mass measure?
3. How can you measure the volume of an object that has an irregular shape?

CRITICAL THINKING

4. **Calculate** What is the volume of a box that is 12 cm long, 6 cm wide, and 4 cm high?
5. **Synthesize** What is the relationship between the units of measurement for the volume of a liquid and of a solid object?

CHALLENGE

6. **Infer** Why might a small increase in the dimensions of an object cause a large change in its volume?

ANSWERS

1. Sample answer: air, water, a rock

2. Weight measures the downward pull on an object due to gravity. Mass measures how much matter an object contains.

3. Submerge the object in water and subtract the volume of the water before the object was added from the volume of the water and the object together.

4. 12 cm · 6 cm · 4 cm = 288 cm^3

5. One cubic centimeter of solid is equivalent to one milliliter of liquid.

6. Each of the three dimensions increases, so, for example, doubling the dimensions increases the volume by a multiple of eight.

Ongoing Assessment

Describe how to measure the volume of matter.

Ask: How would you find the volume of a box? of a small rock? *Multiply the box's length by its width and height. Put the small stone in a graduated cylinder of water and determine how much water the rock displaces.*

Teacher Demo

Place a clear glass that is full to its brim with water in a shallow tray. Ask students to predict what will happen if you slowly drop a stone into the glass. Ask them to explain their predictions.

Reinforce the BIG idea

Have students relate the section to the Big Idea.

 Reinforcing Key Concepts, p. 20

1.1 ASSESS & RETEACH

Assess

 Section 1.1 Quiz, p. 3

Reteach

Write the heading *Matter* on the board. Ask a volunteer to define *matter* and write the definition on the board. Add these two subheads under *Matter: Mass* and *Volume*. Ask a volunteer to define *mass* and *volume* and write the definitions on the board. Ask students to describe how each is measured, which tools are used for the measurements, and which units are used.

Technology Resources

Have students visit **ClassZone.com** for reteaching of Key Concepts.

 **CONTENT REVIEW**

 **CONTENT REVIEW CD-ROM**

CHAPTER INVESTIGATION

Focus

PURPOSE To practice measuring mass and volume

OVERVIEW Students will measure the mass and volume of four objects and find the average of three trials for each object. Students will find:

- the volume of a rectangular object by multiplying the object's length by its width and height.
- the volume of an irregular object by subtracting the volume of water in a graduated cylinder without the object from the volume of the water with the object.

Lab Preparation

- Collect small rocks that will fit into graduated cylinders, pennies, sponges, and tissue boxes. You also could substitute a variety of other objects from the classroom. You could ask students to bring pennies from home.
- Prior to the investigation, have students read through the investigation and prepare their data tables. Or you may wish to copy and distribute datasheets and rubrics.

UNIT RESOURCE BOOK, pp. 63–71

SCIENCE TOOLKIT, F15

Lab Management

- Review with students how to use a beam balance and a graduated cylinder. Remind them to read the volume of the water at the bottom of the meniscus in the cylinder.
- If time is a factor, you could reduce the number of objects to be tested.
- Point out that three trials and an average are used because sometimes there are human errors in measuring.

INCLUSION Students with disabilities might need help making accurate measurements of mass and volume.

CHAPTER INVESTIGATION

Mass and Volume

OVERVIEW AND PURPOSE In order for scientists around the world to communicate with one another about calculations in their research, they use a common system of measurement called the metric system. Scientists use the same tools and methods for the measurement of length, mass, and volume. In this investigation you will

- use a ruler, a graduated cylinder, and a balance to measure the mass and the volume of different objects
- determine which method is best for measuring the volume of the objects

MATERIALS
- small rock
- 5 pennies
- rectangular sponge
- tissue box
- beam balance
- large graduated cylinder
- water
- ruler

Procedure

1. Make a data table like the one shown on the sample notebook page.

2. Measure the mass of each object: rock, pennies, sponge, and tissue box. Record each mass.
3. For each object, conduct three trials for mass. Average the trials to find a final mass measurement.
4. Decide how you will find the volume of each object.

For rectangular objects, you will use the following formula:

$$\text{Volume} = \text{length} \cdot \text{width} \cdot \text{height}$$

For irregular objects, you will use the displacement method and the following formula:

$$\text{Volume of object} = \text{volume of water with object} - \text{volume of water without object}$$

INVESTIGATION RESOURCES

CHAPTER INVESTIGATION, Mass and Volume
- Level A, pp. 63–66
- Level B, pp. 67–70
- Level C, p. 71

Advanced students should complete Levels B & C.

Writing a Lab Report, D12–13

Technology Resources

Customize this student lab as needed or look for an alternative. Print rubrics to assess student lab reports.

Lab Generator CD-ROM

5. For each object, you will conduct three trials for measuring volume. Average the trials to find a final volume measurement.

6. For rectangular objects, use metric units for measuring the length, width, and height. Record the measurements in your data table.

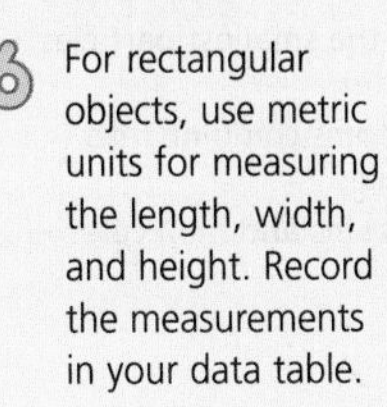

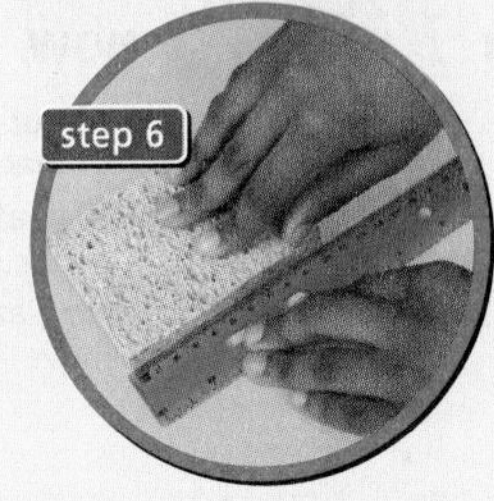

7. For irregular objects, fill the graduated cylinder about half full with water. Record the exact volume of water in the cylinder. **Note:** The surface of the liquid will be curved in the graduated cylinder. Read the volume of the liquid at the bottom of the curve called the meniscus.

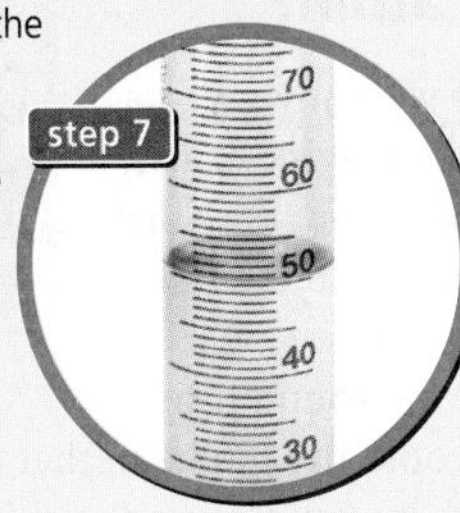

8. Carefully place the object you are measuring into the cylinder. The object must be completely under the water. Record the exact volume of water in the cylinder containing the object by reading the meniscus.

Observe and Analyze

Write It Up

1. **RECORD OBSERVATIONS** Make sure you have filled out your data table completely.
2. **INTERPRET** For each object, explain why you chose a particular method for measuring the volume.

Conclude

1. **IDENTIFY LIMITS** Which sources of error might have affected your measurements?
2. **APPLY** Doctors need to know the mass of a patient before deciding how much of a medication to prescribe. Why is it important to measure each patient's mass before prescribing medicine?
3. **APPLY** Scientists in the United States work closely with scientists in other countries to develop new technology. What are the advantages of having a single system of measurement?

INVESTIGATE Further

CHALLENGE Measuring cups and spoons used in cooking often include both customary and metric units. Convert the measurements in a favorite recipe into metric units. Convert the amounts of solid ingredients to grams, and liquid ingredients to milliliters or liters. If possible, use the new measurements to follow the recipe and prepare the food. Were your conversions accurate?

Mass and Volume

Observe and Analyze

Table 1. Masses of Various Objects

Object	Mass (g)			
	Trial 1	Trial 2	Trial 3	Average
rock				
5 pennies				
sponge				
tissue box				

Table 2. Volumes of Various Objects

Object	Method Used	Volume (cm³ or mL)			
		Trial 1	Trial 2	Trial 3	Average
rock					
5 pennies					
sponge					
tissue box					

Observe and Analyze

1. *SAMPLE DATA Table 1 Masses of Various Objects, Average: rock, 3 g; 5 pennies, 12.33 g; sponge, 8.9 g; tissue box, 293.66 g. Table 2 Volumes of Various Objects, Average: rock, 4 mL; 5 pennies, 2 mL; sponge, 129.4 cm^3; tissue box, 2448 cm^3. See students' tables.*
2. *Students should have chosen the correct formula for measuring the volume of objects with a regular shape (i.e., sponge, tissue box) and the displacement method for measuring the volume of objects with an irregular shape (i.e., rock, pennies).*

Conclude

1. *incorrect readings of volume or length*
2. *Doctors prescribe a dosage based on a patient's body mass. Smaller patients should receive less medicine.*
3. *Scientists should use the same system of measurements so that they can share information without having to convert their results to another measurement system.*

INVESTIGATE Further

CHALLENGE Answers will vary.

Post-Lab Discussion

Have groups compare their measurements and averages. Discuss the importance of doing a number of trials and finding the average. Ask:

- Based on your measurements, why is finding an average measurement important? *Measurements might vary due to small errors in individual measurements. An average measurement is more accurate.*
- Which methods or tools did you use to measure mass and volume? *a beam balance for mass, a formula for the volume of objects with a regular shape, and displacement for the volume of objects with an irregular shape*

1.2 FOCUS

Set Learning Goals

Students will

- Identify the smallest particles of matter.
- Describe how atoms combine into molecules.
- Describe how atoms and molecules move.
- Use modeling to draw conclusions about atoms and their masses.

3-Minute Warm-Up

Display Transparency 4 or copy this exercise on the board:

Decide if these statements are true. If not, correct them.

1. Substances you can see are matter, but substances you cannot see are not matter. *All substances are matter.*
2. Mass and weight are the same thing. *They are not the same.*
3. Formulas can be used to calculate the volume of many solids with a regular shape. *true*

3-Minute Warm-Up, p. T4

1.2 MOTIVATE

THINK ABOUT

PURPOSE To understand the size of atoms

DISCUSS Brainstorm examples of scales in which a length or a size represents a much larger length or size. Ask: If an atom were the size of the round head of the pin in the photograph, how large would the pin be? *It would cover about 90 square miles and be about 80 miles high.*

Ongoing Assessment

CHECK YOUR READING *Answer: Atoms are the smallest basic units of matter. Like building blocks, atoms are the small parts that make up a larger structure.*

KEY CONCEPT

1.2 Matter is made of atoms.

BEFORE, you learned

- Matter has mass
- Matter has volume

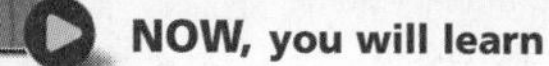
NOW, you will learn

- About the smallest particles of matter
- How atoms combine into molecules
- How atoms and molecules move

VOCABULARY

atom p. 16
molecule p. 18

THINK ABOUT

How small is an atom?

All matter is made up of very tiny particles called atoms. It is hard to imagine exactly how small these particles are. Suppose that each of the particles making up the pin shown in the photograph on the right were actually the size of the round head on the pin. How large would the pin be in that case? If you could stick such a pin in the ground, it would cover about 90 square miles—about one-seventh the area of London, England. It would also be about 80 miles high—almost 15 times the height of Mount Everest.

Atoms are extremely small.

VOCABULARY
Make a four square diagram for *atom* that includes details that will help you remember the term.

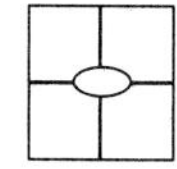

How small can things get? If you break a stone wall into smaller and smaller pieces, you would have a pile of smaller stones. If you could break the smaller stones into the smallest pieces possible, you would have a pile of atoms. An **atom** is the smallest basic unit of matter.

The idea that all matter is made of extremely tiny particles dates back to the fifth century B.C., when Greek philosophers proposed the first atomic theory of matter. All matter, they said, was made of only a few different types of tiny particles called atoms. The different arrangements of atoms explained the differences among the substances that make up the world. Although the modern view of the atom is different from the ancient view, the idea of atoms as basic building blocks has been confirmed. Today scientists have identified more than 100 different types of atoms.

CHECK YOUR READING What are atoms? How are they like building blocks?

A 16 Unit: **Matter and Energy**

RESOURCES FOR DIFFERENTIATED INSTRUCTION

Below Level

UNIT RESOURCE BOOK

- Reading Study Guide A, pp. 23–24
- Decoding Support, p. 58

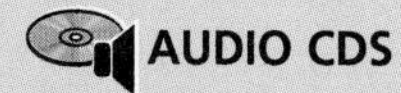
AUDIO CDS

Advanced

UNIT RESOURCE BOOK

Challenge and Extension, p. 29

English Learners

UNIT RESOURCE BOOK

Spanish Reading Study Guide, pp. 27–28

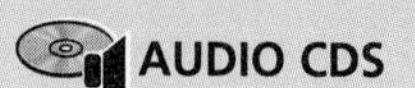
AUDIO CDS

- Audio Readings in Spanish
- Audio Readings (English)

Atoms

It is hard to imagine that visible matter is composed of particles too tiny to see. Although you cannot see an individual atom, you are constantly seeing large collections of them. You are a collection of atoms. So are your textbook, a desk, and all the other matter around you. Matter is not something that contains atoms; matter is atoms. A desk, for example, is a collection of atoms and the empty space between those atoms. Without the atoms, there would be no desk—just empty space.

Atoms are so small that they cannot be seen even with very strong optical microscopes. Try to imagine the size of an atom by considering that a single teaspoonful of water contains approximately 500,000,000,000,000,000,000,000 atoms. Although atoms are extremely small, they do have a mass. The mass of a single teaspoonful of water is about 5 grams. This mass is equal to the mass of all the atoms that the water is made of added together.

READING TIP

The word *atom* comes from the Greek word *atomos*, meaning "indivisible," or "cannot be divided."

INVESTIGATE Mass

How do you measure the mass of an atom?

PROCEDURE

1. Find the mass of the empty beaker. Record your result.
2. Place 10 pennies into the beaker. Find the mass of the beaker with the pennies in it. Record your result.
3. Subtract the mass of the empty beaker from the mass of the beaker with the pennies. Record your result.
4. Divide the difference in mass by 10. Record your result.

WHAT DO YOU THINK?

- What is the mass of one penny? What assumptions do you make when you answer this question?
- How might scientists use a similar process to find the mass of a single atom?

CHALLENGE All pennies may not be the same. After years of use, some pennies may have had some of their metal rubbed away. Also, the materials that make up pennies have changed. Find the individual mass of several pennies and compare the masses. Do all pennies have exactly the same mass?

SKILL FOCUS
Modeling

MATERIALS
- beam balance
- beaker
- 10 pennies

TIME
20 minutes

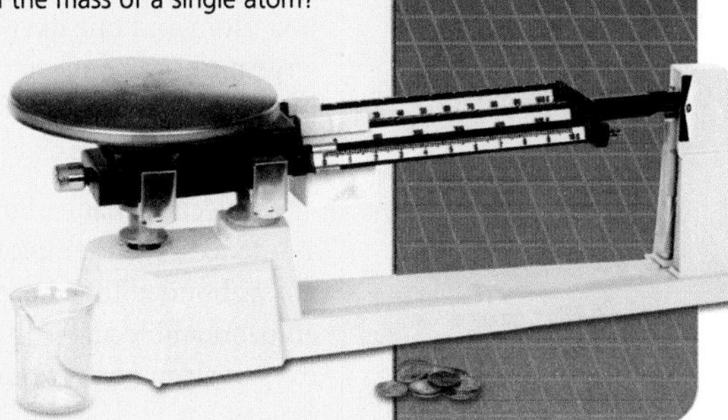

1.2 INSTRUCT

History of Science

The Greek philosophers Leucippus and Democritus developed the first atomic theory in the fifth century B.C. Their view of the atom was different from the modern view of the atom. For example, they thought that objects had different physical properties because their atoms had different shapes. Something that tasted sweet was thought to be made of large, round atoms and sour things made of rough, sharp atoms.

INVESTIGATE Mass

PURPOSE To use modeling to draw conclusions about things that cannot be directly observed

TIPS *20 min.*

- You may want to have students bring pennies from home.
- If necessary, explain that the pennies represent a substance and that each individual penny represents an atom of the substance.

WHAT DO YOU THINK? *About 2.3–2.7 grams; like atoms in a single substance, the pennies are all alike. Scientists could find the mass of a large amount of a substance and divide by the number of atoms to estimate the mass of one atom.*

CHALLENGE *Pennies vary in mass by tiny amounts.*

 Datasheet, Mass, p. 30

Technology Resources

Customize this student lab as needed or look for an alternative. Print rubrics to assess students' lab reports.

 **Lab Generator CD-ROM**

DIFFERENTIATE INSTRUCTION

More Reading Support

A What makes up all matter? *atoms*

English Learners The phrase "hard to imagine" is used several times in this section. Make sure English learners do not interpret the word "hard" as physically hard—like metal is hard. Rather, stress that hard in this context means difficult. Help students understand that some words have more than one meaning.

Advanced Large numbers can be written in scientific notation. Have students learn how to use scientific notation to write large numbers.

 Challenge and Extension, p. 29

Ongoing Assessment

Identify the smallest particles of matter.

Ask: What is the smallest particle that makes up all matter? *atom*

Address Misconceptions

IDENTIFY Ask students to identify what makes up substances besides atoms and molecules. If students do not answer "nothing" or "empty space," they may hold the misconception that atoms and molecules are just some of the parts in a substance, not the substance itself.

CORRECT Explain that atoms or combinations of atoms, such as molecules, are the only matter in a substance. Empty space between atoms is too small to see and has no mass.

REASSESS Ask students what makes up substances. *atoms and combinations of atoms, such as molecules*

Technology Resources

Visit **ClassZone.com** for background on common student misconceptions.

MISCONCEPTION DATABASE

History of Science

Antoine Laurent Lavoisier was a French scientist during the 1700s. Many of his experiments involved breaking compounds into smaller substances and recombining them. He wrote a book that synthesized the chemical knowledge of his time and laid the groundwork for modern atomic theory.

EXPLORE the BIG idea

Revisit "Internet Activity: Scale" on p. 7. Have students compare a substance and its smallest parts.

Ongoing Assessment

Describe how atoms combine into molecules.

Ask: Compare and contrast carbon monoxide molecules with vitamin E molecules. *Both molecules combine atoms, but have different types and numbers of atoms.*

CHECK YOUR READING *Answer: A molecule consists of two or more atoms bonded together.*

Molecules

When two or more atoms bond together, or combine, they make a particle called a **molecule.** A molecule can be made of atoms that are different or atoms that are alike. A molecule of water, for example, is a combination of different atoms—two hydrogen atoms and one oxygen atom. Hydrogen gas molecules are made of the same atom—two hydrogen atoms bonded together.

A molecule is the smallest amount of a substance made of combined atoms that is considered to be that substance. Think about what would happen if you tried to divide water to find its smallest part. Ultimately you would reach a single molecule of water. What would you have if you divided this molecule into its individual atoms of hydrogen and oxygen? If you break up a water molecule, it is no longer water. Instead, you would have hydrogen and oxygen, two different substances.

READING TIP Not all atoms and molecules have color. In this book atoms and molecules are given colors to make them easier to identify.

CHECK YOUR READING How is a molecule related to an atom?

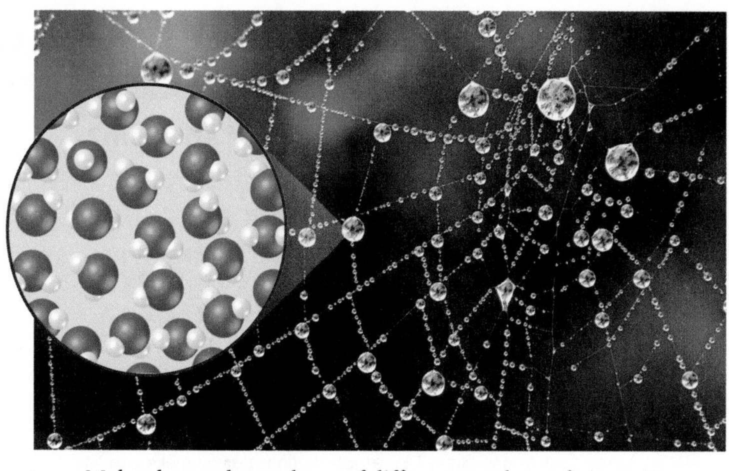

The droplets of water in this spider web are made of water molecules. Each molecule contains two hydrogen atoms (shown in white) and one oxygen atom (shown in red).

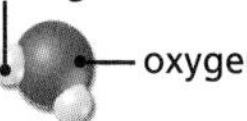

Molecules can be made up of different numbers of atoms. For example, carbon monoxide is a molecule that is composed of one carbon atom and one oxygen atom. Molecules also can be composed of a large number of atoms. The most common type of vitamin E molecule, for example, contains 29 carbon atoms, 50 hydrogen atoms, and 2 oxygen atoms.

Molecules made of different numbers of the same atom are different substances. For example, an oxygen gas molecule is made of two oxygen atoms bonded together. Ozone is also composed of oxygen atoms, but an ozone molecule is three oxygen atoms bonded together. The extra oxygen atom gives ozone properties that are different from those of oxygen gas.

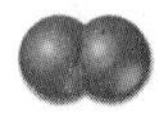
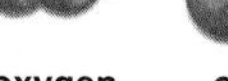

DIFFERENTIATE INSTRUCTION

More Reading Support

B What makes up a molecule? *atoms*

C How do an oxygen atom and an oxygen molecule differ? *O atom = one atom; O_2 molecule = O_2 atoms bond together.*

Below Level Have students draw and label a water molecule and the atoms that are in it.

Inclusion Have tactile learners and students with visual impairments make a model of a water molecule by using two identical gumdrops, a marshmallow, and toothpicks.

This photograph shows the interior of Grand Central Terminal in New York City. Light from the window reflects off dust particles that are being moved by the motion of the molecules in air.

Atoms and molecules are always in motion.

If you have ever looked at a bright beam of sunlight, you may have seen dust particles floating in the air. If you were to watch carefully, you might notice that the dust does not fall toward the floor but instead seems to dart about in all different directions. Molecules in air are constantly moving and hitting the dust particles. Because the molecules are moving in many directions, they collide with the dust particles from different directions. This action causes the darting motion of the dust that you observe.

Atoms and molecules are always in motion. Sometimes this motion is easy to observe, such as when you see evidence of molecules in air bouncing dust particles around. Water molecules move too. When you place a drop of food coloring into water, the motion of the water molecules eventually causes the food coloring to spread throughout the water.

The motion of individual atoms and molecules is hard to observe in solid objects, such as a table. The atoms and molecules in a table cannot move about freely like the ones in water and air. However, the atoms and molecules in a table are constantly moving—by shaking back and forth, or by twisting—even if they stay in the same place.

1.2 Review

KEY CONCEPTS

1. What are atoms?
2. What is the smallest particle of a substance that is still considered to be that substance?
3. Why do dust particles in the air appear to be moving in different directions?

CRITICAL THINKING

4. **Apply** How does tea flavor spread from a tea bag throughout a cup of hot water?
5. **Infer** If a water molecule (H_2O) has two hydrogen atoms and one oxygen atom, how would you describe the makeup of a carbon dioxide molecule (CO_2)?

CHALLENGE

6. **Synthesize** Assume that a water balloon has the same number of water molecules as a helium balloon has helium atoms. If the mass of the water is 4.5 times greater than the mass of the helium, how does the mass of a water molecule compare with the mass of a helium atom?

ANSWERS

1. *the smallest basic unit of matter*
2. *an atom or a molecule*
3. *Constantly moving air molecules collide with the dust particles and bounce them around.*
4. *The motion of the water molecules spreads the tea flavor throughout the water.*
5. *one carbon atom and two oxygen atoms*
6. *The mass of a water molecule is 4.5 times greater.*

Ongoing Assessment

Describe how atoms and molecules move.

Ask: How do the molecules in air, water, and a table move? *Molecules in air and water move about freely. Molecules in a table shake back and forth or twist but stay in the same place.*

Reinforce the BIG idea

Have students relate the section to the Big Idea.

 Reinforcing Key Concepts, p. 31

1.2 ASSESS & RETEACH

Assess

 Section 1.2 Quiz, p. 4

Reteach

Have students write a paragraph that explains what they know about atoms, molecules, and their movement. *Atoms are the smallest particles of matter. They combine to form molecules. Atoms and molecules are always in motion.*

Technology Resources

Have students visit **ClassZone.com** for reteaching of Key Concepts.

 **CONTENT REVIEW**

 **CONTENT REVIEW CD-ROM**

EXTREME SCIENCE
Fun and Motivating Science

Set Learning Goal

To understand how scanning tunneling microscopes make images of atoms

Present the Science

Gerd Binnig and Heinrich Rohrer invented the scanning tunneling microscope in 1981. It is the first microscope that can make three-dimensional images of the atoms on surfaces. Organic molecules, such as DNA, can be attached to a surface and scanned to image their structure.

Discussion Questions

Ask: What does a scanning tunneling microscope measure to make an image of atoms on the surface of a material? *the interaction between the electrically charged needle tip and the nearest atom on the surface of the material*

Ask: What does the series of bumps on the image show? *where the atoms are located*

Ask: Besides making images, what can the tip of a STM needle do? *move atoms around on a surface*

Close

Ask: How do the structures in a scanning tunneling microscope compare with what you can see with your eyes or a regular light microscope? *You cannot see atoms with your eyes or an optical microscope. Scanning tunneling microscopes give you images similar to those of a contour map.*

Technology Resources

Have students visit **ClassZone.com** to find more images from scanning tunneling microscopes

RESOURCE CENTER

EXTREME SCIENCE

LOOKING AT ATOMS

Particles Too Small to See

Atoms are so small that you cannot see them through an ordinary microscope. In fact, millions of them could fit in the period at the end of this sentence. Scientists can make images of atoms, however, using an instrument called a scanning tunneling microscope (STM).

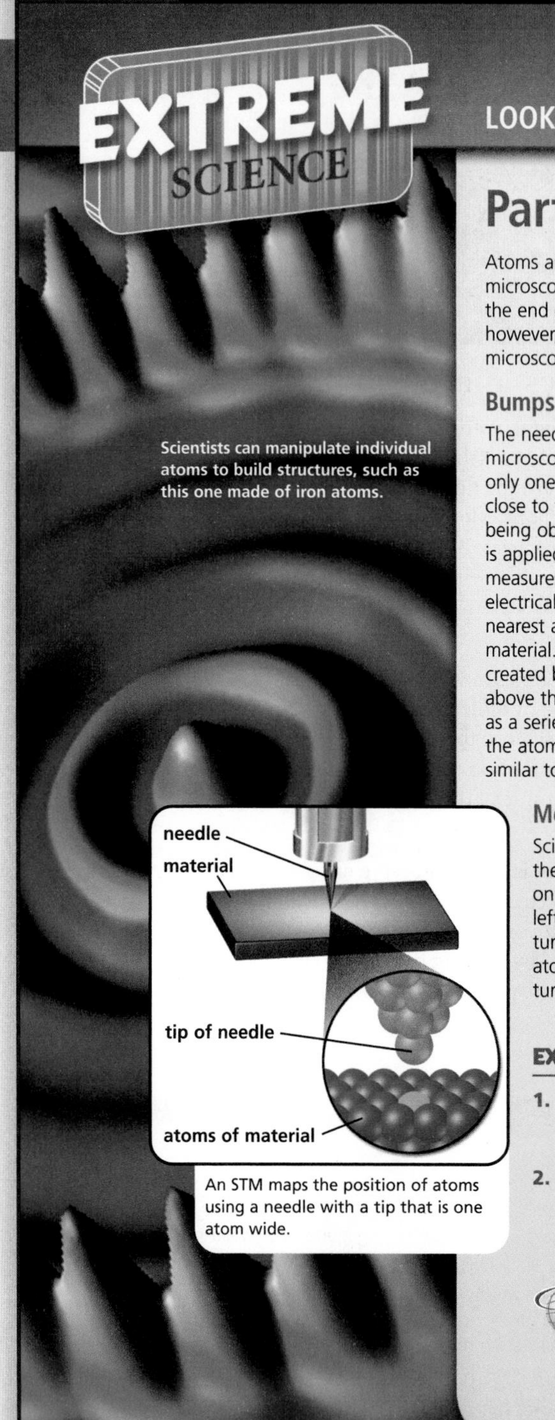

Scientists can manipulate individual atoms to build structures, such as this one made of iron atoms.

An STM maps the position of atoms using a needle with a tip that is one atom wide.

Bumps on a Surface

The needle of the scanning tunneling microscope has a very sharp tip that is only one atom wide. The tip is brought close to the surface of the material being observed, and an electric current is applied to the tip. The microscope measures the interaction between the electrically charged needle tip and the nearest atom on the surface of the material. An image of the surface is created by moving the needle just above the surface. The image appears as a series of bumps that shows where the atoms are located. The result is similar to a contour map.

Moving Atoms

Scientists also can use the tip of the STM needle to move atoms on a surface. The large image at left is an STM image of a structure made by pushing individual atoms into place on a very smooth metal surface. This structure was designed as a corral to trap individual atoms inside.

Tiny Pieces of Matter

- Images of atoms did not exist until 1970.
- Atoms are so small that a single raindrop contains more than 500 billion trillion atoms.
- If each atom were the size of a pea, your fingerprint would be larger than Alaska.
- In the space between stars, matter is so spread out that a volume of one liter contains only about 1000 atoms.

EXPLORE

1. **INFER** Why must the tip of a scanning tunneling microscope be only one atom wide to make an image of atoms on a surface?
2. **CHALLENGE** Find out more about images of atoms on the Internet. How are STM images used in research to design better materials?

RESOURCE CENTER
CLASSZONE.COM
Find more images from scanning tunneling microscopes.

EXPLORE

1. ***INFER*** *The tip needs to get close to one atom to be able to measure the interaction between the atom and the tip.*
2. ***CHALLENGE*** *STM images can be used to help scientists understand structural details and alter the structures for a specific task.*

KEY CONCEPT

Matter combines to form different substances.

BEFORE, you learned

- Matter is made of tiny particles called atoms
- Atoms combine to form molecules

NOW, you will learn

- How pure matter and mixed matter are different
- How atoms and elements are related
- How atoms form compounds

VOCABULARY

element p. 22
compound p. 23
mixture p. 23

EXPLORE Mixed Substances

What happens when substances are mixed?

PROCEDURE

1. Observe and describe a teaspoon of cornstarch and a teaspoon of water.
2. Mix the two substances together in the cup. Observe and describe the result.

WHAT DO YOU THINK?

- After you mixed the substances, could you still see each substance?
- How was the new substance different from the original substances?

MATERIALS

- cornstarch
- water
- small cup
- spoon

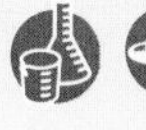

Matter can be pure or mixed.

MAIN IDEA AND DETAILS Continue to organize your notes in a two-column chart as you read.

Matter can be pure, or it can be two or more substances mixed together. Most of the substances you see around you are mixed, although you can't always tell that by looking at them. For example, the air you breathe is a combination of several substances. Wood, paper, steel, and lemonade are all mixed substances.

You might think that the water that you drink from a bottle or from the tap is a pure substance. However, drinking water has minerals dissolved in it and chemicals added to it that you cannot see. Often the difference between pure and mixed substances is apparent only on the atomic or molecular level.

A pure substance has only one type of component. For example, pure water contains only water molecules. Pure silver contains only silver atoms. Coins and jewelry that look like silver are often made of silver in combination with other metals.

RESOURCES FOR DIFFERENTIATED INSTRUCTION

Below Level

UNIT RESOURCE BOOK

- Reading Study Guide A, pp. 34–35
- Decoding Support, p. 58

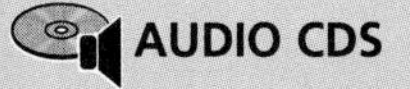

AUDIO CDS

Advanced

UNIT RESOURCE BOOK

Challenge and Extension, p. 40

English Learners

UNIT RESOURCE BOOK

Spanish Reading Study Guide, pp. 38–39

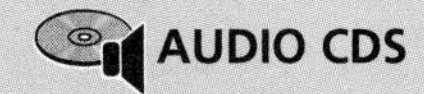

AUDIO CDS

- Audio Readings in Spanish
- Audio Readings (English)

1.3 FOCUS

Set Learning Goals

Students will

- Describe how pure and mixed matter are different.
- Explain how atoms and elements are related.
- Describe how atoms form compounds.
- Observe and compare in an experiment the properties of individual substances with the properties of mixtures.

3-Minute Warm-Up

Display Transparency 5 or copy this exercise on the board:

Match each definition with the correct term.

Definitions

1. the smallest basic unit of matter *c*
2. a particle made when two or more atoms combine *e*
3. anything that has mass and takes up space *a*

Terms

a. matter
b. volume
c. atom
d. weight
e. molecule

 3-Minute Warm-Up, p. T5

1.3 MOTIVATE

EXPLORE Mixed Substances

PURPOSE To introduce students to mixtures

TIP *10 min.* Ask students to clean up any spills immediately.

WHAT DO YOU THINK? *No; after the two substances were mixed, the new substance had a consistency different from the one that either substance had alone.*

1.3 INSTRUCT

Teacher Demo

Many students have the misconception that atoms and molecules have the same properties as the substances they make up. To help students understand that properties they are familiar with, such as color and texture, are not the same as the properties of atoms, try the following:

- Show students pictures or samples of graphite and diamond. Explain that both are pure substances.
- Have students discuss whether or not graphite and diamond are made up of the same atoms or molecules.
- Explain to students that graphite and diamond are both made of carbon atoms. Although the individual atoms all have the same properties, different arrangements of carbon atoms give graphite and diamond very different properties.

Teach from Visuals

To help students interpret the photographs of gold and dry ice along with the diagrams of atoms and molecules, ask:

- Based on the diagram, what does the gold in the photograph consist of? *atoms of gold*
- Based on the diagram, what does the dry ice consist of? *molecules of carbon dioxide, which consist of one carbon atom and two oxygen atoms*
- Is dry ice a pure substance? Why? *Yes; it contains only one kind of molecule.*

Ongoing Assessment

Explain how atoms and elements are related.

Ask: What is the smallest possible amount of an element? *an atom*

CHECK YOUR READING *Answer: An element contains only one type of atom.*

REMINDER

A molecule consists of two or more atoms that are bonded together.

If you could look at the atoms in a bar of pure gold, you would find only gold atoms. If you looked at the atoms in a container of pure water, you would find water molecules, which are a combination of hydrogen and oxygen atoms. Does the presence of two types of atoms mean that water is not really a pure substance after all?

A substance is considered pure if it contains only a single type of atom, such as gold, or a single combination of atoms that are bonded together, such as a water molecule. Because the hydrogen and oxygen atoms are bonded together as molecules, water that has nothing else in it is considered a pure substance.

Elements

One type of pure substance is an element. An **element** is a substance that contains only a single type of atom. The number of atoms is not important as long as all the atoms are of the same type. You cannot separate an element into other substances.

You are probably familiar with many elements, such as silver, oxygen, hydrogen, helium, and aluminum. There are as many elements as there are types of atoms—more than 100. You can see the orderly arrangement of atoms in the element gold, on the left below.

CHECK YOUR READING Why is an element considered to be a pure substance?

The atoms in gold are all the same type of atom. Therefore, gold is an element.

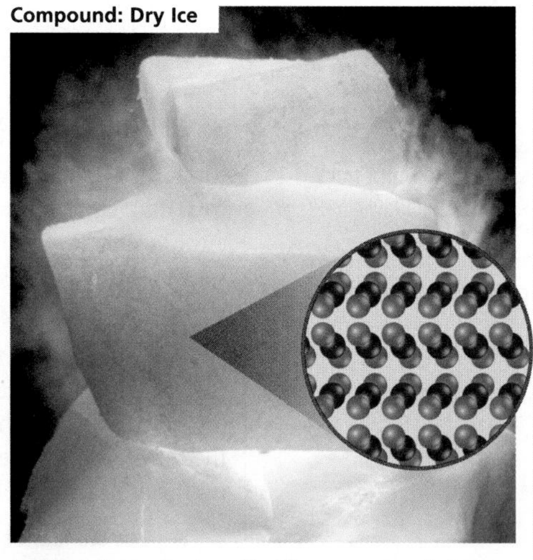

Dry ice is frozen carbon dioxide, a compound. Each molecule is made of one carbon atom and two oxygen atoms.

A 22 Unit: Matter and Energy

DIFFERENTIATE INSTRUCTION

More Reading Support

A Why is water a pure substance even though it contains two types of atoms? *The atoms are bonded in one type of molecule.*

English Learners Help English learners recognize cause-and-effect relationships in sentences that do not follow the *If/then* convention. For example, on p. 22, *then* is implied and not stated; "If you could look at the atoms in a bar of pure gold, you would find only gold atoms." Likewise, the sentence beginning with *because* in the second paragraph conveys a cause-and-effect relationship; the fact that hydrogen and oxygen atoms are bonded together in water molecules makes water a pure substance.

Compounds

A **compound** is a substance that consists of two or more different types of atoms bonded together. A large variety of substances can be made by combining different types of atoms to make different compounds. Some types of compounds are made of molecules, such as water and carbon dioxide, shown on page 22. Other compounds are made of atoms that are bonded together in a different way. Table salt is an example.

A compound can have very different properties from the individual elements that make up that compound. Pure table salt is a common compound that is a combination of sodium and chlorine. Although table salt is safe to eat, the individual elements that go into making it—sodium and chlorine—are poisonous.

What is the relationship between atoms and a compound?

Mixtures

Most of the matter around you is a mixture of different substances. Seawater, for instance, contains water, salt, and other minerals mixed together. Your blood is a mixture of blood cells and plasma. Plasma is also a mixture, made up of water, sugar, fat, protein, salts, and minerals.

A **mixture** is a combination of different substances that remain the same individual substances and can be separated by physical means. For example, if you mix apples, oranges, and bananas to make a fruit salad, you do not change the different fruits into a new kind of fruit. Mixtures do not always contain the same amount of the various substances. For example, depending on how the salad is made, the amount of each type of fruit it contains will vary.

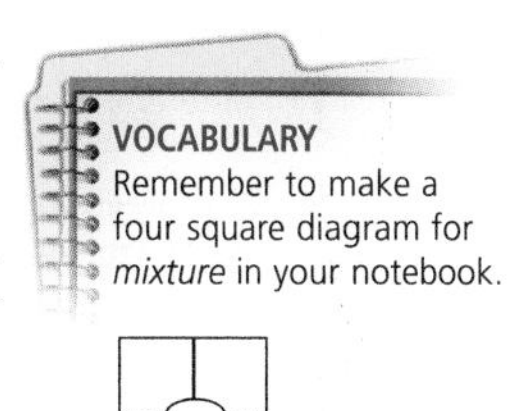

VOCABULARY
Remember to make a four square diagram for *mixture* in your notebook.

APPLY In what ways can a city population be considered a mixture?

DIFFERENTIATE INSTRUCTION

More Reading Support

B What is a combination of substances that remain the same individual substances? *a mixture*

Below Level Have students draw and label a graphic organizer or table that compares atoms, elements, molecules, compounds, and mixtures. Ask students to include a definition and an example of each.

Teach Difficult Concepts

Explain that if you break a pure compound into the smallest parts that are still that substance, you do not have individual atoms or elements. If you break a pure element into the smallest parts that are still that substance, you have atoms of that element. Ask: What do you have if you break a compound into the parts that compose it? *atoms of the elements that compose it*

Integrate the Sciences

The human body and other organisms contain organic compounds and inorganic compounds. Organic compounds always contain the element carbon. Carbon can bond with other elements and with thousands of other carbon atoms to make huge molecules. Proteins and carbohydrates are organic compounds. Most inorganic compounds do not contain carbon and are small. Oxygen, carbon dioxide, and water are some inorganic compounds in the human body.

Ongoing Assessment

Describe how pure and mixed matter are different.

Ask: Are most of the substances around you pure or mixed matter? *mixed*

Ask: What is the difference between a pure substance and a mixed substance? *A pure substance has only one type of atom or molecule, while a mixed substance contains two or more different substances.*

Describe how atoms form compounds.

Ask: What is the lowest number of types of atoms that can bond to form a compound? *two*

CHECK YOUR READING *Answer: A compound is made up of two or more different types of atoms bonded together.*

PHOTO CAPTION Answer: A city population is a mixture of men and women, people of different ages, and people of different ethnic backgrounds.

INVESTIGATE Mixtures

PURPOSE To show that not all liquids behave the same way when mixed with other liquids

TIPS *20 min.*

- You may want to have students bring clear, plastic jars with screw-on lids from home.
- Test tubes with stoppers could be substituted for the jars.

WHAT DO YOU THINK? *Yes; the water changes color. No; the oil floats on the water. The oil floats to the new "top." Some liquids mix easily; others do not.*

CHALLENGE *The soap breaks up the oil into smaller droplets. This helps water mix with the oil.*

Datasheet, Mixtures, p. 41

Technology Resources

Customize this student lab as needed or look for an alternative. Print rubrics to assess student lab reports.

Lab Generator CD-ROM

Ongoing Assessment

Answer: A mixture combines different substances that remain those substances. A compound has different properties than the individual elements that make it up.

INVESTIGATE Mixtures

How well do oil and water mix?

PROCEDURE

1. Add a few drops of food coloring to the water in the beaker. Swirl the water around in the beaker until the water is evenly colored throughout.
2. Pour the colored water from the beaker into the jar until the jar is about one-fourth full.
3. Add the same amount of vegetable oil to the jar. Screw the lid tightly on the jar.
4. Carefully shake the jar several times with your hand over the cover, and then set it on the table. Observe and record what happens to the liquids in the jar.
5. Turn the jar upside down and hold it that way. Observe what happens to the liquids and record your observations.

WHAT DO YOU THINK?

- Does water mix with food coloring? What evidence supports your answer?
- Do water and oil mix? What evidence supports your answer?
- What happened when you turned the jar upside down?
- Based on your observations, what can you infer about the ability of different liquids to mix?

CHALLENGE To clean greasy dishes, you add soap to the dishwater. Try adding soap to your mixture. What does the soap do?

SKILL FOCUS
Inferring

MATERIALS
- food coloring
- beaker of water
- clear jar with screw-on lid
- vegetable oil

TIME
20 minutes

Comparing Mixtures and Compounds

Find out more about mixtures.

Although mixtures and compounds may seem similar, they are very different. Consider how mixtures and compounds compare with each other.

- The substances in mixtures remain the same substances. Compounds are new substances formed by atoms that bond together.
- Mixtures can be separated by physical means. Compounds can be separated only by breaking the bonds between atoms.
- The proportions of different substances in a mixture can vary throughout the mixture or from mixture to mixture. The proportions of different substances in a compound are fixed because the type and number of atoms that make up a basic unit of the compound are always the same.

How is a mixture different from a compound?

DIFFERENTIATE INSTRUCTION

More Reading Support

C If two substances bond together, is the new substance a compound or a mixture? *compound*

Advanced

Challenge and Extension, p. 40

Parts of mixtures can be the same or different throughout.

It is obvious that something is a mixture when you can see the different substances in it. For example, if you scoop up a handful of soil, you might see that it contains dirt, small rocks, leaves, and even insects. You can separate the soil into its different parts.

Exactly what you see depends on what part of the soil you scoop up. One handful of soil might have more pebbles or insects in it than another handful would. There are many mixtures, such as soil, that have different properties in different areas of the mixture. Such a mixture is called a heterogeneous (HEHT-uhr-uh-JEE-nee-uhs) mixture.

In some types of mixtures, however, you cannot see the individual substances. For example, if you mix sugar into a cup of water and stir it well, the sugar seems to disappear. You can tell that the sugar is still there because the water tastes sweet, but you cannot see the sugar or easily separate it out again.

When substances are evenly spread throughout a mixture, you cannot tell one part of the mixture from another part. For instance, one drop of sugar water will be almost exactly like any other drop. Such a mixture is called a homogeneous (HOH-muh-JEE-nee-uhs) mixture. Homogenized milk is processed so that it becomes a homogeneous mixture of water and milk fat. Milk that has not been homogenized will separate—most of the milk fat will float to the top as cream while leaving the rest of the milk low in fat.

READING TIP

The prefix *hetero* means "different," and the prefix *homo* means "same." The Greek root *genos* means "kind."

1.3 Review

KEY CONCEPTS

1. What is the difference between pure and mixed matter?
2. How are atoms and elements related?
3. How are compounds different from mixtures?

CRITICAL THINKING

4. **Infer** What can you infer about the size of sugar particles that are dissolved in a mixture of sugar and water?
5. **Infer** Why is it easier to remove the ice cubes from cold lemonade than it is to remove the sugar?

CHALLENGE

6. **Apply** A unit of sulfuric acid is a molecule of 2 atoms of hydrogen, 1 atom of sulfur, and 4 atoms of oxygen. How many of each type of atom are there in 2 molecules of sulfuric acid?

ANSWERS

1. Pure matter is made of only one type of atom or molecule; mixed matter has two or more substances.

2. An element is composed of only one type of atom.

3. Mixtures combine two or more substances that remain the same substances. Compounds contain two or more types of atoms bonded together. Mixtures can be separated by physical means, but compounds can be separated only by breaking the bond between atoms.

4. The sugar particles are very small.

5. The ice cubes are bigger.

6. 4 hydrogen atoms, 2 sulfur atoms, 8 oxygen atoms

Integrate the Sciences

Crops grow best in soil that is a mixture of sand, silt, clay, and a large amount of humus. This mixture retains water long enough for plants to use it but allows water to drain well enough that the soil doesn't stay too wet. Soil dries out more quickly when it contains more sand. Soil with a larger amount of clay particles doesn't drain as well and is more difficult for plant roots to grow in. Soil with a large amount of humus retains moisture, drains well, and contains nutrients that plants need to grow.

EXPLORE the BIG idea

Revisit "Where Does the Sugar Go?" on p. 7. Have students explain their observations and answer the questions again.

Reinforce the BIG idea

Have students relate the section to the Big Idea.

 Reinforcing Key Concepts, p. 42

1.3 ASSESS & RETEACH

Assess

 Section 1.3 Quiz, p. 5

Reteach

Ask students to compare each pair of terms below and give a definition and examples.

- pure matter and mixed matter
- elements and atoms
- elements and compounds
- mixtures and compounds
- heterogeneous mixtures and homogeneous mixtures

Technology Resources

Have students visit **ClassZone.com** for reteaching of Key Concepts.

 **CONTENT REVIEW**

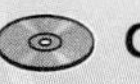 **CONTENT REVIEW CD-ROM**

MATH IN SCIENCE
Math Skills Practice for Science

Set Learning Goal

To make circle graphs to show the amount of different substances in mixtures

Present the Science

Cumin is from the dried seedlike fruit of a cumin plant. Ginger is from the underground stems, or rhizomes, of a ginger plant. Nutmeg is from the seed of a tree. Spice and herb mixtures that vary in terms of amounts of each ingredient include chili powder, curry powder, and poultry seasoning.

Develop Graphing Skills

- Remind students that when they multiply a number by a fraction, they first multiply the number by the numerator of the fraction and then divide by the denominator.
- Remind students that the number of degrees of the individual substances in a circle graph must add up to 360.
- Review how to use a protractor.

DIFFERENTIATION TIP Allow students who have trouble with fine motor skills or are who are visually impaired to round the numbers to whole numbers, which are easier to see on the protractor and easier to draw.

Close

Ask students what they can learn about mixtures from a circle graph and its labels. *how much of different substances make up the mixture*

- Math Support, p. 61
- Math Practice, p. 62

Technology Resources

Students can visit **ClassZone.com** for practice in making circle graphs.

MATH TUTORIAL

MATH in SCIENCE

SKILL: MAKING A CIRCLE GRAPH

A Mixture of Spices

MATH TUTORIAL
CLASSZONE.COM
Click on Math Tutorial for more help with circle graphs.

Two different mixtures of spices may contain the exact same ingredients but have very different flavors. For example, a mixture of cumin, nutmeg, and ginger powder can be made using more cumin than ginger, or it can be made using more ginger than cumin.

One way to show how much of each substance a mixture contains is to use a circle graph. A circle graph is a visual way to show how a quantity is divided into different parts. A circle graph represents quantities as parts of a whole.

Example

Make a circle graph to represent a spice mixture that is 1/2 cumin, 1/3 nutmeg, and 1/6 ginger.

(1) To find the angle measure for each sector of the circle graph, multiply each fraction in your mixture by 360°.

Cumin: $\frac{1}{2} \cdot 360° = 180°$

Nutmeg: $\frac{1}{3} \cdot 360° = 120°$

Ginger: $\frac{1}{6} \cdot 360° = 60°$

(2) Use a compass to draw a circle. Use a protractor to draw the angle for each sector.

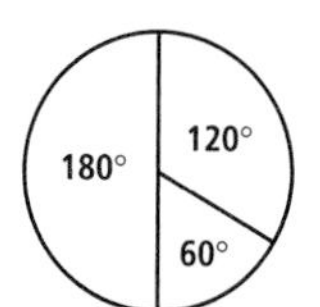

(3) Label each sector and give your graph a title.

ANSWER

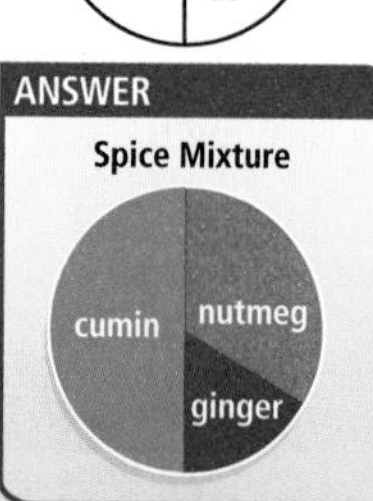

Answer the following questions.

1. Draw a circle graph representing a spice mixture that is 1/2 ginger, 1/4 cumin, and 1/4 crushed red pepper.
2. A jeweler creates a ring that is 3/4 gold, 3/16 silver, and 1/16 copper. Draw a circle graph representing the mixture of metals in the ring.
3. Draw a circle graph representing a mixture that is 1/5 sand, 2/5 water, and 2/5 salt.

CHALLENGE Dry air is a mixture of about 78 percent nitrogen, 21 percent oxygen, and 1 percent other elements. Create a circle graph representing the elements found in air.

A 26 Unit: Matter and Energy

ANSWERS

1. Graphs should show 180° ginger, 90° cumin, and 90° crushed red pepper.

2. Graphs should show 270° gold, 67.5° silver, and 22.5° copper.

3. Graphs should show 72° sand, 144° water, and 144° salt.

CHALLENGE Graphs should show 280.8° nitrogen, 75.6° oxygen, and 3.6° other elements.

KEY CONCEPT

1.4 Matter exists in different physical states.

BEFORE, you learned
- Matter has mass
- Matter is made of atoms
- Atoms and molecules in matter are always moving

NOW, you will learn
- About the different states of matter
- How the different states of matter behave

VOCABULARY
states of matter p. 27
solid p. 28
liquid p. 28
gas p. 28

EXPLORE Solids and Liquids

How do solids and liquids compare?

PROCEDURE

1. Observe the water, ice, and marble. Pick them up and feel them. Can you change their shape? their volume?
2. Record your observations. Compare and contrast each object with the other two.

WHAT DO YOU THINK?
- How are the ice and the water in the cup similar? How are they different?
- How are the ice and the marble similar? How are they different?

MATERIALS
- water in a cup
- ice cube
- marble
- pie tin

Particle arrangement and motion determine the state of matter.

When you put water in a freezer, the water freezes into a solid (ice). When you place an ice cube on a warm plate, the ice melts into liquid water again. If you leave the plate in the sun, the water becomes water vapor. Ice, water, and water vapor are made of exactly the same type of molecule—a molecule of two hydrogen atoms and one oxygen atom. What, then, makes them different?

Ice, water, and water vapor are different states of water. **States of matter** are the different forms in which matter can exist. The three familiar states are solid, liquid, and gas. When a substance changes from one state to another, the molecules in the substance do not change. However, the arrangement of the molecules does change, giving each state of matter its own characteristics.

1.4 FOCUS

Set Learning Goals

Students will
- Describe the different states of matter.
- Describe how the different states of matter behave.
- Experiment with the behavior of different liquids.

3-Minute Warm-Up

Display Transparency 5 or copy this exercise on the board:

Draw and label a diagram that shows the relationship between atoms and molecules. Write a caption for your diagram that explains the relationship. *Diagrams should show at least two atoms bonded together. The atoms may be different or alike. Sample caption: Atoms combine to form a molecule.*

T 3-Minute Warm-Up, p. T5

1.4 MOTIVATE

EXPLORE Solids and Liquids

PURPOSE To compare solids and liquids

TIP *10 min.* Ask students to clean up any spills immediately.

WHAT DO YOU THINK? *The ice and water are both made of water molecules; however, one is solid and the other is liquid. The ice and the marble are both solid; however, the ice can melt and change shape.*

RESOURCES FOR DIFFERENTIATED INSTRUCTION

Below Level

UNIT RESOURCE BOOK
- Reading Study Guide A, pp. 45–46
- Decoding Support, p. 58

AUDIO CDS

Additional INVESTIGATION, Thick and Thin Liquids, A, B, & C, pp. 72–80; Teacher Instructions, pp. 262–263

Advanced

UNIT RESOURCE BOOK
- Challenge and Extension, p. 51
- Challenge Reading, pp. 54–55

English Learners

UNIT RESOURCE BOOK
Spanish Reading Study Guide, pp. 49–50

AUDIO CDS
- Audio Readings in Spanish
- Audio Readings (English)

1.4 INSTRUCT

Develop Critical Thinking

COMPARE Have students compare the three states of matter in a table while listing the characteristics of each. The tables should include at least one example of each state of matter.

Teaching with Technology

If a video camera is available, have students record real-world models of the movements of particles in solids, liquids, and gases. Encourage students to narrate a comparison of the movements they record.

Real World Example

The expansion of water as it freezes often causes damage. Put a can of soda in a freezer, and it will expand enough to break the seal. Pipes carrying water may freeze and burst if temperatures drop too low. Frost in the ground can push posts and pavement out of alignment. Ice expanding in small cracks and in the road base leads to potholes.

Ongoing Assessment

CHECK YOUR READING *Answer: A solid has a fixed shape, but a liquid does not. Particles in a solid are fixed in one place, but particles in a liquid are not fixed in place and can move.*

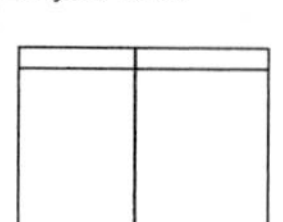

MAIN IDEA AND DETAILS Remember to organize your notes in a two-column chart as you read.

A Solid, liquid, and gas are common states of matter.

A substance can exist as a solid, a liquid, or a gas. The state of a substance depends on the space between its particles and on the way in which the particles move. The illustration on page 29 shows how particles are arranged in the three different states.

1. A **solid** is a substance that has a fixed volume and a fixed shape. In a solid, the particles are close together and usually form a regular pattern. Particles in a solid can vibrate but are fixed in one place. Because each particle is attached to several others, individual particles cannot move from one location to another, and the solid is rigid.
2. A **liquid** has a fixed volume but does not have a fixed shape. Liquids take on the shape of the container they are in. The particles in a liquid are attracted to one another and are close together. However, particles in a liquid are not fixed in place and can move from one place to another.
3. A **gas** has no fixed volume or shape. A gas can take on both the shape and the volume of a container. Gas particles are not close to one another and can move easily in any direction. There is much more space between gas particles than there is between particles in a liquid or a solid. The space between gas particles can increase or decrease with changes in temperature and pressure.

CHECK YOUR READING Describe two differences between a solid and a gas.

B

The particles in a solid are usually closer together than the particles in a liquid. For example, the particles in solid steel are closer together than the particles in molten—or melted—steel. However, water is an important exception. The molecules that make up ice actually have more space between them than the molecules in liquid water do.

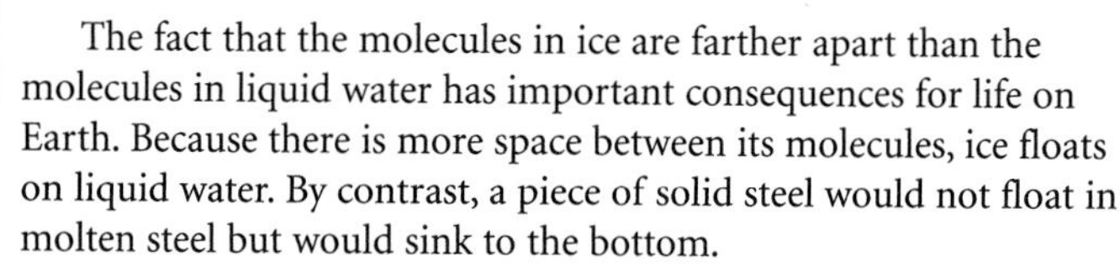

The fact that the molecules in ice are farther apart than the molecules in liquid water has important consequences for life on Earth. Because there is more space between its molecules, ice floats on liquid water. By contrast, a piece of solid steel would not float in molten steel but would sink to the bottom.

Because ice floats, it remains on the surface of rivers and lakes when they freeze. The ice layer helps insulate the water and slow down the freezing process. Animals living in rivers and lakes can survive in the liquid water layer below the ice layer.

DIFFERENTIATE INSTRUCTION

More Reading Support

A What are three common states of matter? *solid, liquid, gas*

B What substance's molecules are farther apart when it is a solid than when it is a liquid? *water*

English Learners English learners may have trouble comprehending certain words and phrases used repeatedly in this section. Tell students that the terms *fixed*, *regular*, *definite*, and *rigid* describe things that do not change.

Use the Chapter Summary page to help students focus on key concepts and vocabulary.

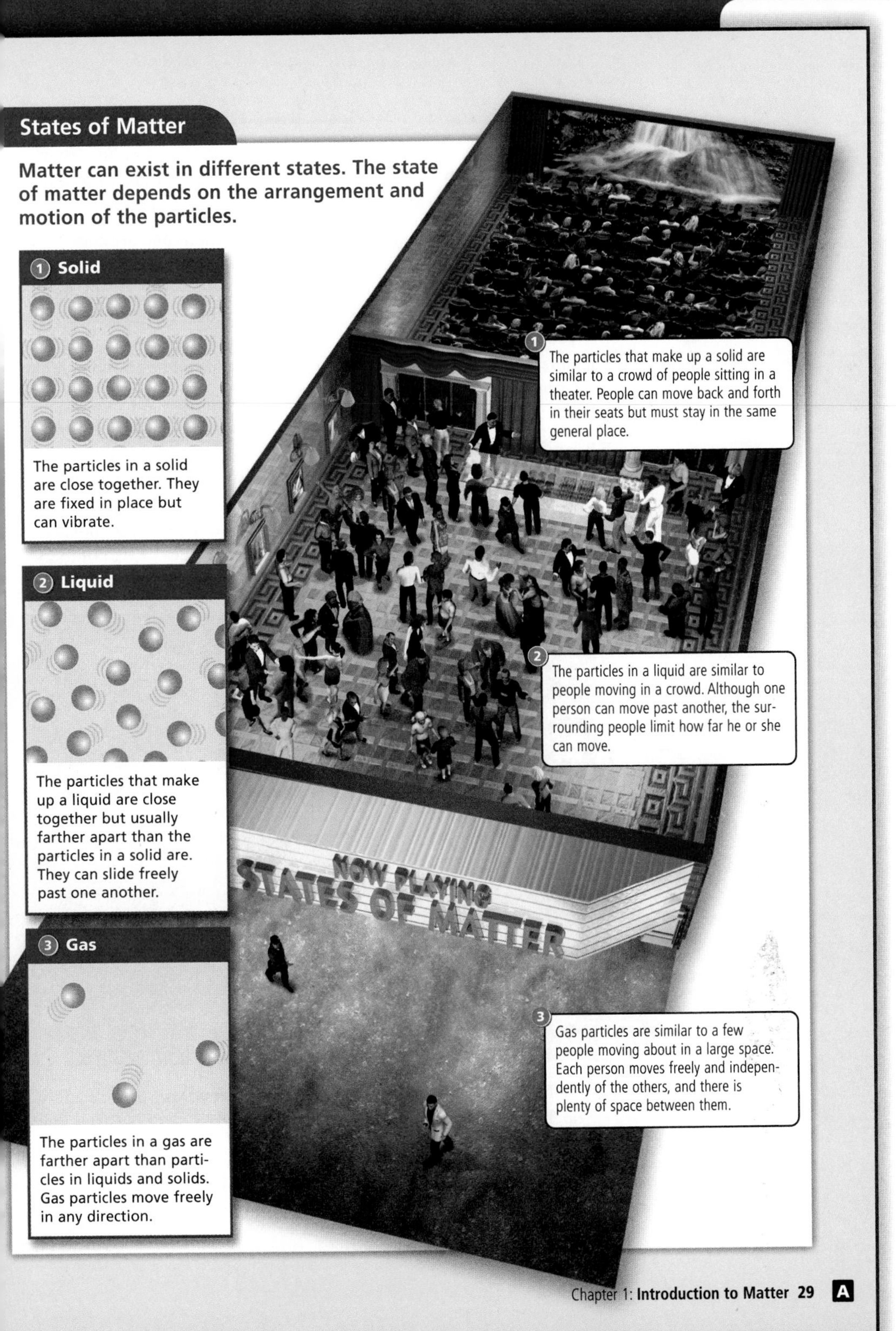

States of Matter
Matter can exist in different states. The state of matter depends on the arrangement and motion of the particles.
1 Solid
The particles in a solid are close together. They are fixed in place but can vibrate.
2 Liquid
The particles that make up a liquid are close together but usually farther apart than the particles in a solid are. They can slide freely past one another.
3 Gas
The particles in a gas are farther apart than particles in liquids and solids. Gas particles move freely in any direction.
1 The particles that make up a solid are similar to a crowd of people sitting in a theater. People can move back and forth in their seats but must stay in the same general place.
2 The particles in a liquid are similar to people moving in a crowd. Although one person can move past another, the surrounding people limit how far he or she can move.
3 Gas particles are similar to a few people moving about in a large space. Each person moves freely and independently of the others, and there is plenty of space between them.
NOW PLAYING
STATES OF MATTER

Chapter 1: **Introduction to Matter** 29 A

Teach from Visuals

To help students interpret the diagrams of the arrangement and motion of particles in three states of matter, ask:

- Which two characteristics determine the state of matter? *arrangement and motion of particles*
- Which state of matter has particles that are farthest apart? *gas*
- Which state of matter has particles that are generally closest together? *solid*
- Which state of matter has particles that can slide past each other but cannot move freely in any direction? *liquid*

 This visual is also available as T6 in the Unit Transparency Book.

Teach Difficult Concepts

Students may have difficulty appreciating how much farther apart gas molecules are compared with the molecules in liquids and solids. On average, gas molecules are spaced about ten times farther apart than molecules in a liquid or solid. Have students model the relative spacing of the three states of matter with their own bodies in a large room.

DIFFERENTIATE INSTRUCTION

Advanced Plasma is a fourth state of matter and makes up at least 99 percent of the universe. The Sun and lightning are made up of plasma. Have students make a list of other examples of where plasma exists. Ask them to identify examples of artificial plasma and natural plasma.

 Challenge and Extension, p. 51

Have students who are interested in quarks, the fundamental building blocks of matter, read the following article.

 Challenge Reading, pp. 54–55

Teach Difficult Concepts

In most cases when a solid is broken, the volume of the pieces will add up to the volume of the original solid. For example, if you saw a block of wood into two pieces, the volume of each piece will add up to the original volume of the block. However, if a solid is divided into small enough pieces, such as when wood is ground into sawdust, the total volume might change.

Teach from Visuals

Have students identify the hexagonal pattern of the water molecules in the visual. Compare the hexagonal pattern with the solid shown in the illustration on p. 29. Point out that it is the relatively large amount of space between the molecules in ice that causes ice to float in water.

Ongoing Assessment

CHECK YOUR READING *Answer: a definite shape and a definite volume*

Solids have a definite volume and shape.

REMINDER
Volume is the amount of space that an object occupies.

A piece of ice, a block of wood, and a ceramic cup are solids. They have shapes that do not change and volumes that can be measured. Any matter that is a solid has a definite shape and a definite volume.

? C

The molecules in a solid are in fixed positions and are close together. Although the molecules can still vibrate, they cannot move from one part of the solid to another part. As a result, a solid does not easily change its shape or its volume. If you force the molecules apart, you can change the shape and the volume of a solid by breaking it into pieces. However, each of those pieces will still be a solid and have its own particular shape and volume.

The particles in some solids, such as ice or table salt, occur in a very regular pattern. The pattern of the water molecules in ice, for example, can be seen when you look at a snowflake like the one shown below. The water molecules in a snowflake are arranged in hexagonal shapes that are layered on top of one another. Because the molecular pattern has six sides, snowflakes form with six sides or six points. Salt also has a regular structure, although it takes a different shape.

The particles in many solids, such as the water molecules in this snowflake, have a regular pattern.

Not all solids have regular shapes in the same way that ice and salt do, however. Some solids, such as plastic or glass, have particles that are not arranged in a regular pattern.

CHECK YOUR READING What two characteristics are needed for a substance to be a solid?

DIFFERENTIATE INSTRUCTION

More Reading Support

C Which state of matter has a definite volume and a definite shape? *solid*

Below Level Point out that different words sometimes refer to the same concept. For example, *arrangement of particles* refers to the space between particles. The words *fixed* and *definite* mean the same thing when referring to the volume and the shape of matter.

Liquids have a definite volume but no definite shape.

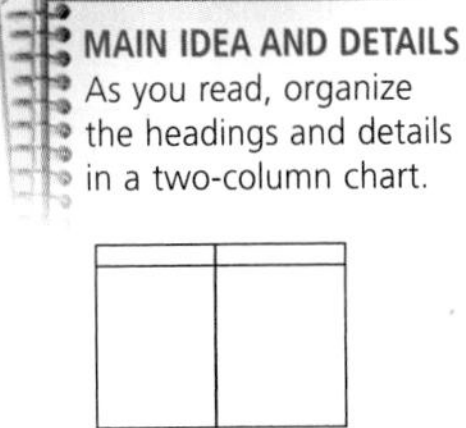

MAIN IDEA AND DETAILS
As you read, organize the headings and details in a two-column chart.

Water, milk, and oil are liquids. A liquid has a definite volume but does not have a definite shape. The volume of a certain amount of oil can be measured, but the shape that the oil takes depends on what container it is in. If the oil is in a tall, thin container, it has a tall, thin shape. If it is in a short, wide container, it has a short, wide shape. Liquids take the shape of their containers.

The molecules in a liquid are close together, but they are not tightly attached to one another as the molecules in a solid are. Instead, molecules in liquids can move independently. As a result, liquids can flow. Instead of having a rigid form, the molecules in a liquid move and fill the bottom of the container they are in.

How is a liquid different from a solid?

INVESTIGATE Liquids

How do different liquids behave?

PROCEDURE

1. Using the graduated cylinder, measure 5 mL of colored water. Add it to the test tube.
2. Measure 5 mL of vegetable oil. Pour the oil into the test tube. Record your observations.
3. Pour a small amount of corn syrup directly into the test tube. Record what happens to all three liquids.
4. Add 10 mL more of colored water to the test tube and record what happens.
5. Add 5 mL more of vegetable oil and record what happens.

WHAT DO YOU THINK?

- How did the layers change as more liquid was added?
- What are some behaviors of each of the liquids in this experiment that can be used to tell them apart?
- What would happen if you changed the order in which you added the liquids?

CHALLENGE Think of a liquid you are familiar with that was not used in this experiment. What do you think would happen if you added that liquid to your test tube? Explain.

SKILL FOCUS
Measuring

MATERIALS
- graduated cylinder
- colored water
- test tube
- test-tube rack
- vegetable oil
- corn syrup

TIME
20 minutes

INVESTIGATE Liquids

PURPOSE To practice measuring liquids and to learn about the behavior of different liquids

TIPS *20 min.*

- Use dark corn syrup so that it cannot be confused with the oil.
- Have students wipe up spills immediately.
- If students have trouble measuring accurately, the results will not be affected.

INCLUSION Students can draw pictures of the layers of liquids each time a liquid is added instead of answering the questions.

WHAT DO YOU THINK? *The layers change positions and get thicker. The oil layer floats on top of the other layers, the corn syrup layer sinks to the bottom, and the water layer is in the middle. Changing the order of adding the liquids would not affect where they settle.*

CHALLENGE *The density of a new liquid will determine whether it combines with one of the existing layers or forms a new layer.*

Datasheet, Liquids, p. 52

Technology Resources

Customize this student lab as needed or look for an alternative. Print rubrics to assess students' lab reports.

Lab Generator CD-ROM

Ongoing Assessment

CHECK YOUR READING *Answer: A liquid has a definite volume but not a definite shape. It takes the shape of its container. A solid has a rigid form, and the molecules are more tightly attached to one another than they are in a liquid.*

DIFFERENTIATE INSTRUCTION

More Reading Support

D What determines the shape of a liquid? *the container it is in*

Additional Investigation To reinforce Section 1.4 learning goals, use the following full-period investigation:

R **Additional INVESTIGATION,** Thick and Thin Liquids, A, B, & C, pp. 72–80, 262–263
(Advanced students should complete levels B and C.)

Advanced Have students investigate the role of density in determining if liquids mix or not. For example, students can create a colored salt solution with the same density as the corn syrup and see if they get the same results.

Teacher Demo

Demonstrate gas pressure by placing a small piece of dry ice in a small container with a lid that can easily pop off. Wear safety goggles and use tongs or gloves; keep students at a safe distance. Dry ice changes from solid carbon dioxide directly to a gas.

Teach from Visuals

To help students understand the diagrams of the helium tanks, ask:

- What happened to the molecules of gas in the container after half of them were removed? *They moved farther apart.*
- What happened to the volume of the gas in the container when half the gas molecules were removed? *It stayed the same.*

Ongoing Assessment

Describe the different states of matter.

Ask: How would you describe the volume and shape of the different states of matter? *A solid has a definite volume and a definite shape. A liquid has a definite volume but no definite shape. A gas has no definite volume or shape.*

Describe how the different states of matter behave.

Ask: How do particles in a solid, liquid, and gas move? *Particles in a solid vibrate in place, particles in a liquid slide over one another, and particles in a gas move about easily in any direction.*

CHECK YOUR READING *Answer: Gas molecules are far apart compared with molecules in a solid or a liquid. The amount of space between gas molecules varies with the size of the container.*

VOCABULARY
Add a four square diagram to your notebook for *gas.*

Gases have no definite volume or shape.

E The air that you breathe, the helium in a balloon, and the neon inside the tube in a neon light are gases. A gas is a substance with no definite volume and no definite shape. Solids and liquids have volumes that do not change easily. If you have a container filled with one liter of a liquid that you pour into a two-liter container, the liquid will occupy only half of the new container. A gas, on the other hand, has a volume that changes to match the volume of its container.

Gas Composition

F The molecules in a gas are very far apart compared with the molecules in a solid or a liquid. The amount of space between the molecules in a gas can change easily. If a rigid container—one that cannot change its shape—has a certain amount of air and more air is pumped in, the volume of the gas does not change. However, there is less space between the molecules than there was before. If the container is opened, the molecules spread out and mix with the air in the atmosphere.

As you saw, gas molecules in a container can be compared to a group of people in a room. If the room is small, there is less space between people. If the room is large, people can spread out so that there is more space between them. When people leave the room, they go in all different directions and mix with all of the other people in the surrounding area.

CHECK YOUR READING Contrast the molecules in a gas with those of a liquid and a solid.

Gas and Volume

The amount of space between gas particles depends on how many particles are in the container.

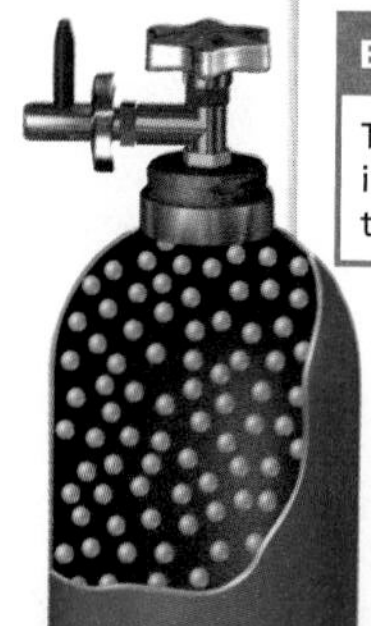

Before Use

The atoms of helium gas are constantly in motion. The atoms are spread throughout the entire tank.

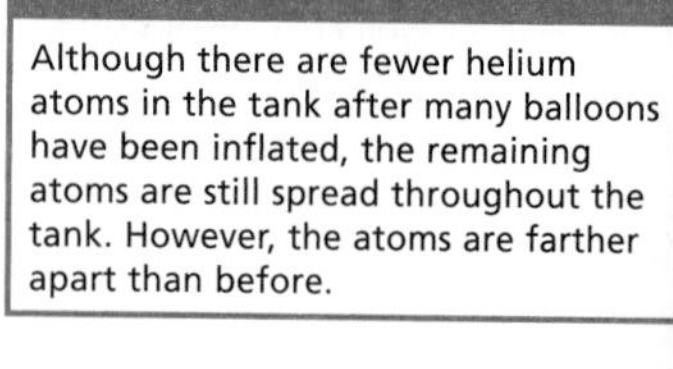

After Use

Although there are fewer helium atoms in the tank after many balloons have been inflated, the remaining atoms are still spread throughout the tank. However, the atoms are farther apart than before.

DIFFERENTIATE INSTRUCTION

More Reading Support

E What state of matter has no definite volume or shape? *gas*

F How easily can the space between gas molecules change? *very easily*

Gas Behavior

Because gas molecules are always in motion, they are continually hitting one another and the sides of any container they may be in. As the molecules bounce off one another and the surfaces of the container, they apply a pressure against the container. You can feel the effects of gas pressure if you pump air into a bicycle tire. The more air you put into the tire, the harder it feels because more gas molecules are pressing the tire outward.

SIMULATION CLASSZONE.COM
Explore the behavior of a gas.

The speed at which gas molecules move depends on the temperature of the gas. Gas molecules move faster at higher temperatures than at lower temperatures. The volume, pressure, and temperature of a gas are related to one another, and changing one can change the others.

Pressure ▲ Volume ▼ Temp. ■

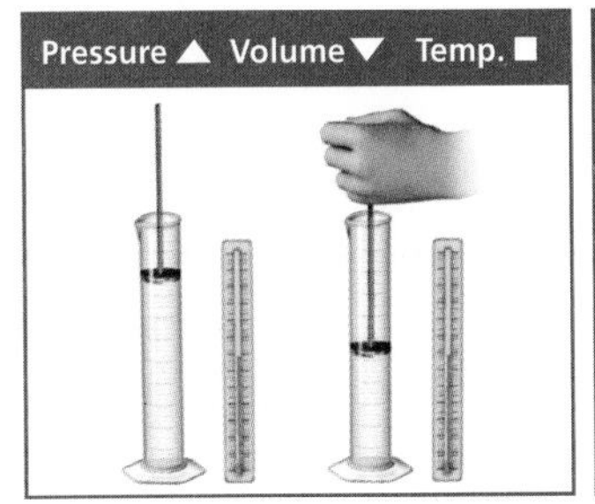

If the temperature of a gas stays the same, increasing the pressure of the gas decreases its volume.

Pressure ▲ Volume ■ Temp. ▲

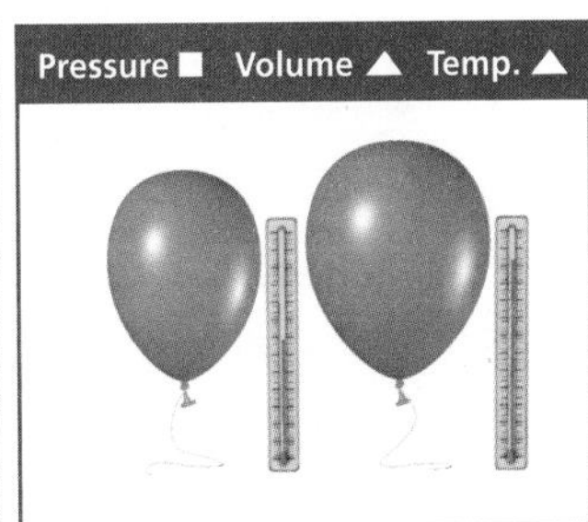

If the volume of a gas stays the same, increasing the temperature of the gas also increases the pressure.

Pressure ■ Volume ▲ Temp. ▲

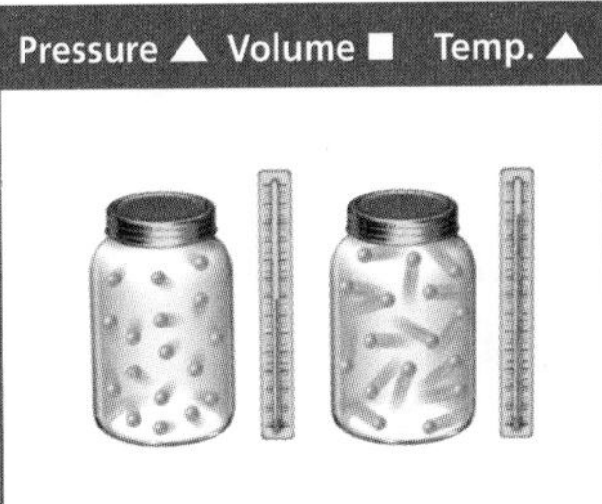

If the pressure of a gas stays the same, increasing the temperature of the gas also increases the volume.

In nature, volume, pressure, and temperature may all be changing at the same time. By studying how gas behaves when one property is kept constant, scientists can predict how gas will behave when all three properties change.

1.4 Review

KEY CONCEPTS

1. What are the characteristics of the three familiar states of matter?
2. How can you change the shape and volume of a liquid?
3. How does gas behave inside a closed container?

CRITICAL THINKING

4. **Infer** What happens to a liquid that is not in a container?
5. **Synthesize** What is the relationship between the temperature and the volume of a gas?

CHALLENGE

6. **Synthesize** Can an oxygen canister ever be half empty? Explain.

ANSWERS

1. Solids have a definite volume and shape, liquids have a definite volume but not a definite shape, and gases have no definite volume or shape.

2. You can change the shape of a liquid by pouring it into a different container with a different shape. You can't change the volume of a liquid unless you remove some of the liquid or add liquid to it.

3. Gas molecules are always in motion; they bounce off one another and the insides of the container.

4. It spreads out.

5. Increasing the temperature of a gas increases its volume.

6. No; it can have half as much oxygen, but the molecules will spread out evenly to fill the canister.

Teach from Visuals

Point out the symbols at the top of each box. Ask students what they mean. To help students understand the diagrams of gas behavior, ask:

- What happens to the volume of a gas if the temperature stays the same and the pressure on the gas increases? *It decreases.*
- What happens to the pressure of a gas if the volume stays the same and the temperature increases? *It increases.*
- What happens to the volume of a gas if the pressure stays the same and the temperature increases? *It increases.*

Reinforce the BIG idea

Have students relate the section to the Big Idea.

 Reinforcing Key Concepts, p. 53

1.4 ASSESS & RETEACH

Assess

 Section 1.4 Quiz, p. 6

Reteach

Have students make a set of flashcards. One side of each card should show a diagram of a characteristic of a solid, liquid, or gas. The other side should name the state of matter that the diagram describes. Students should make enough flashcards to include all the characteristics in the text. Pairs of students can use the cards to test each other's knowledge.

Technology Resources

Have students visit **ClassZone.com** for reteaching of Key Concepts.

 CONTENT REVIEW

 CONTENT REVIEW CD-ROM

BACK TO

the BIG idea

Have students look at the photograph on pp. 6–7. Ask them to list ten objects or substances that represent different states of matter in the photograph. Have them explain what makes each object matter. *Each object has mass and takes up space.*

KEY CONCEPTS SUMMARY

SECTION 1.1
Ask: Which tools are being used to measure mass and volume? *a pan balance to measure mass and a graduated cylinder to measure volume* Ask: Which tool would you use to measure the volume of a box? *a ruler or meter stick*

SECTION 1.2
Ask: How many different kinds of atoms formed the molecule? *two*

SECTION 1.3
Ask: Which picture shows a pure substance? Explain. *The picture on the left of the gold coins shows a pure substance. The coins contain only one substance, the element gold.*

SECTION 1.4
Ask: Which state of matter has a fixed volume but takes the shape of the container it is in? *liquid*

Review Concepts

- Big Idea Flow Chart, p. T1
- Chapter Outline, pp. T7–T8

1 Chapter Review

the BIG idea

Everything that has mass and takes up space is matter.

KEY CONCEPTS SUMMARY

1.1 Matter has mass and volume.

Mass is a measure of how much matter an object contains.

Volume is the measure of the amount of space matter occupies.

VOCABULARY
matter p. 9
mass p. 10
weight p. 11
volume p. 11

1.2 Matter is made of atoms.

An atom is the smallest basic unit of matter. Two or more atoms bonded together form a molecule. Atoms and molecules are always in motion.

VOCABULARY
atom p. 16
molecule p. 18

1.3 Matter combines to form different substances.

Matter can be pure, such as an element (gold), or a compound (water).

Matter can be a mixture. Mixtures contain two or more pure substances.

VOCABULARY
element p. 22
compound p. 23
mixture p. 23

1.4 Matter exists in different physical states.

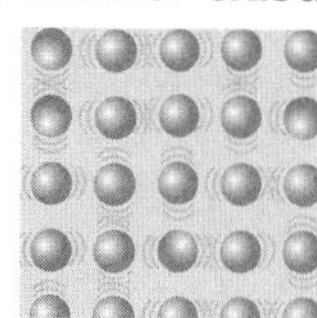

Solids have a fixed volume and a fixed shape.

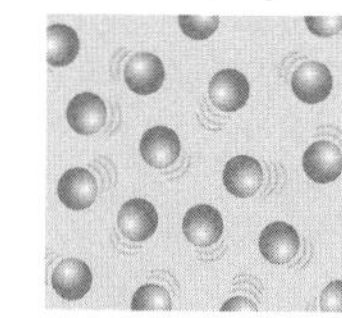

Liquids have a fixed volume but no fixed shape.

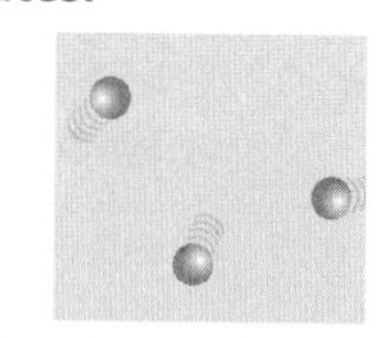

Gases have no fixed volume and no fixed shape.

VOCABULARY
states of matter p. 27
solid p. 28
liquid p. 28
gas p. 28

Technology Resources

Have students visit **ClassZone.com** or use the CD-ROM for a cumulative review of concepts.

 **CONTENT REVIEW**

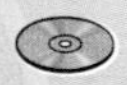 **CONTENT REVIEW CD-ROM**

Engage students in a whole-class interactive review of Key Concepts. Edit content as you wish.

 **POWER PRESENTATIONS**

Reviewing Vocabulary

Copy and complete the chart below. If the right column is blank, give a brief description or definition. If the left column is blank, give the correct term.

Term	Description
1.	the downward pull of gravity on an object
2. liquid	
3.	the smallest basic unit of matter
4. solid	
5.	state of matter with no fixed volume and no fixed shape
6.	a combination of different substances that remain individual substances
7. matter	
8.	a measure of how much matter an object contains
9. element	
10.	a particle made of two or more atoms bonded together
11. compound	

Reviewing Key Concepts

Multiple Choice *Choose the letter of the best answer.*

12. The standard unit for measuring mass is the
- **a.** kilogram
- **b.** gram per cubic centimeter
- **c.** milliliter
- **d.** milliliter per cubic centimeter

13. A unit for measuring the volume of a liquid is the
- **a.** kilogram
- **b.** gram per cubic centimeter
- **c.** milliliter
- **d.** milliliter per cubic centimeter

14. The weight of an object is measured by using a scale that
- **a.** compares the mass of the object with a standard unit of mass
- **b.** shows the amount of space the object occupies
- **c.** indicates how much water is displaced by the object
- **d.** tells how hard the object is pushing or pulling on it

15. To find the volume of a rectangular box,
- **a.** divide the length by the height
- **b.** multiply the length, width, and height
- **c.** subtract the mass from the weight
- **d.** multiply one atom's mass by the total

16. Compounds can be separated only by
- **a.** breaking the atoms into smaller pieces
- **b.** breaking the bonds between the atoms
- **c.** using a magnet to attract certain atoms
- **d.** evaporating the liquid that contains the atoms

17. Whether a substance is a solid, a liquid, or a gas depends on how close its atoms are to one another and
- **a.** the volume of each atom
- **b.** how much matter the atoms have
- **c.** how free the atoms are to move
- **d.** the size of the container

18. A liquid has
- **a.** a fixed volume and a fixed shape
- **b.** no fixed volume and a fixed shape
- **c.** a fixed volume and no fixed shape
- **d.** no fixed volume and no fixed shape

ASSESSMENT RESOURCES

ASSESSMENT BOOK
- Chapter Test A, pp. 7–10
- Chapter Test B, pp. 11–14
- Chapter Test C, pp. 15–18
- Alternative Assessment, pp. 19–20

SPANISH ASSESSMENT BOOK
Spanish Chapter Test, pp. 213–216

Technology Resources

Edit test items and answer choices.

Test Generator CD-ROM

Visit **ClassZone.com** to extend test practice.

Test Practice

Reviewing Vocabulary

1. weight
2. matter with a definite volume but no definite shape
3. atom
4. matter with a definite volume and shape
5. gas
6. mixture
7. anything that has mass and takes up space
8. mass
9. a substance that contains only a single type of atom
10. molecule
11. a substance that results when two or more different types of atoms bond together

Reviewing Key Concepts

12. a
13. c
14. d
15. b
16. b
17. c
18. c

(Answers to items that appear on p. 36)

19. Particles in a solid vibrate in position. Particles in a liquid slide past one another. Particles in a gas move freely in any direction.
20. Molecules in the air are moving, hitting the dust particles, and pushing them from different directions.
21. because three one-dimensional units are multiplied
22. The molecules apply pressure inside the tire, expanding the tire.

(Answers for items 19–22 appear on p. 35.)

Thinking Critically

23. *Matter: wood, metal, water, air; Not Matter: light, sound*
24. *No; you would have carbon and oxygen atoms. A molecule is the smallest part of a compound that is still that compound.*
25. *Like a liquid, sand has a definite volume but takes the shape of its container as sand grains slide around. Unlike a liquid, sand is made up of individual grains that each have a definite volume and a definite shape.*
26. *Many of the gas molecules escape into the air outside the ball.*
27. *Mixtures: substances remain the same, separated by physical means, percentages of substances can vary; Compounds: new substances formed, separated by breaking bonds, percentages of substances fixed; Alike: combinations of substances*
28. *The ball is a solid; the molecules are fixed in place and cannot move easily.*
29. *720,000 cm^3*
30. *2 trips*
31. *The marble displaces the liquid.*
32. *8 mL or 8 cm^3*
33. *50 mL; the volume of the water does not change and the marble is no longer displacing any of it.*

the BIG idea

34. *Answers should categorize items as solids, liquids, and gases.*
35. *Answers might include water (liquid), air (gas), and table (solid).*

UNIT PROJECTS

Give students the appropriate Unit Project worksheets from the URB for their projects. Both directions and rubrics can be used as a guide.

 Unit Projects, pp. 5–10

Short Answer *Answer each of the following questions in a sentence or two.*

19. Describe the movement of particles in a solid, a liquid, and a gas.

20. In bright sunlight, dust particles in the air appear to dart about. What causes this effect?

21. Why is the volume of a rectangular object measured in cubic units?

22. Describe how the molecules in the air behave when you pump air into a bicycle tire.

Thinking Critically

23. CLASSIFY Write the headings *Matter* and *Not Matter* on your paper. Place each of these terms in the correct category: wood, water, metal, air, light, sound.

24. INFER If you could break up a carbon dioxide molecule, would you still have carbon dioxide? Explain your answer.

25. MODEL In what ways is sand in a bowl like a liquid? In what ways is it different?

26. INFER If you cut a hole in a basketball, what happens to the gas inside?

27. COMPARE AND CONTRAST Create a Venn diagram that shows how mixtures and compounds are alike and different.

28. ANALYZE If you place a solid rubber ball into a box, why doesn't the ball change its shape to fit the container?

29. CALCULATE What is the volume of an aquarium that is 120 cm long, 60 cm wide, and 100 cm high?

30. CALCULATE A truck whose bed is 2.5 m long, 1.5 m wide, and 1 m high is delivering sand for a sand-sculpture competition. How many trips must the truck make to deliver 7 cubic meters of sand?

Use the information in the photograph below to answer the next three questions.

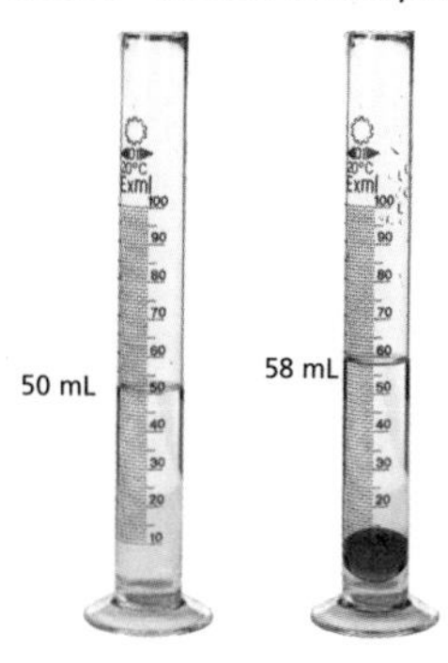

31. INFER One way to find the volume of a marble is by displacement. To determine a marble's volume, add 50 mL of water to a graduated cylinder and place the marble in the cylinder. Why does the water level change when you put the marble in the cylinder?

32. CALCULATE What is the volume of the marble?

33. PREDICT If you carefully removed the marble and let all of the water on it drain back into the cylinder, what would the volume of the water be? Explain.

the BIG idea

34. SYNTHESIZE Look back at the photograph on pages 6–7. Describe the picture in terms of states of matter.

35. WRITE Make a list of all the matter in a two-meter radius around you. Classify each as a solid, liquid, or gas.

UNIT PROJECTS

If you are doing a unit project, make a folder for your project. Include in your folder a list of the resources you will need, the date on which the project is due, and a schedule to track your progress. Begin gathering data.

MONITOR AND RETEACH

If students have trouble applying the concepts in items 29–33, suggest that they review pp. 12–13. Have students make a table that lists standard units of volume in one column and describes the units in the other column.

Students may benefit from summarizing one or more sections of the chapter.

 Summarizing the Chapter, pp. 81–82

Standardized Test Practice

For practice on your state test, go to . . . TEST PRACTICE CLASSZONE.COM

Interpreting Graphs

The graph below shows the changing volume of a gas as it was slowly heated, with the pressure held constant.

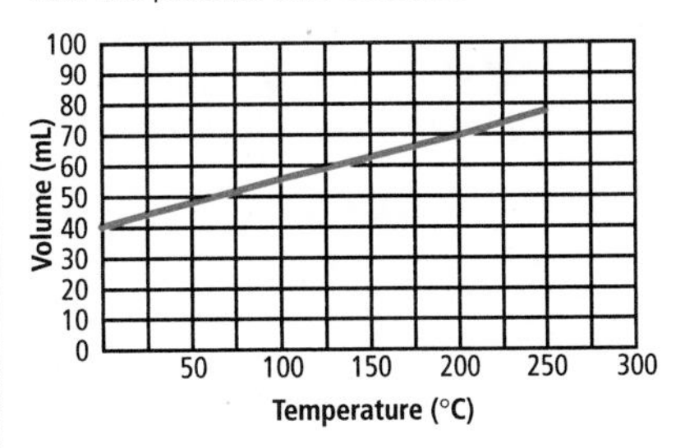

Use the graph to answer the questions.

1. As the temperature of the gas rises, what happens to its volume?
   a. It increases.
   b. It stays the same.
   c. It decreases.
   d. It changes without pattern.
2. What is the volume of the gas at 250°C as compared with the volume at 0°C?
   a. about three times greater
   b. about double
   c. about one-half
   d. about the same
3. What happens to a gas as it is cooled below 0°C?
   a. The volume would increase.
   b. The volume would continue to decrease.
   c. The volume would remain at 40 mL.
   d. A gas cannot be cooled below 0°C.
4. If you raised the temperature of this gas to 300°C, what would be its approximate volume?
   a. 70 mL
   b. 75 mL
   c. 80 mL
   d. 85 mL
5. If the volume of the gas at 0°C was 80 mL instead of 40 mL, what would you expect the volume to be at 200°C?
   a. 35 mL
   b. 70 mL
   c. 80 mL
   d. 140 mL

Extended Response

Answer the two questions below in detail. Include some of the terms from the word box. Underline each term you use in your answer.

gravity	mass	molecule
states of matter	weight	

6. An astronaut's helmet, measured on a balance, has the same number of kilograms on both Earth and the Moon. On a spring scale, though, it registers more newtons on Earth than on the Moon. Why?
7. Explain how water changes as it moves from a solid to a liquid and then to a gas.

METACOGNITIVE ACTIVITY

Have students answer the following questions in their **Science Notebook:**

1. What concept about matter did you find the most challenging to understand?
2. Describe a substance that you think is hard to categorize as a gas, liquid, or solid.
3. What goals have you set for your Unit Project? What is the next step you will complete?

Interpreting Graphs

1. a *3. b* *5. d*
2. b *4. d*

Extended Response

6. RUBRIC
4 points for a response that correctly explains both results and uses the following terms accurately:

- mass
- gravity
- weight

Sample: The balance measures mass, or the amount of matter. Mass is the same on Earth and the moon. The spring scale measures weight, or the force of gravity pulling on an object. Gravity on the moon pulls on the helmet less than gravity on Earth, so the helmet weighs less on the moon.

3 points correctly explains both results and uses two terms correctly
2 points correctly explains both results and uses one term correctly
1 point correctly explains one of the results

7. RUBRIC
4 points for a response that correctly identifies solid, liquid, and gas as states of matter, explains how the arrangement and movement of the molecules change for each state, and uses the following terms accurately:

- states of matter
- molecule

Sample: As water changes from a solid to a liquid and then to a gas, it is changing states of matter. The arrangement of its molecules changes and the movement of the molecules changes. As it changes from a solid to a liquid, its molecules can slide past each other. When water changes to a gas, its molecules can move freely in any direction and are much farther apart.

3 points for a response that correctly identifies solid, liquid, and gas as states of matter, explains how the arrangement and movement of the molecules change for two states of matter, and uses both terms accurately
2 points correctly identifies solid, liquid, and gas as states of matter, explains how the arrangement and movement of the molecules change for one state of matter, and uses both terms accurately
1 point correctly identifies solid, liquid, and gas as states of matter and uses both terms accurately

CHAPTER

Properties of Matter

Physical Science
UNIFYING PRINCIPLES

PRINCIPLE 1
Matter is made of particles too small to see.

PRINCIPLE 2
Matter changes form and moves from place to place.

PRINCIPLE 3
Energy changes from one form to another, but it cannot be created or destroyed.

PRINCIPLE 4
Physical forces affect the movement of all matter on Earth and throughout the universe.

Unit: Matter and Energy
BIG IDEAS

CHAPTER 1
Introduction to Matter
Everything that has mass and takes up space is matter.

CHAPTER 2
Properties of Matter
Matter has properties that can be changed by physical and chemical processes.

CHAPTER 3
Energy
Energy has different forms, but it is always conserved.

CHAPTER 4
Temperature and Heat
Heat is a flow of energy due to temperature differences.

CHAPTER 2
KEY CONCEPTS

SECTION 2.1

Matter has observable properties.
1. Physical properties describe a substance.
2. Chemical properties describe how substances form new substances.

SECTION 2.2

Changes of state are physical changes.
1. Matter can change from one state to another.
2. Solids can become liquids, and liquids can become solids.
3. Liquids can become gases, and gases can become liquids.

SECTION 2.3

Properties are used to identify substances.
1. Substances have characteristic properties.
2. Mixtures can be separated by using the properties of the substances in them.

The Big Idea Flow Chart is available on p. T9 in the **UNIT TRANSPARENCY BOOK.**

Previewing Content

SECTION

Matter has observable properties. pp. 41–49

1. Physical properties describe a substance.
Physical properties of a substance can be observed without changing the identity of the substance. Density, mass, color, size, volume, and texture are examples of physical properties.

- **Density** is the relationship between the mass and the volume of a substance.
- Calculate density by dividing mass by volume, as shown in the sample problem below.

A glass marble has a volume of 5 cm^3 and a mass of 13 g. What is the density of glass?

What do you know? Volume = 5 cm^3, mass = 13 g

What do you want to find out? Density

Write the formula: $D = \frac{m}{V}$

Substitute into the formula: $D = \frac{13\text{ g}}{5\text{ cm}^3}$

Calculate and simplify: $D = 2.6\text{ g/cm}^3$.

Any change in a physical property of a substance is a **physical change.** The identity of the material remains the same during the change. Examples of physical changes include cutting a material, breaking it, and changing its state.

2. Chemical properties describe how substances form new substances.
To observe **chemical properties** in a substance, you must see a **chemical change.**

- To observe the combustibility of a piece of paper, for example, the paper must burn. The products that result from burning the paper differ in identity from the paper itself.
- Signs of a chemical change include the production of an odor, a gas, or a solid and a change in temperature or color.
- Other examples of chemical properties include reactivity, tendency to corrode, and toxicity.

SECTION

Changes of state are physical changes. pp. 50–57

1. Matter can change from one state to another.
Matter has three common states—solid, liquid, and gas. Matter can physically change from one state to another.

- A solid has a fixed volume and a fixed shape.
- A liquid has a fixed volume but assumes the shape of its container.
- Both the volume and shape of a gas depend on the volume and shape of its container.

2. Solids can become liquids, and liquids can become solids.
When a substance **melts,** added energy as heat breaks the tight bonds between particles. This process occurs at a temperature called the **melting point** of the substance. For some substances, the melting point is not a well-defined temperature. **Freezing** is the process by which particles of a liquid lose energy and bond tightly to form a solid. The **freezing point** of a substance is the same as its melting point. While a substance with a well-defined melting point is freezing or melting, the temperature will not change.

3. Liquids can become gases, and gases can become liquids.
During **condensation,** a gas is changed to a liquid. Energy is removed from the gas, and the particles form loose bonds.

- High-energy particles can escape from the surface of a liquid by **evaporation.**
- If energy as heat is added to a liquid, bubbles of gas can form throughout the liquid in a process called **boiling.**
- **Sublimation** is the process by which solids become gases, and deposition is the process by which gases become solids. Sublimation and deposition happen only under certain pressure and temperature conditions.

Common Misconceptions

DENSITY Students may think that *heavy* and *dense* mean the same thing. How heavy something feels depends on its mass. Density, however, depends on both mass and volume. For example, an object can have a low density but still be heavy if the volume is large.

This misconception is addressed on p. 43.

MISCONCEPTION DATABASE
CLASSZONE.COM Background on student misconceptions

EVAPORATION Students may not understand that a liquid does not simply disappear when it evaporates. Instead, evaporation involves a change of state: some of the liquid becomes a gas.

This misconception is addressed on p. 53.

SECTION

Properties are used to identify substances. pp. 58–63

1. Substances have characteristic properties.
The physical and chemical properties of a substance can be used to identify it. Although a substance may share properties with another substance, no two substances have identical sets of properties. The following properties can be used to identify substances because they are the same for every sample of a particular substance:

- density
- heating properties
- solubility
- conductivity
- magnetic properties

2. Mixtures can be separated by using the properties of the substances in them.
Substances can be separated by using differences in physical properties.

- A magnet will separate materials that have magnetic properties from those that don't.
- Filtration can separate solids from liquids and can separate solids that differ in particle size.
- Evaporation can separate a liquid and the substances dissolved in it.

Previewing Labs

Lab Generator CD-ROM
Edit these Pupil Edition labs and generate alternative labs.

EXPLORE the BIG idea

Float or Sink, p. 39 Students mold clay and float it to observe the effect of shape on overall density.	**TIME** 10 minutes **MATERIALS** ball of clay, bowl, water
Hot Chocolate, p. 39 Students melt chocolate, observing the effect of heat on the state of a material.	**TIME** 10 minutes **MATERIALS** four candy-coated chocolates, paper towel
Internet Activity: Physical and Chemical Changes, p. 39 Students are introduced to examples of physical and chemical changes.	**TIME** 20 minutes **MATERIALS** computer with Internet access

SECTION

EXPLORE Physical Properties, p. 41 Students observe physical properties of a material and how they can change.	**TIME** 10 minutes **MATERIALS** rectangular piece of clay
INVESTIGATE Chemical Changes, p. 47 Students observe signs of a chemical change in two solutions.	**TIME** 15 minutes **MATERIALS** 100-mL graduated cylinder, water, 2 clear plastic cups, 2 eyedroppers, iodine solution, cornstarch, spoon, vitamin C tablet

SECTION

CHAPTER INVESTIGATION **Freezing Point,** pp. 56–57 Students freeze melted stearic acid and measure its freezing point.	**TIME** 40 minutes **MATERIAL** large test tube, pure stearic acid, test-tube holder, test-tube rack, wire-loop stirrer, thermometer

SECTION

EXPLORE Identifying Substances, p. 58 Students examine how properties can be used to identify two unknown substances.	**TIME** 10 minutes **MATERIALS** baking soda, baking powder, 2 cups, beaker, water
INVESTIGATE Separating Mixtures, p. 61 Students design a way to separate a mixture of sand, salt, and pepper.	**TIME** 30 minutes **MATERIALS** spoon; mixture of sand, table salt, and pepper; 2 index cards; comb; felt; 100-mL graduated cylinder; water; coffee filter; funnel; small cup; pie tin

Additional INVESTIGATION, Measuring Density, A, B, & C, pp. 133–141; Teacher Instructions, pp. 262–263

Previewing Chapter Resources

CHAPTER 2
Properties of Matter

INTEGRATED TECHNOLOGY

CLASSZONE.COM
- eEdition Plus
- EasyPlanner
- Misconception Database
- Content Review
- Test Practice
- Simulation
- Resource Centers
- Internet Activity: Physical and Chemical Changes
- Math Tutorial

SCILINKS.ORG

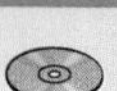
CD-ROMS
- eEdition
- EasyPlanner
- Power Presentations
- Content Review
- Lab Generator
- Test Generator

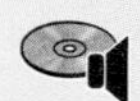
AUDIO CDS
- Audio Readings
- Audio Readings in Spanish

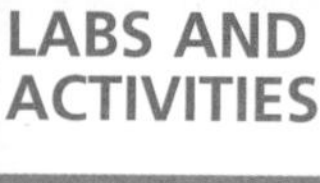
LABS AND ACTIVITIES

EXPLORE the Big Idea, p. 39
- Float or Sink
- Hot Chocolate
- Internet Activity: Physical and Chemical Changes

UNIT RESOURCE BOOK
Unit Projects, pp. 5–10

Lab Generator CD-ROM
Generate customized labs.

SECTION

2.1 Matter has observable properties.
pp. 41–49

Time: 2 periods (1 block)

Lesson Plan, pp. 83–84

Integrated Technology

- **RESOURCE CENTER,** Chemical Properties of Matter
- **MATH TUTORIAL**

UNIT TRANSPARENCY BOOK
- Big Idea Flow Chart, p. T9
- Daily Vocabulary Scaffolding, p. T10
- Note-Taking Model, p. T11
- 3-Minute Warm-Up, p. T12
- "Physical Changes" Visual, p. T14

Labs and Activities

- EXPLORE Physical Properties, p. 41
- INVESTIGATE Chemical Changes, p. 47
- Math in Science, p. 49

UNIT RESOURCE BOOK
- Datasheet, Chemical Changes, p. 92
- Math Support, pp. 120, 122
- Math Practice, pp. 121, 123
- Additional INVESTIGATION, Measuring Density, A, B, & C, pp. 133–141

SECTION

2.2 Changes of state are physical changes.
pp. 50–57

Time: 3 periods (1.5 blocks)

Lesson Plan, pp. 94–95

Integrated Technology

RESOURCE CENTER, Melting Points and Boiling Points

UNIT TRANSPARENCY BOOK
- Daily Vocabulary Scaffolding, p. T10
- 3-Minute Warm-Up, p. T12

Labs and Activities

CHAPTER INVESTIGATION, Freezing Point, pp. 56-57

UNIT RESOURCE BOOK
CHAPTER INVESTIGATION, Freezing Point, pp. 124–132

SECTION

2.3 Properties are used to identify substances.
pp. 58–63

Time: 3 periods (1.5 blocks)

Lesson Plan, pp. 104–105

Integrated Technology

RESOURCE CENTER, Separating Materials from Mixtures

UNIT TRANSPARENCY BOOK
- Big Idea Flow Chart, p. T9
- Daily Vocabulary Scaffolding, p. T10
- 3-Minute Warm-Up, p. T13
- Chapter Outline, p. T15–T16

Labs and Activities

- EXPLORE Identifying Substances, p. 58
- INVESTIGATE Separating Mixtures, p. 61
- Connecting Sciences, p. 63

UNIT RESOURCE BOOK
Datasheet, Separating Mixtures, p. 113

KEY TO ICONS

 CD/CD-ROM
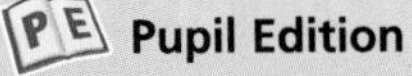 INTERNET
 Pupil Edition
 Teacher Edition
 UNIT RESOURCE BOOK
 UNIT TRANSPARENCY BOOK
 UNIT ASSESSMENT BOOK
 SPANISH ASSESSMENT BOOK
SCIENCE TOOLKIT

READING AND REINFORCEMENT

- Magnet Word, B24–25
- Main Idea Web, C38–39
- Daily Vocabulary Scaffolding, H1–8

UNIT RESOURCE BOOK
- Vocabulary Practice, pp. 117–118
- Decoding Support, p. 119
- Summarizing the Chapter, pp. 142–143

Audio Readings CD
Listen to Pupil Edition.

Audio Readings in Spanish CD
Listen to Pupil Edition in Spanish.

ASSESSMENT

- Chapter Review, pp. 65–66
- Standardized Test Practice, p. 67

UNIT ASSESSMENT BOOK
- Diagnostic Test, pp. 21–22
- Chapter Test, A, B, & C, pp. 26–37
- Alternative Assessment, pp. 38–39

Spanish Chapter Test, pp. 217–220

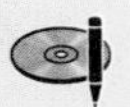

Test Generator CD-ROM
Generate customized tests.

Lab Generator CD-ROM
Rubrics for Labs

STANDARDS

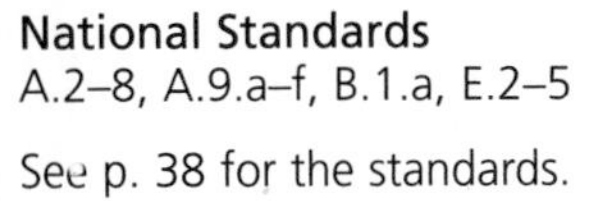

National Standards
A.2–8, A.9.a–f, B.1.a, E.2–5

See p. 38 for the standards.

UNIT RESOURCE BOOK
- Reading Study Guide, A & B, pp. 85–88
- Spanish Reading Study Guide, pp. 89–90
- Challenge and Extension, p. 91
- Reinforcing Key Concepts, p. 93
- Challenge Reading, pp. 115–116

Ongoing Assessment, pp. 41–48

Section 2.1 Review, p. 48

UNIT ASSESSMENT BOOK
Section 2.1 Quiz, p. 23

National Standards
A.2–8, A.9.a–c, A.9.e–f, B.1.a

UNIT RESOURCE BOOK
- Reading Study Guide, A & B, pp. 96–99
- Spanish Reading Study Guide, pp. 100–101
- Challenge and Extension, p. 102
- Reinforcing Key Concepts, p. 103

Ongoing Assessment, pp. 50–55

Section 2.2 Review, p. 55

UNIT ASSESSMENT BOOK
Section 2.2 Quiz, p. 24

National Standards
A.2–7, A.9.a–b, A.9.d–f, B.1.a

UNIT RESOURCE BOOK
- Reading Study Guide, A & B, pp. 106–109
- Spanish Reading Study Guide, pp. 110–111
- Challenge and Extension, p. 112
- Reinforcing Key Concepts, p. 114

Ongoing Assessment, pp. 58–59, 61–62

Section 2.3 Review, p. 62

UNIT ASSESSMENT BOOK
Section 2.3 Quiz, p. 25

National Standards
A.2–7, A.9.a–b, A.9.e–f, E.2–5

Previewing Resources for Differentiated Instruction

CHAPTER INVESTIGATION

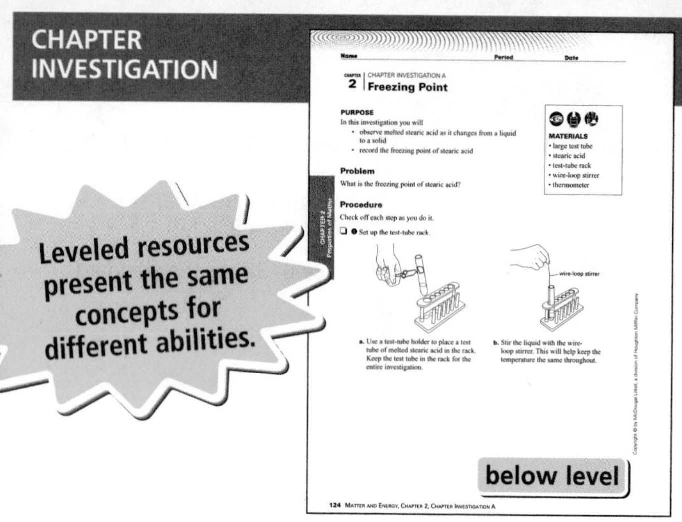

Leveled resources present the same concepts for different abilities.

Name Period Date

CHAPTER 2 | CHAPTER INVESTIGATION A

Freezing Point

PURPOSE
In this investigation you will
- observe melted stearic acid as it changes from a liquid to a solid
- record the freezing point of stearic acid

MATERIALS
- large test tube
- stearic acid
- test-tube rack
- wire-loop stirrer
- thermometer

Problem
What is the freezing point of stearic acid?

Procedure
Check off each step as you do it.

☐ ● Set up the test-tube rack.

a. Use a test-tube holder to place a test tube of melted stearic acid in the rack. Keep the test tube in the rack for the entire investigation.

b. Stir the liquid with the wire-loop stirrer. This will help keep the temperature the same throughout.

124 Matter and Energy, Chapter 2, Chapter Investigation A

below level

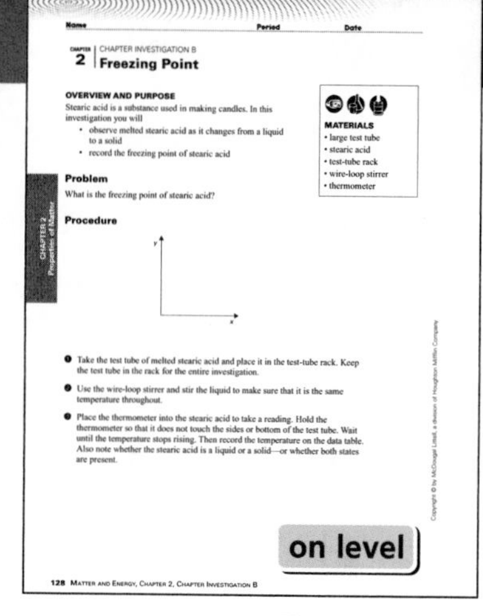

Name Period Date

CHAPTER 2 | CHAPTER INVESTIGATION B

Freezing Point

OVERVIEW AND PURPOSE
Stearic acid is a substance used in making candles. In this investigation you will
- observe melted stearic acid as it changes from a liquid to a solid
- record the freezing point of stearic acid

MATERIALS
- large test tube
- stearic acid
- test-tube rack
- wire-loop stirrer
- thermometer

Problem
What is the freezing point of stearic acid?

Procedure

- Take the test tube of melted stearic acid and place it in the test-tube rack. Keep the test tube in the rack for the entire investigation.
- Use the wire-loop stirrer and stir the liquid to make sure that it is the same temperature throughout.
- Place the thermometer into the stearic acid to take a reading. Hold the thermometer so that it does not touch the sides or bottom of the test tube. Wait until the temperature stops rising. Then record the temperature on the data table. Also note whether the stearic acid is a liquid or a solid—or whether both states are present.

128 Matter and Energy, Chapter 2, Chapter Investigation B

on level

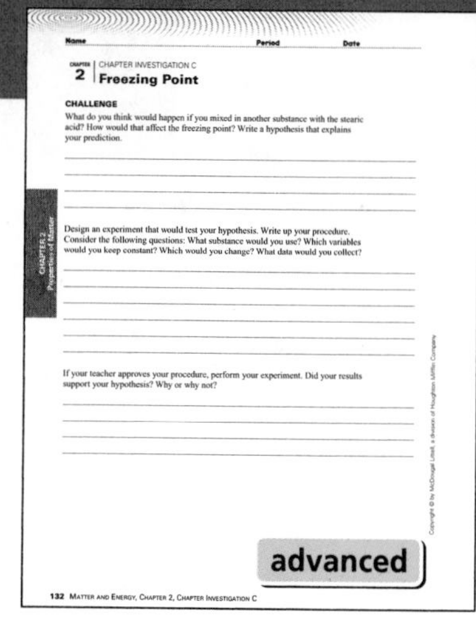

Name Period Date

CHAPTER 2 | CHAPTER INVESTIGATION C

Freezing Point

CHALLENGE
What do you think would happen if you mixed in another substance with the stearic acid? How would that affect the freezing point? Write a hypothesis that explains your prediction.

Design an experiment that would test your hypothesis. Write up your procedure. Consider the following questions: What substance would you use? Which variables would you keep constant? Which would you change? What data would you collect?

If your teacher approves your procedure, perform your experiment. Did your results support your hypothesis? Why or why not?

132 Matter and Energy, Chapter 2, Chapter Investigation C

advanced

UNIT RESOURCE BOOK, pp. 124–127 | pp. 128–131 | pp. 128–132

READING STUDY GUIDE

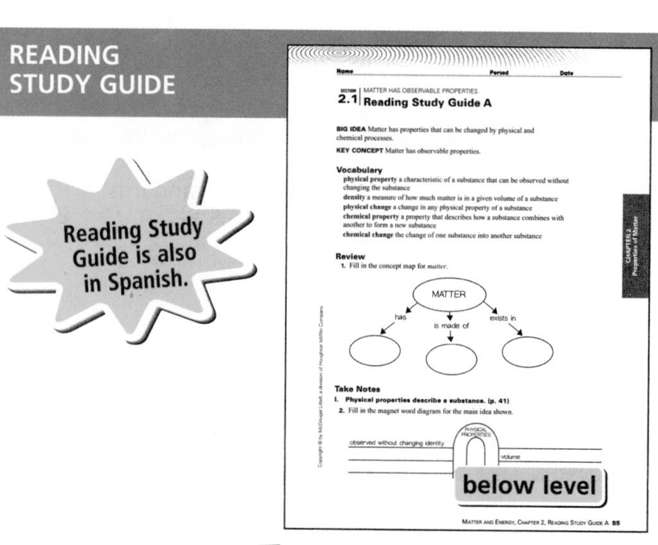

Reading Study Guide is also in Spanish.

Name Period Date

SECTION 2.1 | MATTER HAS OBSERVABLE PROPERTIES

Reading Study Guide A

BIG IDEA Matter has properties that can be changed by physical and chemical processes.

KEY CONCEPT Matter has observable properties.

Vocabulary
physical property a characteristic of a substance that can be observed without changing the substance
density a measure of how much matter is in a given volume of a substance
physical change a change in any physical property of a substance
chemical property a property that describes how a substance combines with another to form a new substance
chemical change the change of one substance into another substance

Review
1. Fill in the concept map for *matter*.

MATTER — has — is made of — exists in

Take Notes
I. Physical properties describe a substance. (p. 41)
2. Fill in the magnet word diagram for the main idea shown.

observed without changing identity — PHYSICAL PROPERTIES — volume

below level

Matter and Energy, Chapter 2, Reading Study Guide A 85

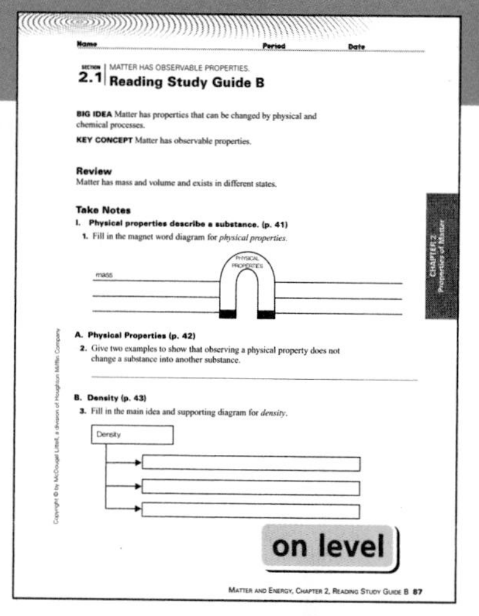

Name Period Date

SECTION 2.1 | MATTER HAS OBSERVABLE PROPERTIES

Reading Study Guide B

BIG IDEA Matter has properties that can be changed by physical and chemical processes.

KEY CONCEPT Matter has observable properties.

Review
Matter has mass and volume and exists in different states.

Take Notes
I. Physical properties describe a substance. (p. 41)
1. Fill in the magnet word diagram for *physical properties*.

mass — PHYSICAL PROPERTIES

A. Physical Properties (p. 42)
2. Give two examples to show that observing a physical property does not change a substance into another substance.

B. Density (p. 43)
3. Fill in the main idea and supporting diagram for *density*.

Density

on level

Matter and Energy, Chapter 2, Reading Study Guide B 87

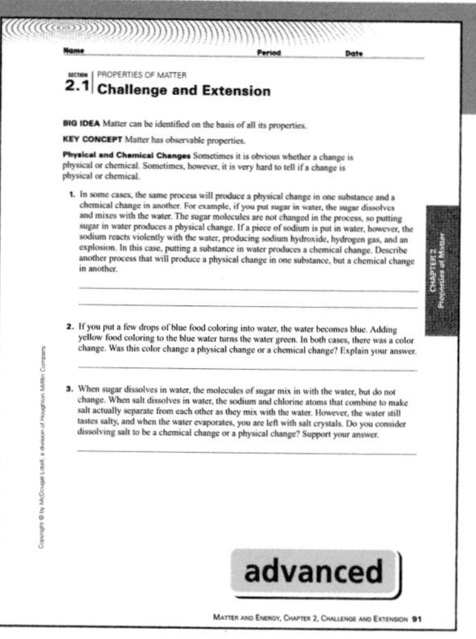

Name Period Date

SECTION 2.1 | PROPERTIES OF MATTER

Challenge and Extension

BIG IDEA Matter can be identified on the basis of all its properties.

KEY CONCEPT Matter has observable properties.

Physical and Chemical Changes Sometimes it is obvious whether a change is physical or chemical. Sometimes, however, it is very hard to tell if a change is physical or chemical.

1. In some cases, the same process will produce a physical change in one substance and a chemical change in another. For example, if you put sugar in water, the sugar dissolves and mixes with the water. The sugar molecules are not changed in the process, so putting sugar in water produces a physical change. If a piece of sodium is put in water, however, the sodium reacts violently with the water, producing sodium hydroxide, hydrogen gas, and an explosion. In this case, putting a substance in water produces a physical change, but a chemical change in another.

2. If you put a few drops of blue food coloring into water, the water becomes blue. Adding yellow food coloring to the blue water turns the water green. In both cases, there was a color change. Was this color change a physical change or a chemical change? Explain your answer.

3. When sugar dissolves in water, the molecules of sugar mix in with the water, but do not change. When salt dissolves in water, the sodium and chlorine atoms that combine to make salt actually separate from each other as they mix with the water. However, the water still tastes salty, and when the water evaporates, you are left with salt crystals. Do you consider dissolving salt to be a chemical change or a physical change? Support your answer.

advanced

Matter and Energy, Chapter 2, Challenge and Extension 91

UNIT RESOURCE BOOK, pp. 85–86 | pp. 87–88 | p. 91

CHAPTER TEST

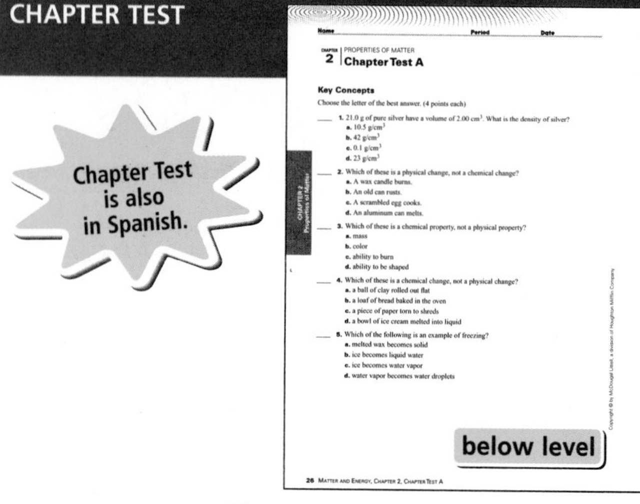

Chapter Test is also in Spanish.

Name Period Date

CHAPTER 2 | PROPERTIES OF MATTER

Chapter Test A

Key Concepts
Choose the letter of the best answer. (4 points each)

___ 1. 21.0 g of pure silver have a volume of 2.00 cm^3. What is the density of silver?
a. 10.5 g/cm^3
b. 42 g/cm^3
c. 0.1 g/cm^3
d. 23 g/cm^3

___ 2. Which of these is a physical change, not a chemical change?
a. A wax candle burns.
b. An old can rusts.
c. A scrambled egg cooks.
d. An aluminum can melts.

___ 3. Which of these is a chemical property, not a physical property?
a. mass
b. color
c. ability to burn
d. ability to be shaped

___ 4. Which of these is a chemical change, not a physical change?
a. a ball of clay rolled out flat
b. a loaf of bread baked in the oven
c. a piece of paper torn to shreds
d. a bowl of ice cream melted into liquid

___ 5. Which of the following is an example of freezing?
a. melted wax becomes solid
b. ice becomes liquid water
c. ice becomes water vapor
d. water vapor becomes water droplets

below level

26 Matter and Energy, Chapter 2, Chapter Test A

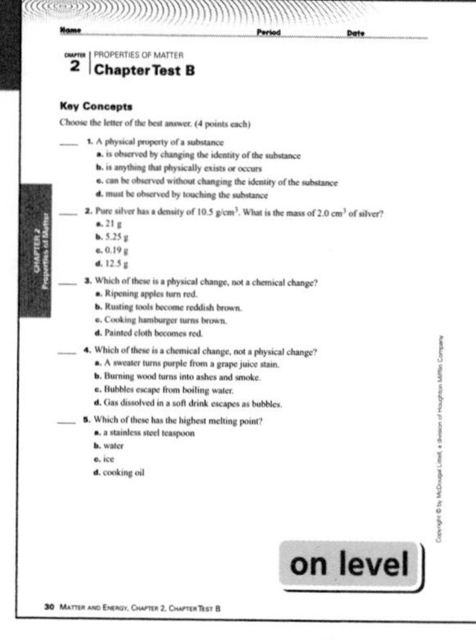

Name Period Date

CHAPTER 2 | PROPERTIES OF MATTER

Chapter Test B

Key Concepts
Choose the letter of the best answer. (4 points each)

___ 1. A physical property of a substance
a. is observed by changing the identity of the substance
b. is anything that physically exists or occurs
c. can be observed without changing the identity of the substance
d. must be observed by touching the substance

___ 2. Pure silver has a density of 10.5 g/cm^3. What is the mass of 2.0 cm^3 of silver?
a. 21 g
b. 5.25 g
c. 0.19 g
d. 12.5 g

___ 3. Which of these is a physical change, not a chemical change?
a. Ripening apples turn red.
b. Rusting tools become reddish brown.
c. Cooking hamburger turns brown.
d. Painted cloth becomes red.

___ 4. Which of these is a chemical change, not a physical change?
a. A sweater turns purple from a grape juice stain.
b. Burning wood turns into ashes and smoke.
c. Bubbles escape from boiling water.
d. Gas dissolved in a soft drink escapes as bubbles.

___ 5. Which of these has the highest melting point?
a. a stainless steel teaspoon
b. water
c. ice
d. cooking oil

on level

30 Matter and Energy, Chapter 2, Chapter Test B

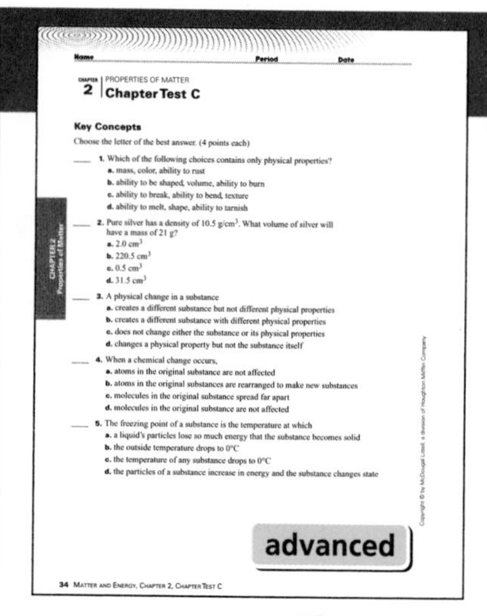

Name Period Date

CHAPTER 2 | PROPERTIES OF MATTER

Chapter Test C

Key Concepts
Choose the letter of the best answer. (4 points each)

___ 1. Which of the following choices contains only physical properties?
a. mass, color, ability to rust
b. ability to be shaped, volume, ability to burn
c. ability to break, ability to bend, texture
d. ability to melt, shape, ability to tarnish

___ 2. Pure silver has a density of 10.5 g/cm^3. What volume of silver will have a mass of 21 g?
a. 2.0 cm^3
b. 220.5 cm^3
c. 0.5 cm^3
d. 31.5 cm^3

___ 3. A physical change in a substance
a. creates a different substance but not different physical properties
b. creates a different substance with different physical properties
c. does not change either the substance or its physical properties
d. changes a physical property but not the substance itself

___ 4. When a chemical change occurs,
a. atoms in the original substance are not affected
b. atoms in the original substances are rearranged to make new substances
c. molecules in the original substance spread far apart
d. molecules in the original substance are not affected

___ 5. The freezing point of a substance is the temperature at which
a. a liquid's particles lose so much energy that the substance becomes solid
b. the outside temperature drops to 0°C
c. the temperature of any substance drops to 0°C
d. the particles of a substance increase in energy and the substance changes state

advanced

34 Matter and Energy, Chapter 2, Chapter Test C

UNIT ASSESSMENT BOOK, pp. 26–29 | pp. 30–33 | pp. 34–37

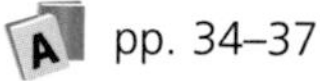

TECHNOLOGY

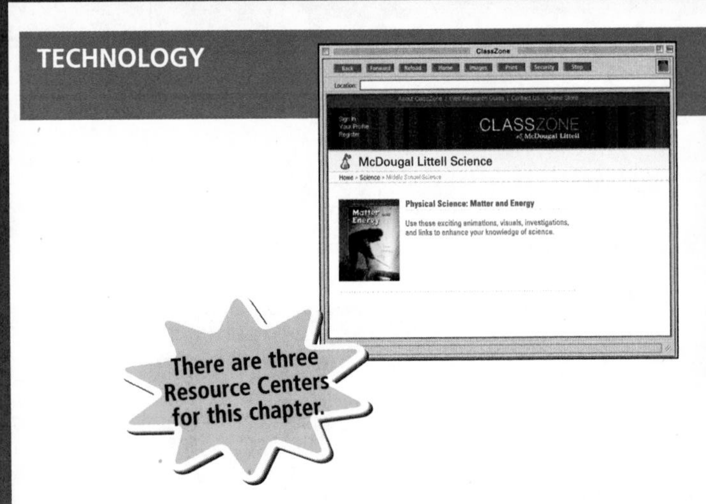

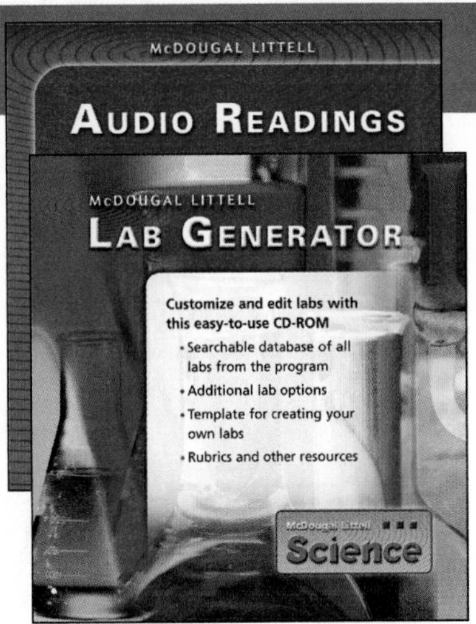

McDOUGAL LITTELL
AUDIO READINGS
McDOUGAL LITTELL
LAB GENERATOR
Customize and edit labs with this easy-to-use CD-ROM
• Searchable database of all labs from the program
• Additional lab options
• Template for creating your own labs
• Rubrics and other resources
McDougal Littell Science

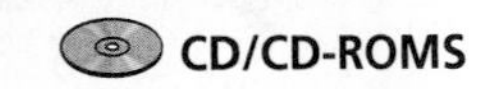

VISUAL CONTENT

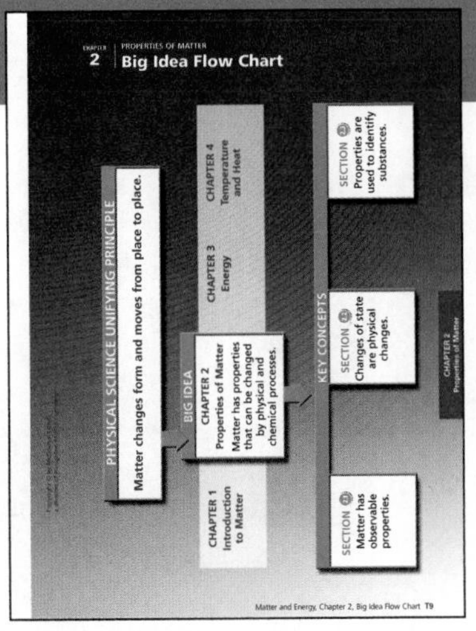

UNIT TRANSPARENCY BOOK, p. 9

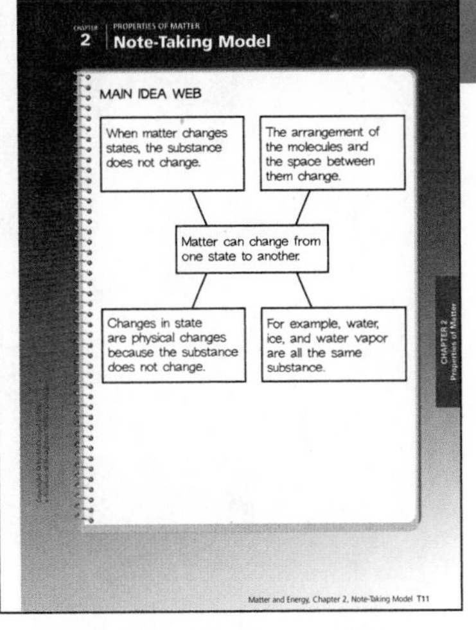

CHAPTER 2 PROPERTIES OF MATTER
Note-Taking Model
MAIN IDEA WEB
When matter changes states, the substance does not change.
The arrangement of the molecules and the space between them change.
Matter can change from one state to another.
Changes in state are physical changes because the substance does not change.
For example, water, ice, and water vapor are all the same substance.
Matter and Energy, Chapter 2, Note-Taking Model T11

p. 11

CHAPTER 2 PROPERTIES OF MATTER
Physical Changes
The process of turning wool into a sweater requires that the wool undergo physical changes.
1 Shearing
2 Spinning
3 Dyeing
Change in shape
Change in volume and texture
Change in color
The final product, a wool sweater, is still wool.
Matter and Energy, Chapter 2, Teaching Visual T14

p. 14

MORE SUPPORT

Reinforcing Key Concepts for each section

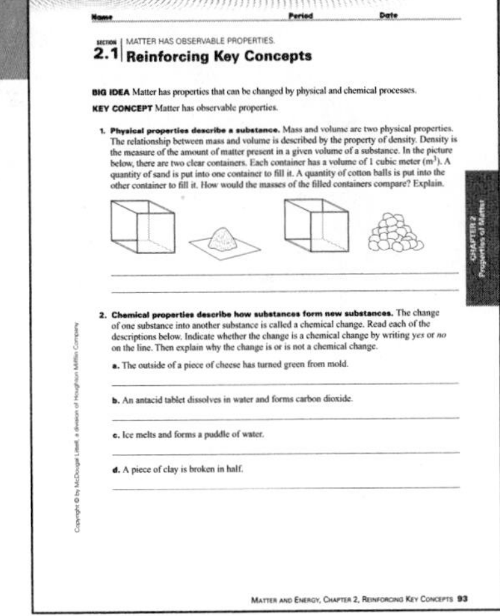

SECTION 2.1 MATTER HAS OBSERVABLE PROPERTIES
Reinforcing Key Concepts

BIG IDEA Matter has properties that can be changed by physical and chemical processes.

KEY CONCEPT Matter has observable properties.

1. **Physical properties describe a substance.**
2. **Chemical properties describe how substances form new substances.**

a. The outside of a piece of cheese has turned green from mold.

b. An antacid tablet dissolves in water and forms carbon dioxide.

c. Ice melts and forms a puddle of water.

d. A piece of clay is broken in half.

MATTER AND ENERGY, CHAPTER 2, REINFORCING KEY CONCEPTS 93

UNIT RESOURCE BOOK, p. 93

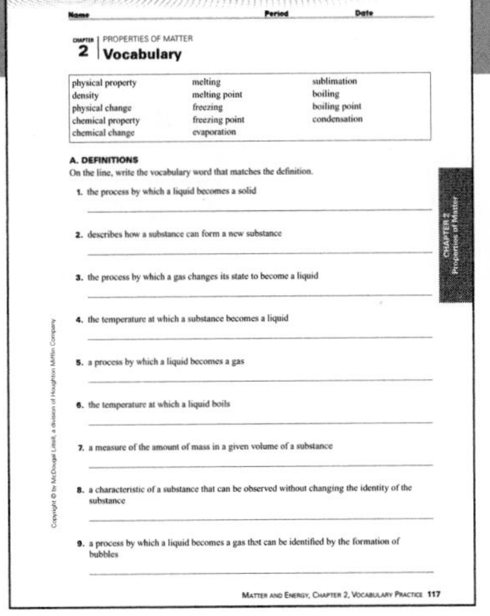

CHAPTER 2 PROPERTIES OF MATTER
Vocabulary

physical property	melting	sublimation
density	melting point	boiling
physical change	freezing	boiling point
chemical property	freezing point	condensation
chemical change	evaporation	

A. DEFINITIONS

On the line, write the vocabulary word that matches the definition.

1. the process by which a liquid becomes a solid
2. describes how a substance can form a new substance
3. the process by which a gas changes its state to become a liquid
4. the temperature at which a substance becomes a liquid
5. a process by which a liquid becomes a gas
6. the temperature at which a liquid boils
7. a measure of the amount of mass in a given volume of a substance
8. a characteristic of a substance that can be observed without changing the identity of the substance
9. a process by which a liquid becomes a gas that can be identified by the formation of bubbles

MATTER AND ENERGY, CHAPTER 2, VOCABULARY PRACTICE 117

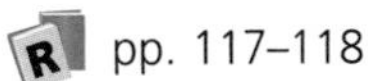

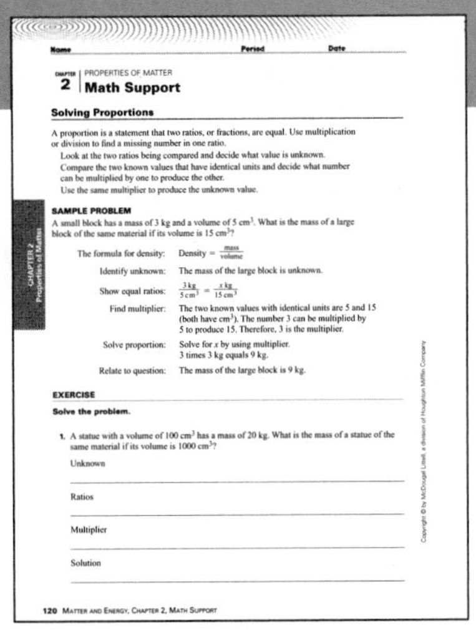

CHAPTER 2 PROPERTIES OF MATTER
Math Support

Solving Proportions

A proportion is a statement that two ratios, or fractions, are equal. Use multiplication or division to find a missing number in one ratio.

SAMPLE PROBLEM

A small block has a mass of 3 kg and a volume of 5 cm³. What is the mass of a large block of the same material if its volume is 15 cm³?

EXERCISE

Solve the problem.

1. A statue with a volume of 100 cm³ has a mass of 20 kg. What is the mass of a statue of the same material if its volume is 1000 cm³?

Unknown

Ratios

Multiplier

Solution

120 MATTER AND ENERGY, CHAPTER 2, MATH SUPPORT

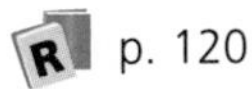

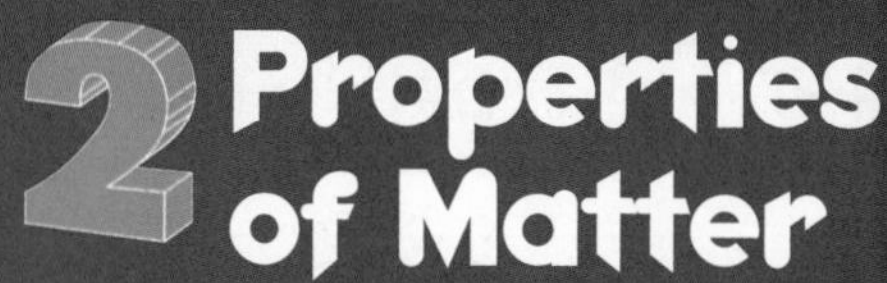

CHAPTER 2 Properties of Matter

INTRODUCE

the BIG idea

Have students look at the photograph of the chef making a sugar sculpture. Discuss how the question in the box relates to the Big Idea:

- How does the sugar in the sculpture compare to granular sugar?
- What is added to the sugar to make it harden?
- In what ways can food change when it is cooked?

National Science Education Standards

Content

B.1.a A substance has characteristic properties, such as density, a boiling point, and solubility, all of which are independent of the amount of the sample. A mixture of substances often can be separated into the original substances using one or more of the characteristic properties.

Process

A.2–8 Design and conduct an investigation; use tools to gather and interpret data; use evidence to describe, predict, explain, model; think critically to make relationships between evidence and explanation; recognize different explanations and predictions; communicate scientific procedures and explanations; use mathematics.

A.9.a–f Understand scientific inquiry by using different investigations, methods, mathematics, technology, explanations based on logic, evidence, and skepticism.

E.2–5 Design, implement, and evaluate a solution or product; communicate technological design.

CHAPTER 2 Properties of Matter

the BIG idea

Matter has properties that can be changed by physical and chemical processes.

Key Concepts

SECTION

2.1 Matter has observable properties.
Learn how to recognize physical and chemical properties.

SECTION

2.2 Changes of state are physical changes.
Learn how energy is related to changes of state.

SECTION

2.3 Properties are used to identify substances.
Learn how the properties of substances can be used to identify them and to separate mixtures.

Internet Preview

CLASSZONE.COM
Chapter 2 online resources: Content Review, Simulation, three Resource Centers, Math Tutorial, Test Practice

What properties could help you identify this sculpture as sugar?

INTERNET PREVIEW

CLASSZONE.COM For student use with the following pages:

Review and Practice
- Content Review, pp. 40, 64
- Math Tutorial: Solving Proportions, p. 49
- Test Practice, p. 67

Activities and Resources
- Internet Activity: Physical/Chemical Changes, p. 39
- Resource Centers: Chemical Properties of Matter, p. 46; Melting Points and Boiling Points, p. 54; Separating Materials from Mixtures, p. 63

Physical Properties of Matter
Code: MDL062

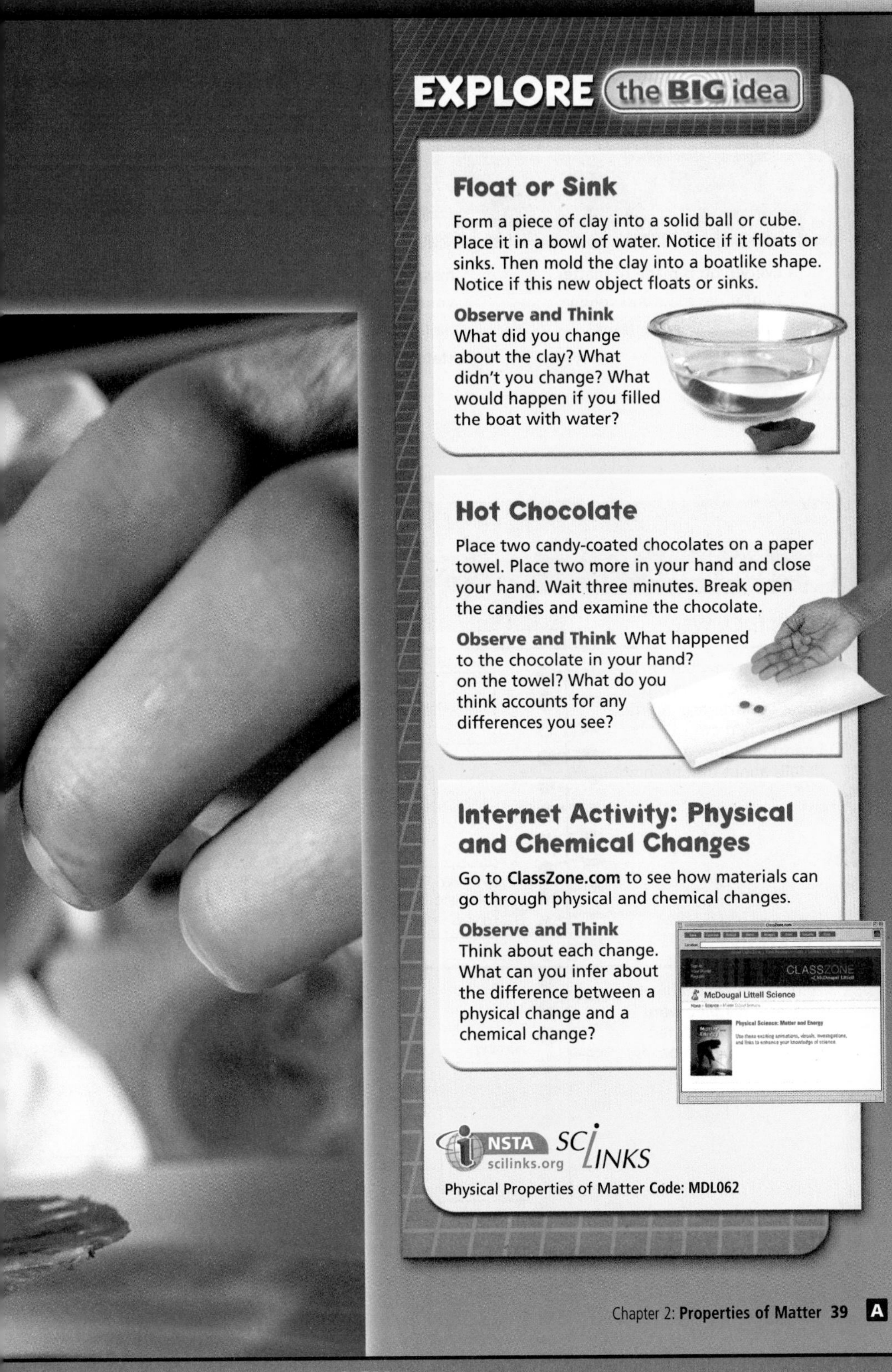
EXPLORE the BIG idea

Float or Sink

Form a piece of clay into a solid ball or cube. Place it in a bowl of water. Notice if it floats or sinks. Then mold the clay into a boatlike shape. Notice if this new object floats or sinks.

Observe and Think What did you change about the clay? What didn't you change? What would happen if you filled the boat with water?

Hot Chocolate

Place two candy-coated chocolates on a paper towel. Place two more in your hand and close your hand. Wait three minutes. Break open the candies and examine the chocolate.

Observe and Think What happened to the chocolate in your hand? on the towel? What do you think accounts for any differences you see?

Internet Activity: Physical and Chemical Changes

Go to **ClassZone.com** to see how materials can go through physical and chemical changes.

Observe and Think Think about each change. What can you infer about the difference between a physical change and a chemical change?

NSTA scilinks.org SCI LINKS

Physical Properties of Matter **Code: MDL062**

TEACHING WITH TECHNOLOGY

Digital Camera While students perform the Chapter Investigation on pp. 56–57, have them photograph of the various steps of the procedure. Have them prepare a slide show of the photographs on a computer.

CBL and Probeware Use a temperature probe instead of a thermometer in the Chapter Investigation, pp. 56–57.

EXPLORE the BIG idea

These inquiry-based activities are appropriate for use at home or as a supplement to classroom instruction.

Float or Sink

PURPOSE To introduce students to the effect of shape on overall density. Students shape clay to explore its density and volume.

TIP *10 min.* Be sure the boatlike shape contains no holes through which water can enter.

Answer: The shape of the clay changed. By creating a shape that held air, the clay boat was able to float. Other properties, such as total amount of clay, mass, weight, and color, didn't change. If the boat were filled with water, it would sink.

REVISIT after p. 43.

Hot Chocolate

PURPOSE To introduce students to the effect of heat on the state of a material. Students apply heat to chocolate and observe that it changes state.

TIP *10 min.* Remind students never to eat anything in a laboratory setting.

Answer: It melted; it stayed the same; an increase in temperature melts chocolate.

REVISIT after p. 51.

Internet Activity: Physical and Chemical Changes

PURPOSE To introduce students to examples of physical and chemical changes.

TIP *20 min.* Encourage students to generalize about what physical changes have in common and what chemical changes have in common.

Answer: The identity of a substance changes during a chemical change but does not during a physical change.

REVISIT after p. 46.

PREPARE

CONCEPT REVIEW

Activate Prior Knowledge

- At the beginning of class, place an ice cube in a transparent cup in the front of the room. Ask students if the ice is matter. *It is.*
- After 15 minutes, ask students if what is in the cup is still matter. *yes*
- Ask them to describe any changes the matter has undergone. *Some of the solid matter is now a liquid.*

TAKING NOTES

Main Idea Web

Point out that different webs can have different numbers of boxes around them. Some might have only two connected to the heading box. Suggest that students limit the number of boxes surrounding the center box to reflect an appropriate level of detail.

Vocabulary Strategy

Discuss why a magnet is a good representation of the central concept.

Vocabulary and Note-Taking Resources

- Vocabulary Practice, pp. 117–118
- Decoding Support, p. 119

- Daily Vocabulary Scaffolding, p. T10
- Note-Taking Model, p. T11

- Magnet Word, B24–25
- Main Idea Web, C38–39
- Daily Vocabulary Scaffolding, H1–8

CHAPTER 2

Getting Ready to Learn

CONCEPT REVIEW

- Everything is made of matter.
- Matter has mass and volume.
- Atoms combine to form molecules.

VOCABULARY REVIEW

mass p. 10
volume p. 11
molecule p. 18
states of matter p. 27

CONTENT REVIEW
CLASSZONE.COM
Review concepts and vocabulary.

TAKING NOTES

MAIN IDEA WEB

Write each new blue heading in a box. Then write notes in boxes around the center box that give important terms and details about that heading.

VOCABULARY STRATEGY

Think about a vocabulary term as a **magnet word** diagram. Write related terms and ideas in boxes around it.

See the Note-Taking Handbook on pages R45–R51.

SCIENCE NOTEBOOK

Physical properties describe a substance.
- color, shape, size, texture, volume, mass
- melting point, boiling point
- density: a measure of the amount of matter in a given volume

CHEMICAL CHANGE
- burning
- rusting
- tarnishing
- change in temperature
- change in color
- formation of bubbles

CHECK READINESS

Administer the Diagnostic Test to determine students' readiness for new science content and requisite math skills.

 Diagnostic Test, pp. 21–22

Technology Resources

Students needing content and math skills should visit **ClassZone.com.**

- CONTENT REVIEW
- MATH TUTORIAL

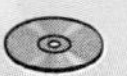 CONTENT REVIEW CD-ROM

KEY CONCEPT

Matter has observable properties.

BEFORE, you learned

- Matter has mass and volume
- Matter is made of atoms
- Matter exists in different states

NOW, you will learn

- About physical and chemical properties
- About physical changes
- About chemical changes

VOCABULARY

physical property p. 41
density p. 43
physical change p. 44
chemical property p. 46
chemical change p. 46

EXPLORE Physical Properties

How can a substance be changed?

PROCEDURE

1. Observe the clay. Note its physical characteristics, such as color, shape, texture, and size.
2. Change the shape of the clay. Note which characteristics changed and which ones stayed the same.

MATERIAL

rectangular piece of clay

WHAT DO YOU THINK?

- How did reshaping the clay change its physical characteristics?
- How were the mass and the volume of the clay affected?

Physical properties describe a substance.

What words would you use to describe a table? a chair? the sandwich you ate for lunch? You would probably say something about the shape, color, and size of each item. Next you might consider whether it is hard or soft, smooth or rough to the touch. Normally, when describing an object, you identify the characteristics of the object that you can observe without changing the identity of the object.

The characteristics of a substance that can be observed without changing the identity of the substance are called **physical properties.** In science, observation can include measuring and handling a substance. All of your senses can be used to detect physical properties. Color, shape, size, texture, volume, and mass are a few of the physical properties you probably have encountered.

VOCABULARY
Make a magnet word diagram in your notebook for *physical property.*

CHECK YOUR READING Describe some of the physical properties of your desk.

2.1 FOCUS

Set Learning Goals

Students will

- Describe physical and chemical properties.
- Give examples of physical changes.
- Explain that chemical changes form new substances.
- Observe signs of chemical change in an experiment.

3-Minute Warm-Up

Display Transparency 12 or copy this exercise on the board:

Describe a common object by naming its distinctive properties. Trade your mystery-object description with a partner's and try to guess what object he or she has described.

3-Minute Warm-Up, p. T12

2.1 MOTIVATE

EXPLORE Physical Properties

PURPOSE To observe the physical properties of a material before and after a physical change

TIP *10 min.* Students can use the same clay they used for the exploration on p. 39.

WHAT DO YOU THINK? *The shape changed, but the color and texture stayed the same. Mass and volume stayed the same.*

Ongoing Assessment

CHECK YOUR READING *Answer: Answers should include color, texture, and shape.*

RESOURCES FOR DIFFERENTIATED INSTRUCTION

Below Level

UNIT RESOURCE BOOK

- Reading Study Guide A, pp. 85–86
- Decoding Support, p. 119

AUDIO CDS

Additional INVESTIGATION, Measuring Density, A, B, & C, pp. 133–141; Teacher Instructions, pp. 262–263

Advanced

UNIT RESOURCE BOOK

- Challenge and Extension, p. 91
- Challenge Reading, pp. 115–116

English Learners

UNIT RESOURCE BOOK

Spanish Reading Study Guide, pp. 89–90

AUDIO CDS

- Audio Readings in Spanish
- Audio Readings (English)

2.1 INSTRUCT

Real World Example

A centrifuge is a machine that separates the components of a liquid on the basis of the physical property of density. It spins a liquid at great speed so that denser materials sink to the bottom of the container that holds the liquid. One liquid whose components are separated in a centrifuge is blood. Blood cells are denser than plasma, the liquid part of blood. As a result, the cells end up at the bottom, with the plasma on top. Why would anyone want to separate blood? Sometimes blood tests are done on only one part of blood, so that part must be isolated. Also, some blood transfusions involve only plasma or cells.

Develop Critical Thinking

APPLY This activity illustrates the concept of density.

- Collect two groups of items. One group should have items of the same mass but different volumes. The second group should have items of the same volume, but different masses.
- Have students order the items in each group according to which they think have the greatest mass. Students should write down and explain their choices without touching or picking up the objects.
- Next, have students pick up the items and adjust their lists based on the new information.
- Ask: How does the relationship between mass and volume depend on the material? *Some materials have a greater mass per unit volume than other materials.*

Ongoing Assessment

Answer: mass, volume; shape, ability to stretch

Answer: texture, color; volume, shape, mass

Physical Properties

A

How do you know which characteristics are physical properties? Just ask yourself whether observing the property involves changing the substance to a different substance. For example, you can stretch a rubber band. Does stretching the rubber band change what it is made of? No. The rubber band is still a rubber band before and after it is stretched. It may look a little different, but it is still a rubber band.

Mass and volume are two physical properties. Measuring these properties does not change the identity of a substance. For example, a lump of clay might have a mass of 200 grams (g) and a volume of 100 cubic centimeters (cm^3). If you were to break the clay in half, you would have two 100 g pieces of clay, each with a volume of 50 cm^3. You can bend and shape the clay too. Even if you were to mold a realistic model of a car out of the clay, it still would be a piece of clay. Although you have changed some of the properties of the object, such as its shape and volume, you have not changed the fact that the substance you are observing is clay.

Because all formulas for volume involve the multiplication of three measurements, volume has a unit that is cubed (such as cm^3).

CHECK YOUR READING Which physical properties listed above are found by taking measurements? Which are not?

Physical Properties

Physical properties of clay—such as volume, mass, color, texture, and shape—can be observed without changing the fact that the substance is clay.

B

Block of Clay

Shaped Clay

READING VISUALS COMPARE AND CONTRAST Which physical properties do the two pieces of clay have in common? Which are different?

DIFFERENTIATE INSTRUCTION

More Reading Support

A If observing a property doesn't change a substance, what kind of property is it? *physical*

B What type of property is texture? *physical*

English Learners The words *would*, *should*, and *could* are not universal in all languages. Help students understand sentences that contain these words. For example, guide them to see that the condition expressed in the following sentence is hypothetical: "If you were to break the clay in half, you would have two 100 g pieces of clay."

Density

The relationship between the mass and the volume of a substance is another important physical property. For any substance, the amount of mass in a unit of volume is constant. For different substances, the amount of mass in a unit of volume may differ. This relationship explains why you can easily lift a shoebox full of feathers but not one filled with pennies, even though both are the same size. A volume of pennies contains more mass than an equal volume of feathers. The relationship between mass and volume is called density.

Density is a measure of the amount of matter present in a given volume of a substance. Density is normally expressed in units of grams per cubic centimeter (g/cm^3). In other words, density is the mass in grams divided by the volume in cubic centimeters.

$$\text{Density} = \frac{\text{mass}}{\text{Volume}} \qquad D = \frac{m}{V}$$

How would you find the density of 200 g of clay with a volume of 100 cm^3? You calculate that the clay has a density of 200 g divided by 100 cm^3, or 2 g/cm^3. If you divide the clay in half and find the density of one piece of clay, it will be 100 g/50 cm^3, or 2 g/cm^3—the same as the original piece. Notice that density is a property of a substance that remains the same no matter how much of the substance you have.

READING TIP

The density of solids is usually measured in grams per cubic centimeter (g/cm^3). The density of liquids is usually measured in grams per milliliter (g/mL). Recall that 1 mL = 1 cm^3.

Calculating Density

Sample Problem

A glass marble has a volume of 5 cm^3 and a mass of 13 g. What is the density of glass?

What do you know? Volume = 5 cm^3, mass = 13 g

What do you want to find out? Density

Write the formula: $D = \frac{m}{V}$

Substitute into the formula: $D = \frac{13\text{ g}}{5\text{ cm}^3}$

Calculate and simplify: $D = 2.6\text{ g/cm}^3$

Check that your units agree: Unit is g/cm^3.
Unit of density is g/cm^3. Units agree.

Answer: $D = 2.6\text{ g/cm}^3$

Practice the Math

1. A lead sinker has a mass of 227 g and a volume of 20 cm^3. What is the density of lead?
2. A glass of milk has a volume of 100 cm^3. If the milk has a mass of 103 g, what is the density of milk?

Address Misconceptions

IDENTIFY Place a ball of modeling clay and a Styrofoam ball in water. Ask why one floats and the other doesn't. If students suggest the clay sinks because it is heavier, they may hold the misconception that *dense* and *heavy* mean the same thing.

CORRECT Remove a small portion of clay that has less mass than the Styrofoam ball. Have students compare the weights of the small portion of clay and the ball. Ask them to predict what will happen if both are put in water again. Demonstrate. Explain that the clay sinks not because it is heavier, but because it is more dense than water.

REASSESS Ask students to discuss how large a Styrofoam ball would have to be before it would sink in water. Students should reach the conclusion that any amount of Styrofoam should float because Styrofoam is less dense than water.

Technology Resources

Visit **ClassZone.com** for background on common student misconceptions.

MISCONCEPTION DATABASE

Develop Algebra Skills

Remind students of the importance of using correct units to determine whether a problem is set up correctly. If correct units are used at each step, the answer should have the correct units.

EXPLORE the BIG idea

Revisit "Float or Sink" on p. 39. Ask students how the density of air affects the average density of the boat. *The average density is decreased because the volume of the boat is filled with air.*

DIFFERENTIATE INSTRUCTION

More Reading Support

C What physical property compares mass and volume? *density*

D How do you calculate density? *Divide mass by volume.*

Additional Investigation To reinforce Section 2.1 learning goals, use the following full-period investigation:

R **Additional INVESTIGATION,** Measuring Density, A, B, & C, pp. 133–141, 262–263
(Advanced students should complete Levels B and C.)

Below Level Have students pan for copper using a mixture of pennies and objects that are less dense, such as buttons. Have students move the pan around so the pennies "sink" to the bottom and the less dense items rise to the top.

Ongoing Assessment

Practice the Math *Answers:*

1. 11.35 g/cm^3

2. 1.03 g/cm^3

- Math Support, p. 120
- Math Practice, p. 121

Integrate the Sciences

There are two types of weathering: physical and chemical. In physical weathering, rock is broken down into smaller pieces. It also undergoes mechanical weathering, a type of physical weathering. Plant-root growth helps break up rocks, as does ice wedging. The latter works as follows: Water seeps into cracks in rocks. When the temperature drops below freezing, the water freezes and expands, applying pressure to the sides of the crack. A series of freezing and thawing cycles will break the rock into small pieces.

In chemical weathering, chemical reactions occur that create new substances. Soil, for example, is formed from decaying organic matter and weathered rock. Rock can weather chemically by the action of acids and other chemicals in the environment.

Teacher Demo

Use a raw egg to show the difference between physical changes and other changes. Have students list each change below and decide whether it is a physical change or not.

- Break a raw egg open and place it in a beaker. *physical*
- Use a fork to beat the egg. *physical*
- Cook the egg over a hot plate. *other*

Ongoing Assessment

CHECK YOUR READING *Answer: It changes a physical property, not the substance itself. For example, cutting paper will change its shape and volume, but will not change the substance of the paper.*

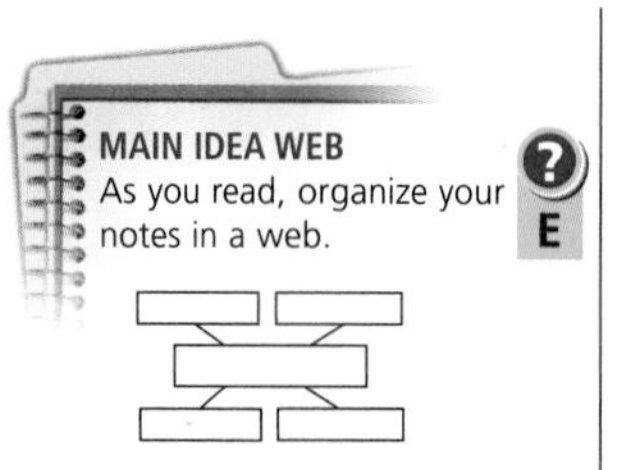

MAIN IDEA WEB
As you read, organize your notes in a web.

Physical Changes

E You have read that a physical property is any property that can be observed without changing the identity of the substance. What then would be a physical change? A **physical change** is a change in any physical property of a substance, not in the substance itself. Breaking a piece of clay in half is a physical change because it changes only the size and shape of the clay. Stretching a rubber band is a physical change because the size of the rubber band changes. The color of the rubber band sometimes can change as well when it is stretched. However, the material that the rubber band is made of does not change. The rubber band is still rubber.

What happens when water changes from a liquid into water vapor or ice? Is this a physical change? Remember to ask yourself what has changed about the material. Ice is a solid and water is a liquid, but both are the same substance—both are composed of H_2O molecules. As you will read in more detail in the next section, a change in a substance's state of matter is a physical change.

CHECK YOUR READING How is a physical change related to a substance's physical properties?

A substance can go through many different physical changes and still remain the same substance. Consider, for example, the changes that happen to the wool that ultimately becomes a sweater.

1. Wool is sheared from the sheep. The wool is then cleaned and placed into a machine that separates the wool fibers from one another. Shearing and separating the fibers are physical changes that change the shape, volume, and texture of the wool.
2. The wool fibers are spun into yarn. Again, the shape and volume of the wool change. The fibers are twisted so that they are packed more closely together and are intertwined with one another.
3. The yarn is dyed. The dye changes the color of the wool, but it does not change the wool into another substance. This type of color change is a physical change.
4. Knitting the yarn into a sweater also does not change the wool into another substance. A wool sweater is still wool, even though it no longer resembles the wool on a sheep.

It can be difficult to determine if a specific change is a physical change or not. Some changes, such as a change in color, also can occur when new substances are formed during the change. When deciding whether a change is a physical change or not, ask yourself whether you have the same substance you started with. If the substance is the same, then the changes it underwent were all physical changes.

DIFFERENTIATE INSTRUCTION

More Reading Support

E Is cutting paper a physical change? *yes*

Advanced Have students select two or three samples of several different substances. Then have them identify properties of each substance—such as mass, volume, and density—and categorize them as "Depends on amount of substance (extensive)" or "Does not depend on amount of substance (intensive)."

Have students who are interested in physical properties of unusual substances read the following article:

Challenge Reading, pp. 115–116

Physical Changes

The process of turning wool into a sweater requires that the wool undergo physical changes. Changes in shape, volume, texture, and color occur as raw wool is turned into a colorful sweater.

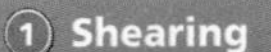

1 Shearing

Preparing the wool produces physical changes. The wool is removed from the sheep and then cleaned before the wool fibers are separated.

2 Spinning

Further physical changes occur as machine twists the wool fibers into a long, thin rope of yarn.

3 Dyeing

Dyeing produces color changes but does not change the basic substance of the wool.

4 The final product, a wool sweater, is still wool.

READING VISUALS How does the yarn in the sweater differ from the wool on the sheep?

DIFFERENTIATE INSTRUCTION

Inclusion Bring a wool sweater and some raw wool to class so that students who are visually impaired can feel and compare the two forms of wool.

Teach Difficult Concepts

Color change produced by covering a material with a pigment, such as paint or dye, is usually considered a physical change. Although many dyes and paints attach to the surface of a material by chemical bonds, these bonds do not change the nature of the original substance. Instead, the bonds cause the pigment to adhere to the material. Ask: How do you think dye colors wool without changing the wool's identity? *It coats the fibers.*

Teach from Visuals

The visual "Physical Changes" is available as T14 in the Unit Transparency Book.

Art Connection

In many cultures, people use dyes to produce textile artwork. Examples include the colorful kente cloth woven in Ghana and the detailed batik images produced in Indonesia and India. The oldest known dyes are indigo, which comes from the indigo plant, and Tyrian purple, which is extracted from snails. Many natural dyes color fabric with no other additive, but some require the use of metal salts, often called mordants, that fix the dye to the fabric. Most dyes used today are synthetic, although some artists still prefer to make their own dyes from natural sources.

Ongoing Assessment

Give examples of physical changes.

Ask: How can you physically change a substance without changing what it is? *by cutting it up, by changing its shape, by changing its color*

READING VISUALS *Answer: The yarn was twisted into strands, dyed, and knit to form the sweater. However, it is still wool.*

Teach Difficult Concepts

Students might have difficulty understanding color as it relates to properties and changes. Color itself is a physical property, because it can be observed without changing the identity of the object. But when color changes, it is usually a sign of a chemical change. Ask students for examples of color changes that indicate a chemical change. To help students understand this concept, try the following demonstration.

Teacher Demo

Fill a small, transparent container halfway with water. Dissolve a crystal of potassium permanganate in the water. Point out that dissolving is a physical change. Add about one gram of sodium hydrogen sulfite to the solution. As it dissolves, the solution loses its color because of a chemical change.

EXPLORE the BIG idea

Revisit "Internet Activity: Physical and Chemical Changes" on p. 39. Have students state the difference between a chemical change and a physical change.

Ongoing Assessment

Describe physical and chemical properties.

Ask: List some physical properties of wood. *hardness, density, color* What chemical properties does wood have? *ability to burn, turns into other substances*

PHOTO CAPTION Answer: When people rub the nose, the oils on their skin remove tarnish, revealing the untarnished bronze. Oils from people's hands also protect the bronze from reacting with the air and forming more tarnish.

Chemical properties describe how substances form new substances.

Learn about the chemical properties of matter.

If you wanted to keep a campfire burning, would you add a piece of wood or a piece of iron? You would add wood, of course, because you know that wood burns but iron does not. Is the ability to burn a physical property of the wood? The ability to burn seems to be quite different from physical properties such as color, density, and shape. More important, after the wood burns, all that is left is a pile of ashes and some new substances in the air. The wood has obviously changed into something else. The ability to burn, therefore, must describe another kind of property that substances have—not a physical property but a chemical property.

Chemical Properties and Changes

Chemical properties describe how substances can form new substances. Combustibility, for example, describes how well an object can burn. Wood burns well and turns into ashes and other substances. Can you think of a chemical property for the metal iron? Especially when left outdoors in wet weather, iron rusts. The ability to rust is a chemical property of iron. The metal silver does not rust, but eventually a darker substance called tarnish forms on its surface. You may have noticed a layer of tarnish on some silver spoons or jewelry.

INFER The bust of Abraham Lincoln is made of bronze. Why is the nose a different color from the rest of the head?

The chemical properties of copper cause it to become a blue-green color when it is exposed to air. A famous example of tarnished copper is the Statue of Liberty. The chemical properties of bronze are different. Some bronze objects tarnish to a dark brown color, like the bust of Abraham Lincoln in the photograph on the left.

Chemical properties can be identified by the changes they produce. The change of one substance into another substance is called a **chemical change.** A piece of wood burning, an iron fence rusting, and a silver spoon tarnishing are all examples of chemical changes. A chemical change affects the substances involved in the change. During a chemical change, combinations of atoms in the original substances are rearranged to make new substances. For example, when rust forms on iron, the iron atoms combine with oxygen atoms in the air to form a new substance that is made of both iron and oxygen.

A chemical change is also involved when an antacid tablet is dropped into a glass of water. As the tablet dissolves, bubbles of gas appear. The water and the substances in the tablet react to form new substances. One of these substances is carbon dioxide gas, which forms the bubbles that you see.

DIFFERENTIATE INSTRUCTION

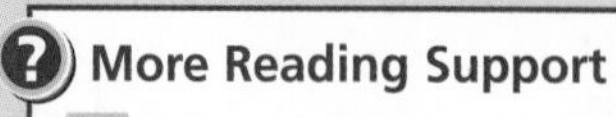

More Reading Support

F What happens to atoms during a chemical change? *They are rearranged.*

Advanced Have students place an old discolored penny in a jar containing a few milliliters of household ammonia. Have them close the jar quickly to avoid breathing the vapors. After 30 minutes, check the penny and ask them to explain what happened in terms of a chemical change. *The ammonia chemically changed the tarnish on the penny. Evidence of this change is the color change of the penny—it became brighter—and the color change of the solution—it turned blue.*

Challenge and Extension, p. 91

Not all chemical changes are as destructive as burning, rusting, or tarnishing. Chemical changes are also involved in cooking. When you boil an egg, for example, the substances in the raw egg change into new substances as energy is added to the egg. When you eat the egg, further chemical changes take place as your body digests the egg. The process forms new molecules that your body then can use to function.

CHECK YOUR READING Give three examples of chemical changes.

The only true indication of a chemical change is that a new substance has been formed. Sometimes, however, it is difficult to tell whether new substances have been formed or not. In many cases you have to judge which type of change has occurred only on the basis of your observations of the change and your previous experience. However, some common signs can suggest that a chemical change has occurred. You can use these signs to guide you as you try to classify a change that you are observing.

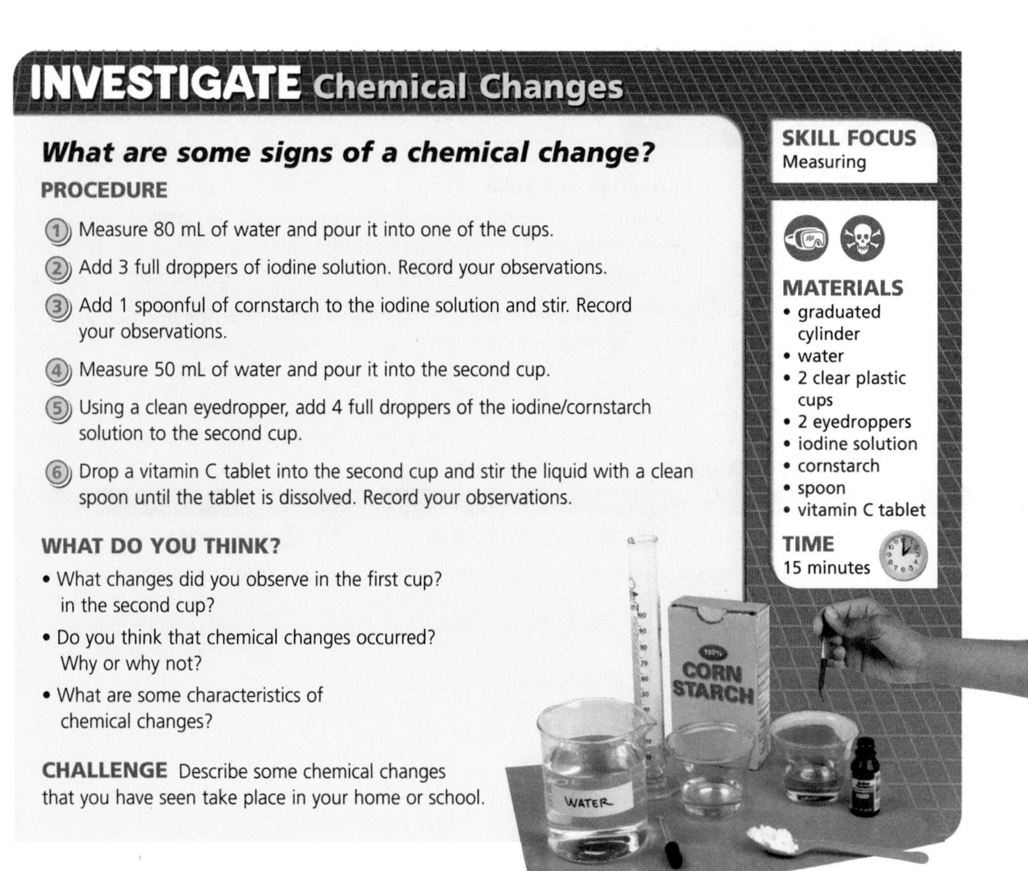

INVESTIGATE Chemical Changes

What are some signs of a chemical change?

PROCEDURE

1. Measure 80 mL of water and pour it into one of the cups.
2. Add 3 full droppers of iodine solution. Record your observations.
3. Add 1 spoonful of cornstarch to the iodine solution and stir. Record your observations.
4. Measure 50 mL of water and pour it into the second cup.
5. Using a clean eyedropper, add 4 full droppers of the iodine/cornstarch solution to the second cup.
6. Drop a vitamin C tablet into the second cup and stir the liquid with a clean spoon until the tablet is dissolved. Record your observations.

WHAT DO YOU THINK?

- What changes did you observe in the first cup? in the second cup?
- Do you think that chemical changes occurred? Why or why not?
- What are some characteristics of chemical changes?

CHALLENGE Describe some chemical changes that you have seen take place in your home or school.

SKILL FOCUS
Measuring

MATERIALS
- graduated cylinder
- water
- 2 clear plastic cups
- 2 eyedroppers
- iodine solution
- cornstarch
- spoon
- vitamin C tablet

TIME
15 minutes

DIFFERENTIATE INSTRUCTION

More Reading Support

G What type of change occurs when an egg cooks? *chemical*

H What is an indication of chemical change? *A new substance forms.*

English Learners Offer synonyms and simplifications for word combinations that may confuse an English learner. For example, "a few" (p. 41) means *some*, "go through" (p. 44) means *experience* or *undergo*, and "most likely" (p. 48) means *probably*.

INVESTIGATE Chemical Changes

PURPOSE To observe signs of a chemical change

TIPS *15 min.*

- Use Lugol's solution as a source of iodine. Add enough iodine to make the solution yellow.
- If vitamin C tablets are not available, use orange juice.

WHAT DO YOU THINK? *The color changed from orange to blue-black in the first cup. In the second cup, the color changed and bubbles formed. Yes; new substances formed, as evidenced by the color changes and bubbles. Some signs of a chemical change are a change in color and the formation of bubbles.*

CHALLENGE *Sample answer: food cooking, tarnish forming on door-knobs, and fuel burning for heat*

Datasheet, Chemical Changes, p. 92

Technology Resources

Customize this student lab as needed or look for an alternative. Print rubrics to assess student lab reports.

Lab Generator CD-ROM

Metacognitive Strategy

Ask students to write a paragraph focusing on questions that occurred to them during the investigation. Have them underline the questions they could answer after completing the activity.

Integrate the Sciences

After a thunderstorm, the air often smells different. As lightning passes through oxygen gas, the oxygen chemically changes to ozone. The results of this chemical reaction can be detected as a change in odor. Ozone, however, is unstable and soon forms oxygen gas again.

Ongoing Assessment

Sample answer: burning, rusting, tarnishing

Ongoing Assessment

Explain that chemical changes form new substances.

Ask: If you mix water and antacid and they react chemically, what substances are you likely to have after the reaction? *new chemicals and maybe a little left-over water or antacid*

Sample answer: change in odor, temperature, and color

Reinforce the BIG idea

Have students relate the section to the Big Idea.

 Reinforcing Key Concepts, p. 93

2.1 ASSESS & RETEACH

Assess

 Section 2.1 Quiz, p. 23

Reteach

Light a previously unburned candle and have students observe it. Ask students to list signs of a chemical change in the burning candle. *Students should notice a change in odor, a rise in temperature, formation of a gas, and a change in wick color.*

Technology Resources

Have students visit **ClassZone.com** for reteaching of Key Concepts.

 CONTENT REVIEW

 CONTENT REVIEW CD-ROM

Signs of a Chemical Change

You may not be able to see that any new substances have formed during a change. Below are some signs that a chemical change may have occurred. If you observe two or more of these signs during a change, you most likely are observing a chemical change.

Carbon dioxide bubbles form as substances in the tablet react with water.

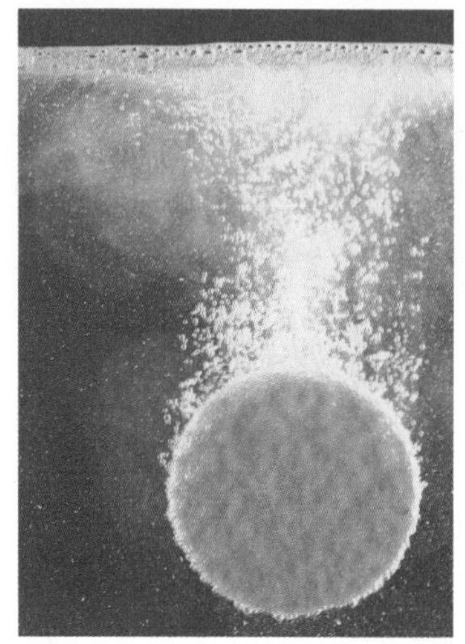

Production of an Odor Some chemical changes produce new smells. The chemical change that occurs when an egg is rotting produces the smell of sulfur. If you go outdoors after a thunderstorm, you may detect an unusual odor in the air. The odor is an indication that lightning has caused a chemical change in the air.

Change in Temperature Chemical changes often are accompanied by a change in temperature. You may have noticed that the temperature is higher near logs burning in a campfire.

Change in Color A change in color is often an indication of a chemical change. For example, fruit may change color when it ripens.

Formation of Bubbles When an antacid tablet makes contact with water, it begins to bubble. The formation of gas bubbles is another indicator that a chemical change may have occurred.

Formation of a Solid When two liquids are combined, a solid called a precipitate can form. The shells of animals such as clams and mussels are precipitates. They are the result of a chemical change involving substances in seawater combining with substances from the creatures.

 Give three signs of chemical changes. Describe one that you have seen recently.

2.1 Review

KEY CONCEPTS

1. What effect does observing a substance's physical properties have on the substance?
2. Describe how a physical property such as mass or texture can change without causing a change in the substance.
3. Explain why burning is a chemical change in wood.

CRITICAL THINKING

4. **Synthesize** Why does the density of a substance remain the same for different amounts of the substance?
5. **Calculate** What is the density of a block of wood with a mass of 120 g and a volume of 200 cm^3?

CHALLENGE

6. **Infer** Iron can rust when it is exposed to oxygen. What method could be used to prevent iron from rusting?

ANSWERS

1. The identity of the substance does not change.

2. Sample answer: Some mass can be removed from the object, but the object still has the same identity.

3. The products of the change are no longer wood.

4. As mass increases or decreases, so does volume. The relationship between mass and volume remains the same.

5. $D = m/V = 120\ g/200\ cm^3 = 0.6\ g/cm^3$

6. Paint the iron so that it does not come into contact with the oxygen.

MATH in SCIENCE

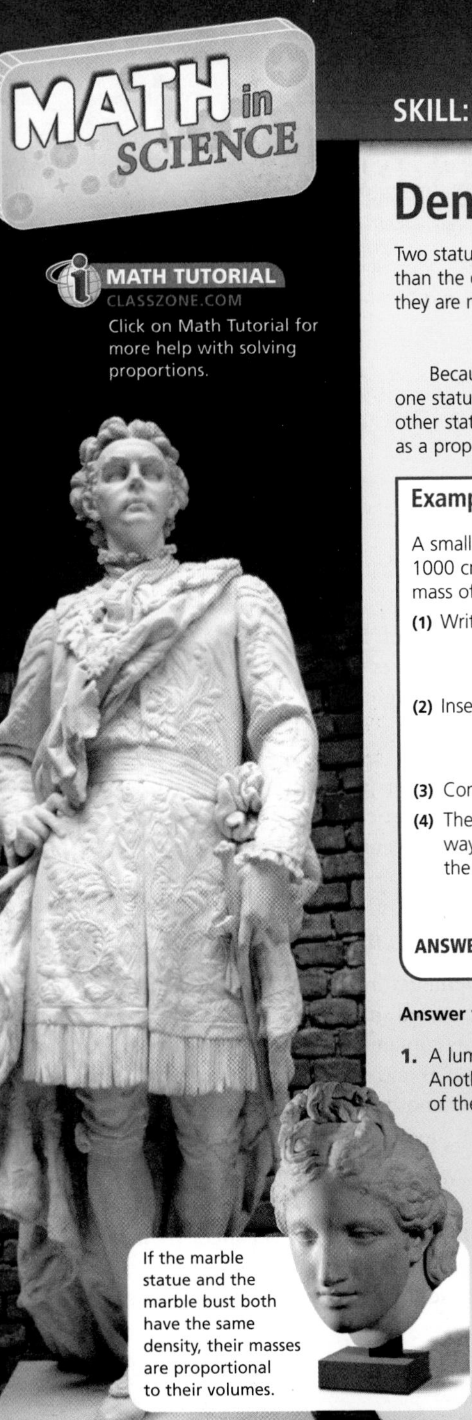

MATH TUTORIAL
CLASSZONE.COM
Click on Math Tutorial for more help with solving proportions.

SKILL: SOLVING PROPORTIONS

Density of Materials

Two statues are made of the same type of marble. One is larger than the other. However, they both have the same density because they are made of the same material. Recall the formula for density.

$$\text{Density} = \frac{\text{mass}}{\text{Volume}}$$

Because the density is the same, you know that the mass of one statue divided by its volume is the same as the mass of the other statue divided by its volume. You can set this up and solve it as a proportion.

Example

A small marble statue has a mass of 2.5 kg and a volume of 1000 cm^3. A large marble statue with the same density has a mass of 10 kg. What is the volume of the large statue?

(1) Write the information as an equation showing the proportion.

$$\frac{\text{mass of small statue}}{\text{volume of small statue}} = \frac{\text{mass of large statue}}{\text{volume of large statue}}$$

(2) Insert the known values into your equation.

$$\frac{2.5 \text{ kg}}{1000 \text{ cm}^3} = \frac{10 \text{ kg}}{\text{volume of large statue}}$$

(3) Compare the numerators: 10 kg is 4 times greater than 2.5 kg.

(4) The denominators of the fractions are related in the same way. Therefore, the volume of the large statue is 4 times the volume of the small one.

$$\text{volume of large statue} = 4 \cdot 1000 \text{ cm}^3 = 4000 \text{ cm}^3$$

ANSWER The volume of the large statue is 4000 cm^3.

Answer the following questions.

1. A lump of gold has a volume of 10 cm^3 and a mass of 193 g. Another lump of gold has a mass of 96.5 g. What is the volume of the second lump of gold?
2. A carpenter saws a wooden beam into two pieces. One piece has a mass of 600 g and a volume of 1000 cm^3. What is the mass of the second piece if its volume is 250 cm^3?
3. A 200 mL bottle is completely filled with cooking oil. The oil has a mass of 180 g. If 150 mL of the oil is poured into a pot, what is the mass of the poured oil?

CHALLENGE You have two spheres made of the same material. One has a diameter that is twice as large as the other. How do their masses compare?

If the marble statue and the marble bust both have the same density, their masses are proportional to their volumes.

ANSWERS

1. 193 g · 0.5 = 96.5 g; 10 cm^3 · 0.5 = 5 cm^3

2. $\frac{1000 \text{ cm}^3}{4} = 250 \text{ cm}^3$; $\frac{600 \text{ g}}{4} = 150 \text{ g}$

3. 200 mL · 0.75 = 150 mL; 180 g · 0.75 = 135 g

CHALLENGE *The formula for the volume of a sphere is $4/3\ \pi r^3$. If the diameter doubles, the radius also doubles. Thus, the volume is 2^3, or 8, times greater. The density is the same for both, so if the volume is 8 times greater, the mass will also be 8 times greater.*

MATH IN SCIENCE

Math Skills Practice for Science

Set Learning Goal

To solve proportions in the context of density

Present the Science

Density can be reliably used for identifying many solids because it remains the same for any-sized sample of the solid. The density of a mineral is a key characteristic in identifying it.

Develop Algebra Skills

- Remind students that a proportion is two ratios in fraction form with an equal sign between them.
- If three of the four parts of a proportion are known, multiplication and division can be used to find the unknown quantity.

Close

Ask: A 450-gram sample of nickel has a volume of 50 cm^3. A sample of iron has a mass of 300 grams. Can you use this information to set up a proportion to find the volume of iron? Explain. *No; the densities of the elements differ, so the ratios are not equal to each other.*

- Math Support, p. 122
- Math Practice, p. 123

Technology Resources

Students can visit **ClassZone.com** for practice in solving proportions.

MATH TUTORIAL

2.2 FOCUS

Set Learning Goals

Students will

- Describe how liquids can become solids, and solids can become liquids.
- Explain how liquids can become gases, and gases can become liquids.
- Determine how energy is related to changes in state.

3-Minute Warm-Up

Display Transparency 12 or copy this exercise on the board:

Decide if the statements are true. If not, correct them.

1. Density is a chemical property. *physical property*
2. Cutting a carrot in half is a physical change. *true*
3. Silver tarnishing is an example of a physical change. *chemical change*

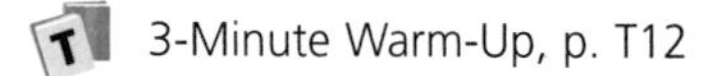

2.2 MOTIVATE

THINK ABOUT

PURPOSE To understand how liquid water forms from water vapor

DISCUSS How does the formation of dew compare with the formation of water on the outside of a glass of ice water? *In both cases, liquid water forms from water vapor in the air.*

Ongoing Assessment

CHECK YOUR READING *Answer: The identity of the substance does not change.*

KEY CONCEPTS

2.2 Changes of state are physical changes.

BEFORE, you learned

- Substances have physical and chemical properties
- Physical changes do not change a substance into a new substance
- Chemical changes result in new substances

NOW, you will learn

- How liquids can become solids, and solids can become liquids
- How liquids can become gases, and gases can become liquids
- How energy is related to changes of state

VOCABULARY

melting p. 51
melting point p. 51
freezing p. 52
freezing point p. 52
evaporation p. 53
sublimation p. 53
boiling p. 54
boiling point p. 54
condensation p. 55

THINK ABOUT

Where does dew come from?

On a cool morning, droplets of dew cover the grass. Where does this water come from? You might think it had rained recently. However, dew forms even if it has not rained. Air is made of a mixture of different gases, including water vapor. Some of the water vapor condenses—or becomes a liquid—on the cool grass and forms drops of liquid water.

MAIN IDEA WEB
Remember to place each blue heading in a box. Add details around it to form a web.

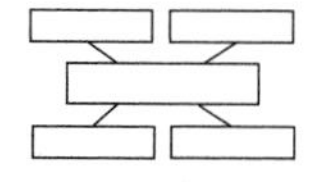

Matter can change from one state to another.

Matter is commonly found in three states: solid, liquid, and gas. A solid has a fixed volume and a fixed shape. A liquid also has a fixed volume but takes the shape of its container. A gas has neither a fixed volume nor a fixed shape. Matter always exists in one of these states, but it can change from one state to another.

When matter changes from one state to another, the substance itself does not change. Water, ice, and water vapor are all the same basic substance. As water turns into ice or water vapor, the water molecules themselves do not change. What changes are the arrangement of the molecules and the amount of space between them. Changes in state are physical changes because changes in state do not change the basic substance.

Why is a change in state a physical change rather than a chemical change?

RESOURCES FOR DIFFERENTIATED INSTRUCTION

Below Level

UNIT RESOURCE BOOK

- Reading Study Guide A, pp. 96–97
- Decoding Support, p. 119

AUDIO CDS

Advanced

UNIT RESOURCE BOOK

Challenge and Extension, p. 102

English Learners

UNIT RESOURCE BOOK

Spanish Reading Study Guide, pp. 100–101

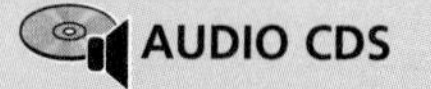

- Audio Readings in Spanish
- Audio Readings (English)

Solids can become liquids, and liquids can become solids.

If you leave an ice cube on a kitchen counter, it changes to the liquid form of water. Water changes to the solid form of water, ice, when it is placed in a freezer. In a similar way, if a bar of iron is heated to a high enough temperature, it will become liquid iron. As the liquid iron cools, it becomes solid iron again.

Melting

Melting is the process by which a solid becomes a liquid. Different solids melt at different temperatures. The lowest temperature at which a substance begins to melt is called its **melting point.** Although the melting point of ice is 0°C (32°F), iron must be heated to a much higher temperature before it will melt.

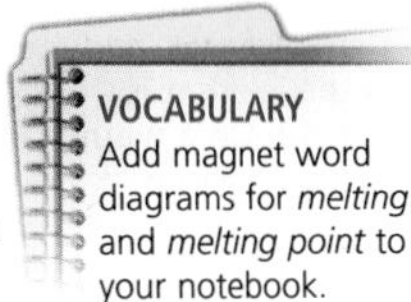

VOCABULARY
Add magnet word diagrams for *melting* and *melting point* to your notebook.

Remember that particles are always in motion, even in a solid. Because the particles in a solid are bound together, they do not move from place to place—but they do vibrate. As a solid heats up, its particles gain energy and vibrate faster. If the vibrations are fast enough, the particles break loose and slide past one another. In other words, the solid melts and becomes a liquid.

Some substances have a well-defined melting point. If you are melting ice, for example, you can predict that when the temperature reaches 0°C, the ice will start to melt. Substances with an orderly structure start melting when they reach a specific temperature.

Melting a Solid

Steel melts at very high temperatures. Liquid steel can be poured into molds to form the beams that are used in bridges like the one shown on the left.

READING VISUALS What would happen to the steel in this bridge if it became as hot as the steel in the bucket?

2.2 INSTRUCT

Real World Example

Fuses and circuit breakers control the amount of electric current in a wire so that it does not overheat. A fuse contains a small piece of metal with a low melting point. When the metal in the fuse becomes hot enough to melt, the circuit breaks, and current no longer flows through the wire. Circuit breakers contain a piece of metal that bends when it becomes hot. The bending opens the circuit.

EXPLORE the BIG idea

Revisit "Hot Chocolate" on p. 39. Have students explain the source of the energy that caused the change of state.

Teach from Visuals

To help students interpret the photographs of solid and liquid steel, ask: What generalization can you make about the melting points of structural materials? *They must be high enough that they will not melt under normal conditions.*

Ongoing Assessment

Describe how liquids can become solids, and solids can become liquids.

Ask: What process does iron ore undergo when it is liquefied to remove the impurities? *melting*

Answer: It would melt, and the bridge would collapse.

DIFFERENTIATE INSTRUCTION

More Reading Support

A What happens to ice at 0°C? *It melts.*

B How do the particles in a liquid differ from those in a solid? *The particles in a liquid can move past each other.*

English Learners English learners may think the sentence "What changes are the arrangement of the molecules and the amount of space between them," (p. 50) is a question because the phrase *what changes* begins the sentence.

Advanced Have students use a high school chemistry book to investigate phase diagrams. Have them write a paragraph that explains the meaning of each area of a phase diagram.

R Challenge and Extension, p. 102

Teach Difficult Concepts

Students might think that melting and dissolving are the same process because in both processes a solid ends up as a liquid. Point out that melting involves changing state and dissolving doesn't. To help students understand the difference, you might try the following demonstration.

Teacher Demo

Take two sugar cubes. Wrap one tightly in plastic wrap. Drop both into a cup of warm water and stir. Discuss the results. If students suggest that the unwrapped cube melted, ask: Were both cubes at the same temperature? *yes* If the wrapped sugar stays in the warm water, will it become a liquid? *no* Explain that the melting point of sugar is about 185°C (365°F). The unwrapped sugar did not melt but broke up into pieces too small to see and then mixed with the water.

Ongoing Assessment

Answer: Particles move fast enough to slide past one another.

Answer: They are the same temperature.

Other substances, such as plastic and chocolate, do not have a well-defined melting point. Chocolate becomes soft when the temperature is high enough, but it still maintains its shape. Eventually, the chocolate becomes a liquid, but there is no specific temperature at which you can say the change happened. Instead, the melting happens gradually over a range of temperatures.

Describe the movement of molecules in a substance that is at its melting point.

Icicles grow as water drips down them, freezes, and sticks to the ice that is already there. On a warm day, the frozen icicles melt again.

Freezing

READING TIP

On the Celsius temperature scale, under normal conditions, water freezes at 0°C and boils at 100°C. On the Fahrenheit scale, water freezes at 32°F and boils at 212°F.

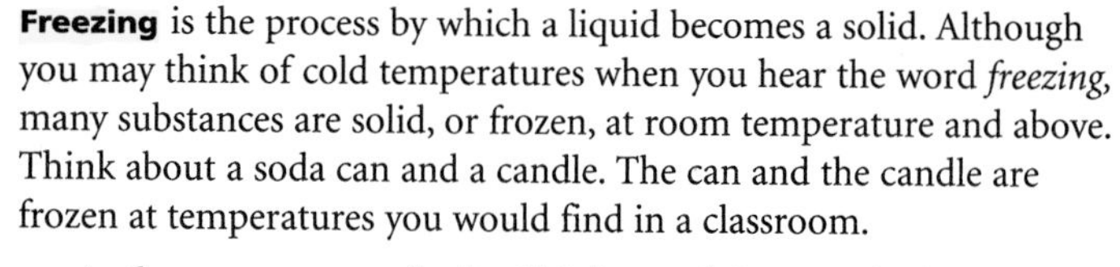

Freezing is the process by which a liquid becomes a solid. Although you may think of cold temperatures when you hear the word *freezing,* many substances are solid, or frozen, at room temperature and above. Think about a soda can and a candle. The can and the candle are frozen at temperatures you would find in a classroom.

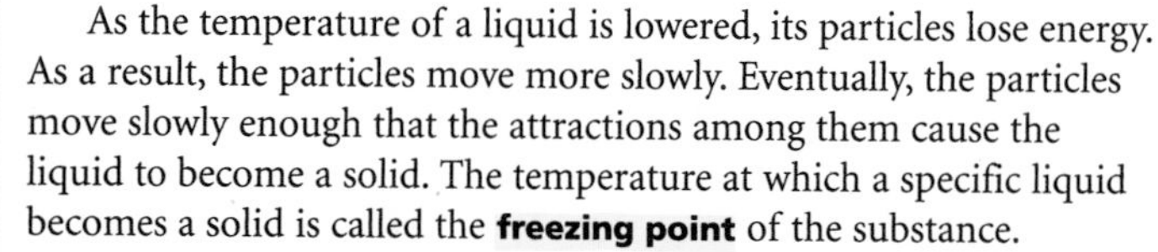

As the temperature of a liquid is lowered, its particles lose energy. As a result, the particles move more slowly. Eventually, the particles move slowly enough that the attractions among them cause the liquid to become a solid. The temperature at which a specific liquid becomes a solid is called the **freezing point** of the substance.

D

The freezing point of a substance is the same as that substance's melting point. At this particular temperature, the substance can exist as either a solid or a liquid. At temperatures below the freezing/melting point, the substance is a solid. At temperatures above the freezing/melting point, the substance is a liquid.

What is the relationship between a substance's melting point and freezing point?

DIFFERENTIATE INSTRUCTION

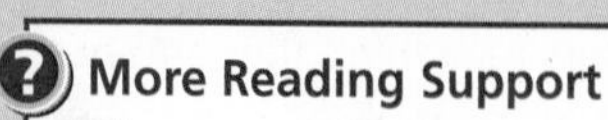

More Reading Support

C How does a liquid becomes a solid? *It freezes.*

D What state is a substance in if its temperature is slightly above its freezing point? *liquid*

Alternative Assessment Have students put together a bulletin board about the three states of matter and the processes that change matter from one state to another. They can use drawings, pictures from magazines, concept maps, or other means of displaying information.

Liquids can become gases, and gases can become liquids.

Suppose you spill water on a picnic table on a warm day. You might notice that the water eventually disappears from the table. What has happened to the water molecules? The liquid water has become water vapor, a gas. The water vapor mixes with the surrounding air. At the same picnic, you might also notice that a cold can of soda has beads of water forming on it. The water vapor in the air has become the liquid water found on the soda can.

Evaporation

E **Evaporation** is a process by which a liquid becomes a gas. It usually occurs at the surface of a liquid. Although all particles in a liquid move, they do not all move at the same speed. Some particles move faster than others. The fastest moving particles at the surface of the liquid can break away from the liquid and escape to become gas particles.

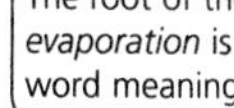

READING TIP: The root of the word *evaporation* is *vapor,* a Latin word meaning "steam."

As the temperature increases, the energy in the liquid increases. More particles can escape from the surface of the liquid. As a result, the liquid evaporates more quickly. This is why spilled water will evaporate faster in hot weather than in cold weather.

CHECK YOUR READING Describe the movement of particles in a liquid as it evaporates.

F It is interesting to note that under certain conditions, solids can lose particles through a process similar to evaporation. When a solid changes directly to a gas, the process is called **sublimation.** You may have seen dry ice being used in a cooler to keep foods cold. Dry ice is frozen carbon dioxide that sublimates in normal atmospheric conditions.

During evaporation, fast-moving particles escape from the surface of a liquid and become gas particles.

Address Misconceptions

IDENTIFY Ask: What happens to water when it evaporates? If students answer, "It disappears," they hold the misconception that a substance disappears when it evaporates.

CORRECT Add room-temperature or slightly warm water to a transparent, round container until it is about one-fourth full. Cover the top with a plate. Add ice cubes to the plate. Near the end of class, carefully remove the plate and observe its underside. There should be drops of water on it.

REASSESS Ask: Where did the water that formed on the bottom of the plate come from? *As the liquid water evaporated, it formed water vapor. The water vapor cooled and changed back into liquid water on the plate.*

Technology Resources

Visit **ClassZone.com** for background on common student misconceptions.

MISCONCEPTION DATABASE

Ongoing Assessment

CHECK YOUR READING *Answer: Particles move past one another at different speeds. The particles near the surface that have the greatest speeds (most kinetic energy) escape from the liquid and form vapor.*

DIFFERENTIATE INSTRUCTION

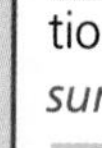

More Reading Support

E Where does evaporation usually occur? *at the surface of a liquid*

F By what process does a solid change directly to a gas? *sublimation*

Inclusion Have students place a few drops of rubbing alcohol on their arm. Ask them to describe how it feels. Tell them that as the particles of alcohol escape from the liquid, the overall energy decreases, so their skin temperature also decreases.

Real World Example

Two problems in a car's fuel line involve changes in state. Fuel-line freeze-up occurs when water in the fuel tank freezes during cold weather. The resulting ice can block the fuel line, and gasoline cannot reach the engine. To avoid this problem, car owners add materials to the gas tank that absorb the water in gasoline. The second problem, vapor lock, occurs during warmer weather. At a hot spot in the gasoline line, gasoline can vaporize, forming a pocket of gas. The engine stalls because the fuel pump in the car is designed to pump a liquid, not a gas.

History of Science

In 1846, Norbert Rillieux patented a vacuum evaporator to crystallize sugar from sugar-cane juice. The first pan in the evaporator held heated cane juice. The vapors from this pan heated the juice in the next pan, and so forth, with the last pan being connected to a condenser that removed water from the system. The whole system worked in a partial vacuum. Decreasing the pressure from pan to pan lowered the boiling point, so less energy was needed for water molecules to escape. Rillieux's vacuum evaporator has been adapted for use in processing sugar beets, soap, and glue.

Ongoing Assessment

Explain how liquids can become gases, and gases can become liquids.

Name two processes by which a liquid becomes a gas. *evaporation, boiling*

Explore melting points and boiling points.

Boiling

Boiling is another process by which a liquid becomes a gas. Unlike evaporation, boiling produces bubbles. If you heat a pot of water on the stove, you will notice that after a while tiny bubbles begin to form. These bubbles contain dissolved air that is escaping from the liquid. As you continue to heat the water, large bubbles suddenly form and rise to the surface. These bubbles contain energetic water molecules that have escaped from the liquid water to form a gas. This process is boiling.

Boiling can occur only when the liquid reaches a certain temperature, called the **boiling point** of the liquid. Liquids evaporate over a wide range of temperatures. Boiling, however, occurs at a specific temperature for each liquid. Water, for example, has a boiling point of 100°C (212°F) at normal atmospheric pressure.

G

In the mountains, water boils at a temperature lower than 100°C. For example, in Leadville, Colorado, which has an elevation of 3094 m (10,152 ft) above sea level, water boils at 89°C (192°F). This happens because at high elevations the air pressure is much lower than at sea level. Because less pressure is pushing down on the surface of the water, bubbles can form inside the liquid at a lower temperature. Less energetic water molecules are needed to expand the bubbles under these conditions. The lower boiling point of water means that foods cooked in water, such as pasta, require a longer time to prepare.

Different substances boil at different temperatures. Helium, which is a gas at room temperature, boils at –270°C (–454°F). Aluminum, on the other hand, boils at 2519°C (4566°F). This fact explains why some substances usually are found as gases but others are not.

Boiling

Bubbles of vapor form inside the boiling water.

DIFFERENTIATE INSTRUCTION

More Reading Support

G How does the boiling point of a liquid at a high elevation differ from the boiling point of that liquid at sea level? *It is lower.*

Below Level Have students plan a demonstration of the difference in particle motion in evaporation and boiling. They can use the figures on pp. 53 and 54 to help them. Allow them to use other students or appropriate objects to represent particles.

Tiny droplets of water form on a window as water vapor from the air condenses into liquid water.

Condensation

The process by which a gas changes its state to become a liquid is called **condensation.** You probably have seen an example of condensation when you enjoyed a cold drink on a warm day. The beads of water that formed on the glass or can were water vapor that condensed from the surrounding air.

The cold can or glass cooled the air surrounding it. When you cool a gas, it loses energy. As the particles move more slowly, the attractions among them cause droplets of liquid to form. Condensed water often forms when warm air containing water vapor comes into contact with a cold surface, such as a glass of ice or ground that has cooled during the night.

As with evaporation, condensation can occur over a wide range of temperatures. Like the particles in liquids, the individual particles in a gas are moving at many different speeds. Slowly moving particles near the cool surface condense as they lose energy. The faster moving particles also slow down but continue to move too fast to stick to the other particles in the liquid that is forming. However, if you cool a gas to a temperature below its boiling point, almost all of the gas will condense.

READING TIP

The root of the word *condensation* is *condense,* which comes from a Latin word meaning "to thicken."

2.2 Review

KEY CONCEPTS

1. Describe three ways in which matter can change from one state to another.
2. Compare and contrast the processes of evaporation and condensation.
3. How does adding energy to matter by heating it affect the energy of its particles?

CRITICAL THINKING

4. **Synthesize** Explain how water can exist as both a solid and a liquid at 0°C.
5. **Apply** Explain how a pat of butter at room temperature can be considered to be frozen.

CHALLENGE

6. **Infer** You know that water vapor condenses from air when the air temperature is lowered. Should it be possible to condense oxygen from air? What would have to happen?

ANSWERS

1. any three: melting, freezing, evaporation, sublimation, boiling, condensation

2. During evaporation, particles in a liquid have enough energy to escape the liquid and become a gas. During condensation, gas particles cool enough to slow down and form attractions to other particles, becoming a liquid.

3. It increases the energy of the particles.

4. That temperature is both the freezing point and the melting point of water. Both processes are occurring, so both states can exist.

5. The butter is solid.

6. Yes; oxygen could condense from air if the temperature were lowered to oxygen's boiling point.

Ongoing Assessment

Determine how energy is related to changes in state.

Ask: If you remove energy from a liquid, will it become a solid or a gas? *a solid*

Reinforce 

Have students relate the section to the Big Idea.

 Reinforcing Key Concepts, p. 103

2.2 ASSESS & RETEACH

Assess

 Section 2.2 Quiz, p. 24

Reteach

Have pairs of students label index cards *condensation, melting, freezing, evaporation, sublimation, solid, liquid,* and *gas.* If you wish to include the gas to solid process, include an index card labeled *deposition*. Otherwise, tell students to ignore the gas to solid transition. Have one student lay down side-by-side two of the cards that list states of matter. Have the other student lay down any cards that name the processes that change matter from the state on the left to the state on the right.

Technology Resources

Have students visit **ClassZone.com** for reteaching of Key Concepts.

 **CONTENT REVIEW**

 **CONTENT REVIEW CD-ROM**

CHAPTER INVESTIGATION

Focus

PURPOSE To observe the change in state from a liquid to a solid and to measure the freezing point of a substance

OVERVIEW Students will use a thermometer to measure the temperature at which liquid stearic acid solidifies. Students will find the following:

- Pure stearic acid starts to harden at about 69°C. Some brands may be mixtures with a melting point of 55°C.
- The temperature remains the same until the stearic acid becomes completely solid.

Lab Preparation

- Prepare the wire-loop stirrers. The loop must fit around the thermometer and into the test tube.
- Obtain stearic acid, or stearin, which is available in most craft stores.
- Melt the stearic acid before class. Keep it liquid by immersing the container in hot water.
- Have students read pp. 56–57 and make data tables in advance or copy and distribute datasheets and rubrics.

UNIT RESOURCE BOOK, pp. 124–132

SCIENCE TOOLKIT, F15

Lab Management

- Have students stir with an up-and-down motion so as not to disturb the thermometer.
- To clean the test tubes, place them upright in hot water to melt the acid. Carefully remove the thermometers and stirrers and wipe them clean. Pour the acid into a metal can and wipe the test tubes clean. Cool the can completely before discarding it.

SAFETY Students should wear safety goggles throughout the activity and use care in handling the warm stearic acid.

Teaching with Technology

If probeware is available, use temperature probes instead of thermometers.

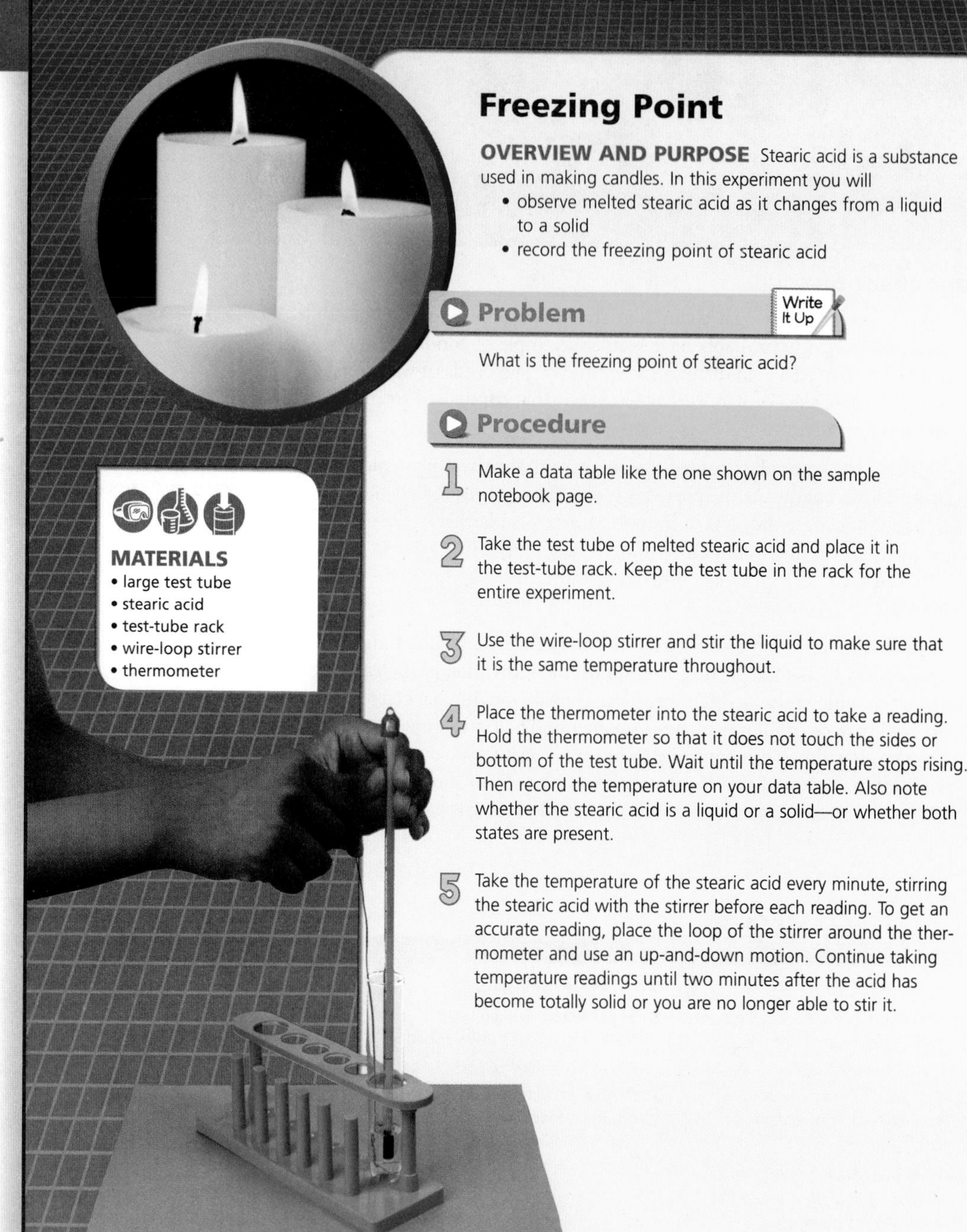

Freezing Point

OVERVIEW AND PURPOSE Stearic acid is a substance used in making candles. In this experiment you will
- observe melted stearic acid as it changes from a liquid to a solid
- record the freezing point of stearic acid

Problem

Write It Up

What is the freezing point of stearic acid?

Procedure

1. Make a data table like the one shown on the sample notebook page.
2. Take the test tube of melted stearic acid and place it in the test-tube rack. Keep the test tube in the rack for the entire experiment.
3. Use the wire-loop stirrer and stir the liquid to make sure that it is the same temperature throughout.
4. Place the thermometer into the stearic acid to take a reading. Hold the thermometer so that it does not touch the sides or bottom of the test tube. Wait until the temperature stops rising. Then record the temperature on your data table. Also note whether the stearic acid is a liquid or a solid—or whether both states are present.
5. Take the temperature of the stearic acid every minute, stirring the stearic acid with the stirrer before each reading. To get an accurate reading, place the loop of the stirrer around the thermometer and use an up-and-down motion. Continue taking temperature readings until two minutes after the acid has become totally solid or you are no longer able to stir it.

MATERIALS
- large test tube
- stearic acid
- test-tube rack
- wire-loop stirrer
- thermometer

A 56 Unit: Matter and Energy

INVESTIGATION RESOURCES

CHAPTER INVESTIGATION, Freezing Point
- Level A, pp. 124–127
- Level B, pp. 128–131
- Level C, p. 132

Advanced students should complete Levels B & C.

Writing a Lab Report, D12–13

Technology Resources

Customize this student lab as needed or look for an alternative. Print rubrics to assess student lab reports.

Lab Generator CD-ROM

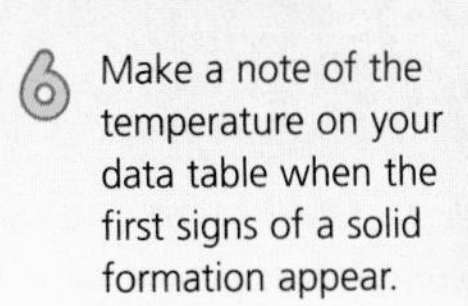

6. Make a note of the temperature on your data table when the first signs of a solid formation appear.

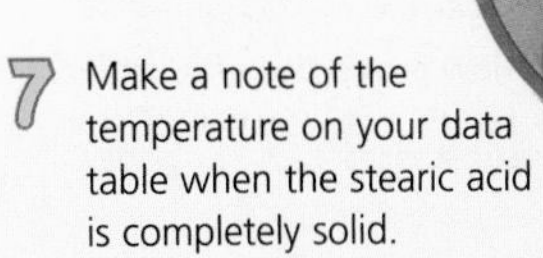

7. Make a note of the temperature on your data table when the stearic acid is completely solid.

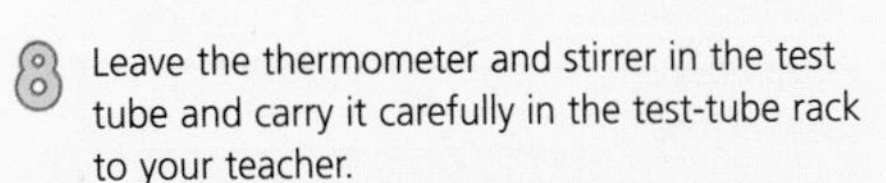

8. Leave the thermometer and stirrer in the test tube and carry it carefully in the test-tube rack to your teacher.

Observe and Analyze

1. **RECORD OBSERVATIONS** Make a line graph showing the freezing curve of stearic acid. Label the vertical axis **Temperature** and the horizontal axis **Time.**
2. **RECORD OBSERVATIONS** Label your graph to show when the stearic acid was a liquid, when it was a solid, and when it was present in both states.
3. **ANALYZE** Explain how your graph tells you the freezing point of stearic acid.

Conclude

1. **INTERPRET** Answer the question in the problem.
2. **IDENTIFY** How does the freezing point of stearic acid compare with the freezing point of water?
3. **INFER** What happened to the energy of the molecules as the stearic acid changed from a liquid to a solid?
4. **APPLY** Candle makers add stearic acid to the paraffin from which they make candles. The stearic acid lowers the melting point of the paraffin. How does that improve the candles?
5. **APPLY** Why do you think stearic acid is used as an ingredient in bar soaps but not in liquid soaps?

INVESTIGATE Further

CHALLENGE What do you think would happen if you mixed in another substance with the stearic acid? How would that affect the freezing point? What experiment would you perform to find the answer?

Freezing Point

Problem What is the freezing point of stearic acid?

Observe and Analyze

Table 1. Freezing Point of Stearic Acid

Time (min)	Temperature (°C)	Liquid	Solid	Both
0.0				
1.0				
2.0				
3.0				
4.0				
5.0				
6.0				
7.0				

Observe and Analyze

1. *See students' graphs. Sample data: 0 min, 73°C; 1 min, 66°C; 2 min, 60°C; 3 min, 57°C; 4 min, 55°C; 5 min, 55°C; 6 min, 55°C; 7 min, 55°C; 8 min, 53°C; 9 min, 52°C*
2. *The slope to the left of the flat part of the line on the graph indicates when the acid was liquid. The slope to the right indicates when the acid was solid. The flat line indicates when the acid was both a solid and a liquid.*
3. *The flat part of the line indicates that both states exist, so the temperature indicated by the flat part of the line is the freezing point.*

Conclude

1. *The freezing point of stearic acid is about 69°C (or 55°C for some brands).*
2. *The freezing point of stearic acid is higher.*
3. *It decreased.*
4. *Answer should match students' freezing point in question 1. Melting and freezing point are the same.*
5. *It keeps bar soaps from liquefying in hot water. Liquid soap is supposed to be liquid, even at room temperature.*

INVESTIGATE Further

CHALLENGE Adding impurities tends to lower the freezing point of a mixture.

Post-Lab Discussion

Commercial stearic acid, such as the acid used in this lab, is usually a mixture of stearic acid and palmitic acid. The freezing point of palmitic acid is about 63°C. Ask: How do you think using a mixture might affect your results? *Freezing point will change. Freezing point may not be a well-defined temperature.*

2.3 FOCUS

Set Learning Goals

Students will

- Describe how properties can help you identify substances.
- Explain how properties of substances can be used to separate substances.
- Design an experiment to separate a mixture.

3-Minute Warm-Up

Display Transparency 13 or copy this exercise on the board:

Create an events-chain concept map that shows the formation of a liquid from a solid, then a gas from a liquid. Use arrows to show energy being added or released. *Map should show a material going from a solid to a liquid to a gas. Arrows between states should show that energy is added or released.*

3-Minute Warm-Up, p. 13

2.3 MOTIVATE

EXPLORE Identifying Substances

PURPOSE To use properties to identify a substance

TIP *10 min.* Remind students that both chemical and physical properties can be used for identification.

WHAT DO YOU THINK? *When water was added to substance A, the substance dissolved but did not change identity. When water was added to substance B, a new substance formed during the chemical change that occurred. Substance B is baking powder.*

Ongoing Assessment

CHECK YOUR READING *Answer: Compare the properties of the unknown substance with the properties of known substances.*

KEY CONCEPT

2.3 Properties are used to identify substances.

BEFORE, you learned

- Matter can change from one state to another
- Changes in state require energy changes

NOW, you will learn

- How properties can help you identify substances
- How properties of substances can be used to separate substances

EXPLORE Identifying Substances

How can properties help you identify a substance?

PROCEDURE

1. Place some of substance A into one cup and some of substance B into the other cup. Label the cups.
2. Carefully add some water to each cup. Observe and record what happens.

WHAT DO YOU THINK?

- Which result was a physical change? a chemical change? Explain.
- The substances are baking soda and baking powder. Baking powder and water produce carbon dioxide gas. Which substance is baking powder?

MATERIALS

- substance A
- substance B
- 2 cups
- water

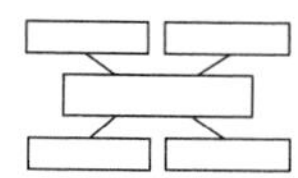
MAIN IDEA WEB
As you read, place each blue heading in a box. Add details around it to form a web.

Substances have characteristic properties.

You often use the properties of a substance to identify it. For example, when you reach into your pocket, you can tell the difference between a ticket stub and a folded piece of tissue because one is stiff and smooth and the other is soft. You can identify nickels, dimes, and quarters without looking at them by feeling their shapes and comparing their sizes. To tell the difference between a nickel and a subway token, however, you might have to use another property, such as color. Texture, shape, and color are physical properties that you use all the time to identify and sort objects.

CHECK YOUR READING How can physical properties be used to identify a substance?

RESOURCES FOR DIFFERENTIATED INSTRUCTION

Below Level

UNIT RESOURCE BOOK

- Reading Study Guide A, pp. 106–107
- Decoding Support, p. 119

AUDIO CDS

Advanced

UNIT RESOURCE BOOK

Challenge and Extension, p. 112

English Learners

UNIT RESOURCE BOOK

Spanish Reading Study Guide, pp. 110–111

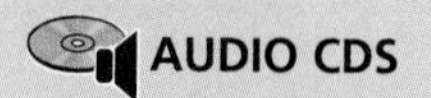
AUDIO CDS

- Audio Readings in Spanish
- Audio Readings (English)

Identifying Unknown Substances

Suppose you have a glass of an unknown liquid that you want to identify. It looks like milk, but you cannot be sure. How could you determine what it is? Of course, you would not taste an unknown substance, but there are many properties other than taste that you could use to identify the substance safely.

To proceed scientifically, you could measure several properties of the unknown liquid and compare them with the properties of known substances. You might observe and measure such properties as color, odor, texture, density, boiling point, and freezing point. A few of these properties might be enough to tell you that your white liquid is glue rather than milk.

To determine the difference among several colorless liquids, scientists would use additional tests. Their tests, however, would rely on the same idea of measuring and comparing the properties of an unknown with something that is already known.

Properties Used for Identifying Substances

You are already familiar with the most common physical properties of matter. Some of these properties, such as mass and volume, depend upon the specific object in question. You cannot use mass to tell one substance from another because two very different objects can have the same mass—a kilogram of feathers has the same mass as a kilogram of peanut butter, for example.

Other properties, such as density, can be used to identify substances. They do not vary from one sample of the same substance to another. For example, you could see a difference between a kilogram of liquid soap and a kilogram of honey by measuring their densities.

The physical properties described below can be used to identify a substance.

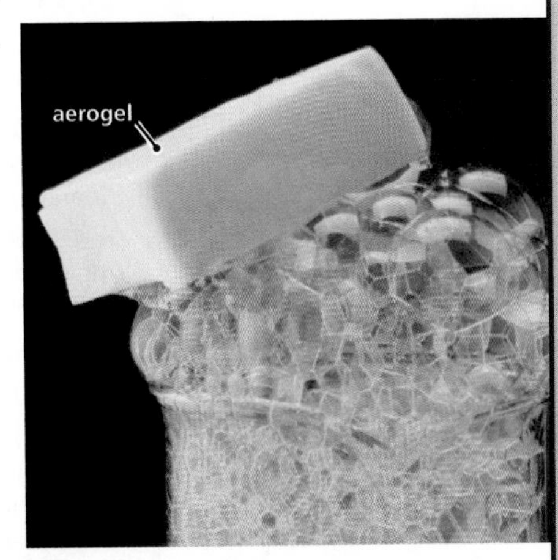

Aerogel, an extremely lightweight material used in the space program, has such a low density that it can float on soap bubbles.

Density The densities of wood, plastic, and steel are all different. Scientists already have determined the densities of many substances. As a result, you can conveniently compare the density of an unknown substance with the densities of known substances. Finding any matching densities will give you information about the possible identity of the unknown substance. However, it is possible for two different substances to have the same density. In that case, in order to identify the substance positively, you would need additional data.

Why can't you identify a substance on the basis of density alone?

2.3 INSTRUCT

Integrate the Sciences

Every element has an identifiable spectrum for emission and absorption of light. Analysis of this physical property of stars can tell a lot about their identity and history. Each star has a unique spectrum that can be interpreted to tell the star's temperature, surface gravity, density, atmospheric conditions, and surface chemistry.

Health Connection

After age 35, a person's bone mass begins to decrease. If bone mass falls below a certain level, the person is diagnosed with osteoporosis. This is a disease that affects the strength and density of bones. Doctors use specialized screening tests to measure a person's bone density. These tests can detect osteoporosis before a fracture occurs. The most accurate test is called dual-energy x-ray absorptiometry, or DEXA. It uses very low doses of radiation to measure the density of the bones of the spine and hip.

Ongoing Assessment

Describe how properties can help you identify substances.

Ask: How can density be used to identify a substance? *If two unknowns have the same density, they might be the same substance. Each substance has a characteristic density.*

CHECK YOUR READING *Answer: More than one substance can have the same density.*

DIFFERENTIATE INSTRUCTION

More Reading Support

A How can you identify a substance by its properties? *Compare the properties of the unknown substance with the properties of known substances.*

English Learners English learners may be unfamiliar with ticket stubs and subway tokens, mentioned on p. 58. Some students may have trouble with abstract and hypothetical questions. For example, the first Synthesize question in this section's review asks students to imagine ("suppose") a situation and then draw conclusions. Help students by modeling expected answers to questions such as these.

Teach from Visuals

The fibers in the photograph contain spaces filled with air.

- Ask: Why do you think the fibers contain spaces filled with air? *Air is not a good conductor of heat, so the air helps insulate the clothing.*
- The fibers themselves act as insulators, as does the air. Ask: If the fibers insulate anyway, why are the spaces for the air desirable? *Air is less dense than the material that makes up the fiber, so using the spaces reduces the weight of the clothing.*

Develop Critical Thinking

ANALYZE Water that is suitable for drinking and other purposes is limited worldwide. One possible source of potable water is salt water. Ask:

- What properties of water and salt might be used to separate them? *Their boiling and melting points are quite different.*
- Could you use a filter to separate water and salt? Explain. *No; a solution is a homogeneous mixture, and all the particles are about the same size.*
- If you boiled or evaporated the water from salt water, how could you reclaim the fresh water that forms? *The water vapor could be collected, cooled, and condensed into liquid water.*

These fibers act as heat insulators to keep the inside of the sleeping bag warm.

Heating Properties Substances respond to heating in different ways. Some warm up very quickly, and others take a long while to increase in temperature. This property is important in selecting materials for different uses. Aluminum and iron are good materials for making pots and pans because they conduct heat well. Various materials used in household insulation are poor heat conductors. Therefore, these insulators are used to keep warm air inside a home on a cold day. You can measure the rate at which a substance conducts heat and compare that rate with the heat conduction rates of other substances. **? B**

READING TIP
The root of the word *solubility* is the Latin word *solvere*, which means "to loosen."

Solubility Solubility is a measure of how much of a substance dissolves in a given volume of a liquid. Sugar and dirt, for instance, have very different solubilities in water. If you put a spoonful of sugar into a cup of water and stir, the sugar dissolves in the water very rapidly. If you put a spoonful of dirt into water and stir, most of the dirt settles to the bottom as soon as you stop stirring. **? C**

Electric Properties Some substances conduct electricity better than others. This means that they allow electric charge to move through them easily. Copper wire is used to carry electricity because it is a good conductor. Materials that do not conduct easily, such as rubber and plastics, are used to block the flow of charge. With the proper equipment, scientists can test the electric conductivity of an unknown substance.

Iron filings are attracted by the magnet. The wood chips, however, are not.

Magnetic Properties Some substances are attracted to magnets, but others are not. You can use a magnet to pick up a paper clip but not a plastic button or a wooden match. The elements iron, cobalt, and nickel are magnetic—meaning they respond to magnets—but copper, aluminum, and zinc are not. Steel, which contains iron, is also magnetic.

DIFFERENTIATE INSTRUCTION

More Reading Support

B Does insulation heat up slowly or quickly? *slowly*

C What do you call the ability of a substance to dissolve in a liquid? *solubility*

English Learners Clarify for English learners that the phrase *used to* (p. 60) explains the function of materials, and does not mean formerly, as students familiar with the phrase *use to* might think. Also caution them against reading "up" as a literal direction in the phrasal verb *break up* on p. 62.

Mixtures can be separated by using the properties of the substances in them.

Suppose you have a bag of cans that you want to recycle. The recycling center accepts only aluminum cans. You know that some of your cans contain steel. You would probably find it difficult to tell aluminum cans from steel ones just by looking at them. How could you separate the cans? Aluminum and steel may look similar, but they have different magnetic properties. You could use a magnet to test each can. If the magnet sticks to the can, the can contains steel. Recycling centers often use magnets to separate aluminum cans from steel cans.

Some mixtures contain solids mixed with liquids. A filter can be used to separate the solid from the liquid. One example of this is a tea bag. The paper filter allows the liquid water to mix with the tea, because water molecules are small enough to pass through the filter. The large pieces of tea, however, cannot pass through the filter and remain inside the tea bag.

INVESTIGATE Separating Mixtures

How can a mixture of sand, salt, and pepper be separated?

Scientists often have to isolate a single substance from a mixture. Use your knowledge of the properties of sand, salt, and pepper to design a method for separating each of these substances from the mixture.

PROCEDURE

1. Examine the mixture and the materials provided. Design a procedure for separating the different substances in your mixture. Carefully consider the order in which you will try each step.
2. Write up your procedure. Explain why you chose the steps you did for each substance.
3. Carry out your procedure.

WHAT DO YOU THINK?

- Was your procedure successful? How would you modify your procedure if you were to perform the separation again?
- How does knowing the properties of matter help you separate the substances in mixtures?

SKILL FOCUS
Designing experiments

MATERIALS
- mixture of sand, salt, and pepper
- 2 index cards
- comb
- felt
- graduated cylinder
- spoon
- water
- coffee filter
- funnel
- small cup
- pie tin

TIME
30 minutes

61 A

INVESTIGATE Separating Mixtures

PURPOSE To design an experiment to separate different substances

TIPS *30 min.*

- Iron filings can be part of the mixture, with a magnet available to use for separation.
- Use coarse salt so it is easily seen.
- Use plenty of pepper in the mixture.
- Students might need to be told that they can use static electricity and the comb to separate materials.

WHAT DO YOU THINK? *You can use your knowledge to create situations where one property will separate one substance from the others.*

Datasheet, Separating Mixtures, p. 113

Technology Resources

Customize this student lab as needed or look for an alternative. Print rubrics to assess student lab reports.

Lab Generator CD-ROM

Ongoing Assessment

Explain how properties of substances can be used to separate substances.

Ask: How could small pebbles be separated from larger gravel? *You could use a sieve with holes large enough for the pebbles to pass through.*

DIFFERENTIATE INSTRUCTION

More Reading Support

D What metal will a magnet attract? *steel*

E What method can be used to separate a solid from a liquid? *using a filter*

Advanced Have students create a poster or bulletin board that shows how relative boiling points are used to separate different petroleum products by the process of fractional distillation. Diagrams and explanations of this process can be found in most chemistry books or on the Internet.

Challenge and Extension, p. 112

Ongoing Assessment

CHECK YOUR READING *Sample answer: making water safe to drink, separating aluminum cans from those containing steel*

Reinforce the BIG idea

Have students relate the section to the Big Idea.

 Reinforcing Key Concepts, p. 114

2.3 ASSESS & RETEACH

Assess

 Section 2.3 Quiz, p. 25

Reteach

List on the board as the head of a column each of the properties described on pp. 59–60. Then have students brainstorm sets of two objects. Ask them to explain how the properties could be used to compare the objects. For example, if paper and copper are compared, copper is denser, a conductor of heat and electricity, insoluble in water, and nonmagnetic; paper is less dense, an insulator, insoluble in water, and nonmagnetic. Point out that not all properties can be used to identify each item. For example, both paper and copper are insoluble in water and nonmagnetic.

Technology Resources

Have students visit **ClassZone.com** for reteaching of Key Concepts.

 **CONTENT REVIEW**

 **CONTENT REVIEW CD-ROM**

This water-treatment plant separates harmful substances from the water.

Some mixtures are more difficult to separate than others. For example, if you stir sugar into water, the sugar dissolves and breaks up into individual molecules that are too tiny to filter out. In this case, you can take advantage of the fact that water is a liquid and will evaporate from an open dish. Sugar, however, does not evaporate. The mixture can be heated to speed the evaporation of the water, leaving the sugar behind.

There are many important reasons for separating substances. One reason is to make a substance safe to consume, such as drinking water. In order to produce drinking water, workers at a water-treatment plant must separate many of the substances that are mixed in with the water.

The process in water-treatment plants generally includes these steps:

- First, a chemical is added to the water that causes the larger particles to stick together. They settle to the bottom of the water, where they can be removed.
- Next, the water is run through a series of special molecular filters. Each filter removes smaller particles than the one before.
- Finally, another chemical, chlorine, is added to disinfect the water and make it safe to drink.

Water-treatment plants use the properties of the substances found in water to produce the clean water that flows from your tap.

 What are two situations in which separating substances is useful?

2.3 Review

KEY CONCEPTS

1. How can properties help you distinguish one substance from another?
2. What are two physical properties that can help you identify a substance?
3. How can understanding properties help you separate substances from a mixture?

CRITICAL THINKING

4. **Apply** Why might an archaeologist digging in ancient ruins sift dirt through a screen?
5. **Synthesize** Suppose you had a mixture of iron pellets, pebbles, and small wood spheres, all of which were about the same size. How would you separate this mixture?

CHALLENGE

6. **Synthesize** You have two solid substances that look the same. What measurements would you take and which tests would you perform to determine whether they actually are the same?

ANSWERS

1. You can notice the physical properties of the two substances that are different.

2. Sample answer: density and color

3. Properties that differ from substance to substance can be used to separate them.

4. Artifacts are likely to be bigger than soil particles, and the screen would catch the artifacts.

5. Pull out the iron with a magnet. Add water to float the wood, and skim it from the top. The pebbles would be left.

6. Sample answer: Compare their conductivity and density.

CONNECTING SCIENCES

PHYSICAL SCIENCE AND EARTH SCIENCE

Separating Minerals

A few minerals, such as rock salt, occur in large deposits that can be mined in a form that is ready to use. Most minerals, however, are combined with other materials, so they need to be separated from the mixtures of which they are a part. Scientists and miners use the differences in physical properties to analyze samples and to separate the materials removed from a mine.

Workers can identify garnets in a mine because their physical properties are different from the physical properties of their surroundings.

Appearance

Gemstones are prized because of their obvious physical properties, such as color, shininess, and hardness. Particularly valuable minerals, such as diamonds and emeralds, are often located by digging underground and noting the differences between the gemstone and the surrounding dirt and rock.

Density

When gold deposits wash into a streambed, tiny particles of gold mix with the sand. It is hard to separate them by appearance because the pieces are so small. In the 1800s, as prospectors swirled this sand around in a pan, the lighter particles of sand washed away with the water. The denser gold particles collected in the bottom of the pan. Some modern gold mines use the same principle in machines that handle tons of material, washing away the lighter dirt and rock to leave bits of gold.

Magnetism

Machines called magnetic separators divide a mixture into magnetic and nonmagnetic materials. In order to separate iron from other materials, rocks are crushed and carried past a strong magnet. Particles that contain iron are drawn toward the magnet and fall into one bin, while the nonmagnetic materials fall into another bin.

Melting Point

Thousands of years ago, people discovered that when some minerals are placed in a very hot fire, metals—such as copper, tin, and zinc—can be separated from the rock around them. When the ores reach a certain temperature, the metal melts and can be collected as a liquid.

EXPLORE

1. **INFER** At a copper ore mine in Chile, one of the world's largest magnets is used to remove pieces of iron from the ore. What can you infer about the copper ore?
2. **CHALLENGE** Electrostatic precipitators are important tools for protecting the environment from pollution. Use the Internet to learn how they are used in power plants and other factories that burn fuels.

Find out more about separating materials from mixtures.

EXPLORE

1. ***INFER*** *Copper ore has no magnetic properties.*
2. ***CHALLENGE*** *Electrostatic precipitators use electricity to charge particles that are products of burning fuels. These charged particles are attracted to a charged plate, so they aren't released into the environment.*

CONNECTING SCIENCES
Integration of Sciences

Set Learning Goal

To learn how the physical properties of minerals are used to sort them from other materials

Present the Science

Chemical properties are useful in separating a mineral from compounds. For example, iron ore can be concentrated by physical means, but iron almost always exists in compounds in nature. Chemical changes are necessary to produce metallic iron from its compounds. When you heat iron ore and charcoal, oxygen in the ore combines with carbon in the charcoal to form carbon dioxide. Carbon dioxide and carbon monoxide are released, leaving metallic iron behind.

Discussion Question

Tell students that a deposit of silver is mixed in with sulfur. Silver has a density of 10.5 g/cm^3 and a melting point of 962°C. Sulfur has a density of 2.07 g/cm^3 and a melting point of 115°C. Ask: How can these physical properties be used to separate silver and sulfur? *Heat the mixture until the sulfur melts. Because it is denser, the solid silver will fall to the bottom and can be removed.*

Close

Tell the class that the melting point of iron is 1538°C and of nickel is 1455°C. Ask: Why wouldn't melting point be a good physical property to use to separate nickel and iron? *The melting points are too close together.*

Technology Resources

Have students visit **ClassZone.com** to find more about separating materials from mixtures.

RESOURCE CENTER

BACK TO

the BIG idea

Have students look at the photograph on pp. 38–39. Ask them to use the photograph to summarize what they have learned about cooking in relation to properties and changes. Have them include examples. *Sample answer: Cooking changes the identity of the substances being cooked, so cooking involves chemical changes. A cake baking, an egg frying, and bread toasting are examples. All have different physical and chemical properties after they have been cooked.*

KEY CONCEPTS SUMMARY

SECTION 2.1
Ask: Which of the photos shows a substance that can be identified as a metal because it has luster? *the one on the right*

SECTION 2.2
Ask: What do the arrows in the diagrams indicate? *the direction of the process*

Ask: What two processes represent what happens to a Popsicle left outside on a warm day? *melting and evaporation*

SECTION 2.3
Ask: Which of the physical properties is being used in the photograph to separate substances? *magnetic properties*

Review Concepts

- Big Idea Flow Chart, p. T9
- Chapter Outline, pp. T15–T16

2 Chapter Review

the BIG idea

Matter has properties that can be changed by physical and chemical processes.

KEY CONCEPTS SUMMARY

2.1 Matter has observable properties.

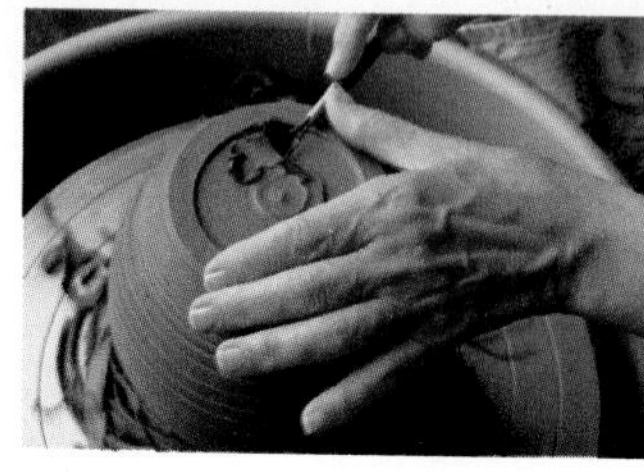

- Physical properties can be observed without changing the substance.
- Physical changes can change some physical properties but do not change the substance.
- Chemical properties describe how substances form new substances.
- Chemical changes create new substances.

VOCABULARY
physical property p. 41
density p. 43
physical change p. 44
chemical property p. 46
chemical change p. 46

2.2 Changes of states are physical changes.

Matter is commonly found in three states: solid, liquid, and gas.

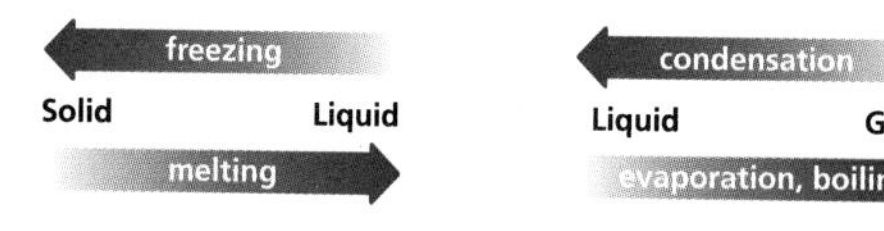

VOCABULARY
melting p. 51
melting point p. 51
freezing p. 52
freezing point p. 52
evaporation p. 53
sublimation p. 53
boiling p. 54
boiling point p. 54
condensation p. 55

2.3 Properties are used to identify substances.

Physical properties that can be used to identify substances include:

- density
- heating properties
- solubility
- electric properties
- magnetic properties

Mixtures can be separated by using the properties of the substances they contain.

Technology Resources

Have students visit **ClassZone.com** or use the CD-ROM for a cumulative review of concepts.

 **CONTENT REVIEW**

 **CONTENT REVIEW CD-ROM**

Engage students in a whole-class interactive review of Key Concepts. Edit content as you wish.

 **POWER PRESENTATIONS**

Reviewing Vocabulary

Describe how the terms in the following sets of terms are related.

1. physical property, physical change
2. chemical property, chemical change
3. density, matter
4. melting, melting point, freezing point
5. boiling, boiling point, liquid
6. evaporation, condensation
7. sublimation, solid

Reviewing Key Concepts

Multiple Choice *Choose the letter of the best answer.*

8. Color, shape, size, and texture are
   a. physical properties
   b. chemical properties
   c. physical changes
   d. chemical changes
9. Density describes the relationship between a substance's
   a. matter and mass
   b. mass and volume
   c. volume and area
   d. temperature and mass
10. Dissolving sugar in water is an example of a
    a. physical change
    b. chemical change
    c. change in state
    d. pressure change
11. An electric current can be used to decompose, or break down, water into oxygen gas and hydrogen gas. This is an example of a
    a. physical change
    b. chemical change
    c. change in state
    d. pressure change
12. The formation of rust on iron is a chemical change because
    a. the color and shape have changed
    b. the mass and volume have changed
    c. the substance remains the same
    d. a new substance has been formed
13. The process by which a solid becomes a liquid is called
    a. boiling
    b. freezing
    c. melting
    d. evaporating
14. The process by which a liquid becomes a solid is called
    a. boiling
    b. freezing
    c. melting
    d. evaporating
15. Two processes by which a liquid can become a gas are
    a. evaporation and boiling
    b. melting and freezing
    c. sublimation and condensation
    d. evaporation and condensation

Short Answer *Answer each of the following questions in a sentence or two.*

16. When a sculptor shapes marble to make a statue, is this a physical or a chemical change? Explain your answer.
17. Describe and identify various physical changes that water can undergo.
18. Why does dew often form on grass on a cool morning, even if there has been no rain?
19. Describe the difference between evaporation and boiling in terms of the movement of the liquid's particles in each case.
20. What effect does altitude have on the boiling point of water?

ASSESSMENT RESOURCES

UNIT ASSESSMENT BOOK
- Chapter Test A, pp. 26–29
- Chapter Test B, pp. 30–33
- Chapter Test C, pp. 34–37
- Alternative Assessment, pp. 38–39

SPANISH ASSESSMENT BOOK
Spanish Chapter Test, pp. 217–220

Technology Resources

Edit test items and answer choices.

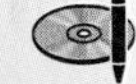 **Test Generator CD-ROM**

Visit **ClassZone.com** to extend test practice.

 **Test Practice**

Reviewing Vocabulary

1. A physical change is a change in any physical property of a substance.

2. A chemical property describes how a substance can form a new substance during a chemical change.

3. Density is a measure of the amount of matter in a given volume of a substance.

4. A substance melts at its melting point, which is the same temperature as its freezing point.

5. A liquid boils, or changes to a gas, at its boiling point.

6. Evaporation is a process by which a liquid changes to a gas, and condensation is the reverse process.

7. Sublimation is the process by which a solid changes directly to a gas.

Reviewing Key Concepts

8. a

9. b

10. a

11. b

12. d

13. c

14. b

15. a

16. The change is physical. The shape changes, but the material does not change identity.

17. Liquid water can freeze into ice or evaporate or boil into water vapor. Solid water can melt into liquid water or sublimate into water vapor. Water vapor can condense into liquid water.

18. Water vapor from the air cools and condenses.

19. During evaporation, only a few liquid particles have enough energy to escape the surface of a liquid and become a gas. During boiling, many particles with enough energy form bubbles of gas throughout the liquid.

20. It lowers the boiling point because atmospheric pressure is lower.

Thinking Critically

21. *From materials originally in the milk, a new substance (lactic acid) forms.*
22. *The shavings and the sharpened pencil still have the identity of the original materials.*
23. *Water is a liquid and thus has ability to flow; spaghetti, which is a solid, is too large to go through the holes.*
24. *It is spoiled.*
25. *Its size and volume change, and the substances dissolved and suspended in it become more concentrated. The water is still liquid, but the solution gets denser.*
26. *Yes. Water can evaporate over a range of temperatures. It will evaporate more slowly, however.*
27. *They would remain behind as solids.*
28. *If the object is only one substance, it is silver. However, if it is a mixture, it could be a combination of elements that have densities greater than and less than 10.5 g/cm^3.*
29. *With a density of 8.91 g/cm^3, nickel would fall between iron and copper.*

the BIG idea

30. *The sugar undergoes physical changes as it is melted and shaped. The sugar undergoes chemical changes when it is heated, changing in color and flavor.*
31. *Students should mention which properties of the matter they might research. Sources of information may include science books and the Internet.*

UNIT PROJECTS

Collect schedules, materials lists, and questions. Be sure dates and materials are obtainable, and questions are focused.

R Unit Projects, pp. 5–10

Thinking Critically

21. ANALYZE Whole milk is a mixture. When bacteria in the milk digest part of the mixture, changes occur. Lactic acid is produced, and the milk tastes sour. Explain why this process is a chemical change.

22. INFER Sharpening a pencil leaves behind pencil shavings. Why is sharpening a pencil a physical change instead of a chemical change?

23. ANALYZE Dumping cooked spaghetti and water into a colander separates the two substances because the liquid water can run through the holes in the colander but the solid spaghetti cannot. Explain how this is an example of separating a mixture based on the physical properties of its components.

24. INFER The density of water is 1.0 g/mL. Anything with a density less than 1.0 g/mL will float in water. The density of a fresh egg is about 1.2 g/mL. The density of a spoiled egg is about 0.9 g/mL. If you place an egg in water and it floats, what does that tell you about the egg?

Use the photograph below to answer the next three questions.

25. COMPARE Which physical properties of the puddle change as the water evaporates? Which physical properties remain the same?

26. ANALYZE Can water evaporate from this puddle on a cold day? Explain your answer.

27. PREDICT What would happen to any minerals and salts in the water if the water completely evaporated?

Use the chart below to answer the next two questions.

Densities Measured at 20°C

Material	Density (g/cm^3)
gold	19.3
lead	11.3
silver	10.5
copper	9.0
iron	7.9

28. PREDICT Suppose you measure the mass and the volume of a shiny metal object and find that its density is 10.5 g/mL. Could you make a reasonable guess as to what material the object is made of? What factor or factors might affect your guess?

29. CALCULATE A solid nickel bar has a mass of 2.75 kg and a volume of 308.71 cm^3. Between which two materials would nickel fall on the chart?

the BIG idea

30. PREDICT Look again at the photograph on pages 38–39. The chef has melted sugar to make a sculpture. Describe how the sugar has changed in terms of its physical and chemical properties. Predict what will happen to the sculpture over time.

31. RESEARCH Think of a question you have about the properties of matter that is still unanswered. For example, there may be a specific type of matter about which you are curious. What information do you need in order to answer your question? How might you find the information?

UNIT PROJECTS

Check your schedule for your unit project. How are you doing? Be sure that you have placed data or notes from your research in your project folder.

MONITOR AND RETEACH

If students have trouble applying the concepts in items 28 and 29, have them determine the densities of several common objects in the classroom and compare them. They can create a similar table, using the densities of these materials. Students can benefit from holding equal volumes of these materials in their hands and comparing how "heavy" they are.

Students may benefit from summarizing one or more sections of the chapter.

 Summarizing the Chapter, pp. 142–143

A 66 Unit: Matter and Energy

Standardized Test Practice

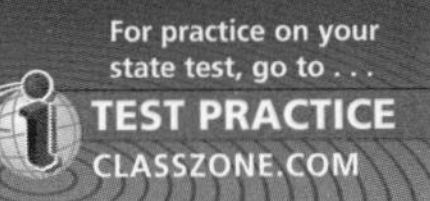

Analyzing Experiments

Read the following description of an experiment together with the chart. Then answer the questions that follow.

Archimedes was a Greek mathematician and scientist who lived in the third century B.C. He figured out that any object placed in a liquid displaced a volume of that liquid equal to its own volume. He used this knowledge to solve a problem.

The king of Syracuse had been given a crown of gold. But he was not sure whether the crown was pure gold. Archimedes solved the king's problem by testing the crown's density.

He immersed the crown in water and measured the volume of water it displaced. Archimedes compared the amount of water displaced by the crown with the amount of water displaced by a bar of pure gold with the same mass. The comparison told him whether the crown was all gold or a mixture of gold and another element.

Element	Density (g/cm^3)
copper	8.96
gold	19.30
iron	7.86
lead	11.34
silver	10.50
tin	7.31

1. Which problem was Archimedes trying to solve?
- **a.** what the density of gold was
- **b.** what the crown was made of
- **c.** what the mass of the crown was
- **d.** how much water the crown displaced

2. Archimedes used the method that he did because a crown has an irregular shape and the volume of such an object cannot be measured in any other way. Which one of the following objects would also require this method?
- **a.** a square wooden box
- **b.** a cylindrical tin can
- **c.** a small bronze statue
- **d.** a rectangular piece of glass

3. Suppose Archimedes found that the crown had a mass of 772 grams and displaced 40 milliliters of water. Using the formula $D = m/V$, what would you determine the crown to be made of?
- **a.** pure gold
- **b.** half gold and half another element
- **c.** some other element with gold plating
- **d.** cannot be determined from the data

4. Using the formula, compare how much water a gold crown would displace if it had a mass of 579 grams.
- **a.** 10 mL
- **b.** 20 mL
- **c.** 30 mL
- **d.** 193 mL

5. If you had crowns made of each element in the chart that were the same mass, which would displace more water than a gold crown of that mass?
- **a.** all
- **b.** lead only
- **c.** tin only
- **d.** none

Extended Response

Answer the two questions below in detail.

6. What is the difference between a physical change and a chemical change? Include examples of each type in your explanation.

7. Why does someone cooking spaghetti at a high elevation need to boil it longer than someone cooking spaghetti at a lower elevation?

Analyzing Experiments

1. b *2. c* *3. a* *4. c* *5. a*

Extended Response

6. RUBRIC

4 points for a response that correctly answers the question, gives at least two examples of each change, and uses the following terms accurately:
- physical change
- chemical change

Sample: Physical changes, such as a change in shape or state, do not change the identity of the material. Chemical changes, such as burning or tarnishing, change the identity of the original material.

3 points correctly answers the question, gives one example of each change, and uses both terms accurately
2 points correctly answers the question, but without examples, and uses both terms accurately
1 point correctly answers the question but without providing examples or using terms accurately

7. RUBRIC

4 points for a response that correctly answers the question and uses the following terms accurately:
- air pressure
- boiling point
- temperature

Sample: The air pressure is less at higher elevations. The energy of the particles doesn't have to be as high to escape, so the boiling point of the water will be at a lower temperature. A lower temperature means a longer cooking time.

3 points correctly answers the question and uses two terms accurately
2 points correctly answers the question and uses one term accurately
1 point correctly answers the question but fails to use any of the terms accurately

METACOGNITIVE ACTIVITY

Have students answer the following questions in their **Science Notebook:**
1. Describe a chemical change that is important in your life.
2. What questions do you still have about physical and chemical properties?
3. How have you solved a problem while working on your Unit Project?

CHAPTER 3 Energy

Physical Science
UNIFYING PRINCIPLES

PRINCIPLE 1
Matter is made of particles too small to see.

PRINCIPLE 2
Matter changes form and moves from place to place.

PRINCIPLE 3
Energy changes from one form to another, but it cannot be created or destroyed.

PRINCIPLE 4
Physical forces affect the movement of all matter on Earth and throughout the universe.

Unit: Matter and Energy
BIG IDEAS

CHAPTER 1
Introduction to Matter
Everything that has mass and takes up space is matter.

CHAPTER 2
Properties of Matter
Matter has properties that can be changed by physical and chemical processes.

CHAPTER 3
Energy
Energy has different forms, but it is always conserved.

CHAPTER 4
Temperature and Heat
Heat is a flow of energy due to temperature differences.

CHAPTER 3
KEY CONCEPTS

SECTION 3.1

Energy exists in different forms.

1. Different forms of energy have different uses.
2. Kinetic energy and potential energy are the two general types of energy.

SECTION 3.2

Energy can change forms but is never lost.

1. Energy changes forms.
2. Energy is always conserved.
3. Energy conversions may produce unwanted forms of energy.

SECTION 3.3

Technology improves the ways people use energy.

1. Technology improves energy conversions.
2. Technology improves the use of energy resources.

The Big Idea Flow Chart is available on p. T17 in the **UNIT TRANSPARENCY BOOK.**

Previewing Content

SECTION

Energy exists in different forms. pp. 71–77

1. Different forms of energy have different uses.
Energy is the ability to cause a change. Different forms of energy cause different changes to occur.

- Mechanical energy involves the position and motion of objects. Mechanical energy is a combination of potential energy and kinetic energy; it may be either or both.
- Sound energy is energy associated with a transfer of vibrations through a solid, liquid, or gas.
- Chemical energy is energy that is stored in the chemical composition of matter due to the atoms, bonds, and arrangement of atoms in substances.
- Thermal energy is the total amount of energy within an object due to the motion of all of the object's particles.
- Electromagnetic energy is energy in electromagnetic waves, including visible light, ultraviolet light, x rays, and microwaves. Electromagnetic energy can travel through a vacuum.
- Nuclear energy holds atomic nuclei together.

2. Kinetic energy and potential energy are the two general types of energy.
Kinetic energy (KE) is the energy of motion. The amount of kinetic energy that any object has depends on its mass and speed. An increase in speed causes a much larger increase in kinetic energy than does an increase in mass.
Potential energy (PE) is energy that is stored in an object as a result of its position, shape, or chemical composition.

- Gravitational potential energy is due to an object's position above Earth's surface. Gravitational potential energy is related to an object's mass and its height above the ground.
- Elastic potential energy is due to position and shape in an object being compressed or flexed. Examples include a compressed spring or a stretched rubber band. Not every object that is compressed will contain elastic potential energy, for example, aluminum foil crumpled into a ball.
- Chemical potential energy is due to a substance's chemical composition—the atoms and bonds contained within the substance. Different substances contain different amounts of chemical potential energy. Examples include energy stored in fossil fuels and in molecules of foods.

SECTION

Energy can change forms but is never lost. pp. 78–85

1. Energy changes forms.
Energy can be converted from one form to another. Often, energy must change forms in order for it to be useful. Many energy transformations occur between potential and kinetic energy. A ski jumper at the top of a slope has gravitational potential energy, which is converted into kinetic energy as the ski jumper moves down the slope. The ski jumper can regain potential energy through the kinetic energy of a chairlift that carries the jumper back up the hill. When gasoline is burned in a car's engine, the chemical potential energy of the fuel is converted into the car's motion, and energy released as heat from the car's engine is the kinetic energy of particle motion.

2. Energy is always conserved.
The **law of conservation of energy** states that energy is neither created nor destroyed. When it appears that energy has been lost, it has simply changed form or been transferred to another object. In the soccer ball photograph below, the soccer ball's kinetic energy decreases, but the energy is converted into sound and heat. As a result, the total amount of energy never changes.

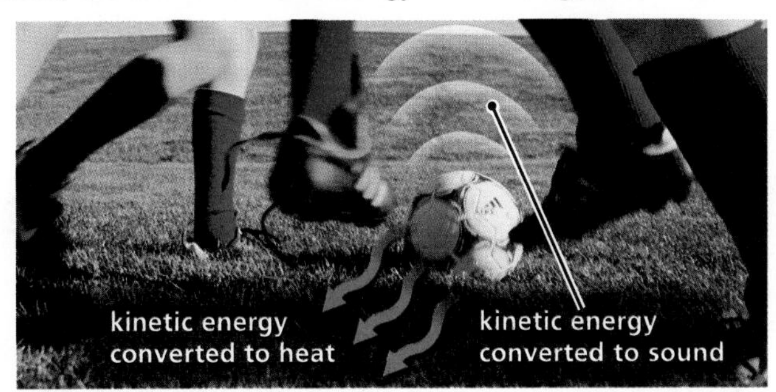

3. Energy conversions may produce unwanted forms of energy.
When energy changes forms, the total amount of energy does not change, but some of the energy may convert to unusable or unwanted forms. **Energy efficiency** is a measure of usable energy after an energy conversion. The more energy-efficient the energy conversion, the more energy is changed into the desired form.

Common Misconceptions

MISCONCEPTION DATABASE
CLASSZONE.COM Background on student misconceptions

ENERGY AND MATTER Students may think that everything that exists is matter, including heat, light, and electricity. Matter has mass and takes up space, whereas energy does not.

 This misconception is addressed on p. 72.

ENERGY AT REST Many students might think that objects at rest do not have any energy. Objects that are not moving do possess different forms of energy, such as gravitational potential energy and chemical potential energy.

 This misconception is addressed on p. 75.

Previewing Content

SECTION

Technology improves the ways people use energy. pp. 86–91

1. Technology improves energy conversions.
Because most energy conversions are very inefficient, an important goal of technology is to improve energy efficiency.
- LEDs convert almost all the electricity they use into light.
- Hybrid cars, which use both a gasoline engine and electrical energy from batteries, are more efficient than conventional gasoline-powered cars.

2. Technology improves the use of energy resources.
Fossil fuels, the most commonly used energy source, are a non-renewable resource. A major goal of technology research is a more efficient usage of other energy sources.
- Solar cells convert sunlight to electrical energy. Solar energy is available in unlimited amounts, is quiet and clean, and is nonpolluting. It is inefficient, however, and the materials used to make solar cells are expensive.
- Windmills are used to convert the kinetic energy of wind into electrical energy. Like solar energy, wind energy is an inexhaustible source of energy that is nonpolluting, but there are limitations to the usefulness of wind power. It takes a large number of windmills to produce enough electrical energy to make a windfarm economically viable. Also, wind power is limited to regions of the country where wind is relatively constant.

Common Misconceptions

MISCONCEPTION DATABASE
CLASSZONE.COM Background on student misconceptions

CONSERVATION OF ENERGY Students may think that "conservation of energy" means that energy should be conserved; this misconception arises due to different uses of the word *conservation*. In terms of the law of conservation of energy, conservation means that the total amount of energy in the universe does not change.

TE This misconception is addressed on p. 82.

Previewing Labs

Lab Generator CD-ROM
Edit these Pupil Edition labs and generate alternative labs.

EXPLORE the BIG idea

A Penny for Your Energy, p. 69 Students explore the transfer of energy from a warm object to a cold object.	**TIME** 10 minutes **MATERIALS** cold glass bottle, cooking oil, coin
Hot Dog! p. 69 Students use a solar-energy collector to cook a hot dog.	**TIME** 40 minutes **MATERIALS** cardboard, aluminum foil, wooden skewer, hot dog, 2 corks
Internet Activity: Energy, p. 69 Students investigate the relationship between potential and kinetic energy.	**TIME** 20 minutes **MATERIALS** computer with Internet access

SECTION  3.1

EXPLORE Energy, p. 71 Students observe that all objects, even when stationary, have energy.	**TIME** 10 minutes **MATERIALS** large plastic bowl, sand, pebble, rock
INVESTIGATE Potential Energy, p. 75 Students design an experiment to change the amount of potential energy an object has.	**TIME** 30 minutes **MATERIALS** model car, meter stick, weights, balance, tape, cardboard, books

SECTION  3.2

CHAPTER INVESTIGATION Energy Conversions, pp. 84–85 Students investigate the amount of energy stored in different kinds of food by constructing a simple calorimeter and burning food samples.	**TIME** 40 minutes **MATERIALS** can opener, empty aluminum can, dowel rod, tap water, graduated cylinder, ring stand with ring, thermometer, aluminum pie plate, aluminum foil, large paper clip, cork, modeling clay, croutons, caramel rice cakes, balance, wooden matches

SECTION  3.3

EXPLORE Solar Cells, p. 86 Students investigate the size of a solar cell needed to provide electrical energy for a solar calculator.	**TIME** 10 minutes **MATERIALS** solar calculator without backup battery, ruler, index card
INVESTIGATE Solar Energy, p. 89 Students observe how the color of a solar-energy collector affects the amount of energy collected.	**TIME** 20 minutes **MATERIALS** 2 plastic cups, white and black plastic to cover cups, 2 rubber bands, scissors, 2 thermometers, stopwatch, aluminum foil

 **Additional INVESTIGATION,** Build a Roller Coaster, A, B, & C, pp. 192–200; Teacher Instructions, pp. 262–263

Previewing Chapter Resources

CHAPTER 3
Energy

INTEGRATED TECHNOLOGY

CLASSZONE.COM
- eEdition Plus
- EasyPlanner Plus
- Misconception Database
- Content Review
- Test Practice
- Simulation
- Visualization
- Resource Centers
- Internet Activity: Energy
- Math Tutorial

CD-ROMS
- eEdition
- EasyPlanner
- Power Presentations
- Content Review
- Lab Generator
- Test Generator

AUDIO CDS
- Audio Readings
- Audio Readings in Spanish

SCILINKS.ORG

LABS AND ACTIVITIES

EXPLORE the Big Idea, p. 69
- A Penny for Your Energy
- Hot Dog!
- Internet Activity: Energy

UNIT RESOURCE BOOK
Unit Projects, pp. 5–10

Lab Generator CD-ROM
Generate customized labs.

SECTION

3.1 Energy exists in different forms.
pp. 71–77

Time: 2 periods (1 block)

Lesson Plan, pp. 144–145

Integrated Technology

RESOURCE CENTERS, Kinetic Energy and Potential Energy; Electric Cars

UNIT TRANSPARENCY BOOK
- Big Idea Flow Chart, p. T17
- Daily Vocabulary Scaffolding, p. T18
- Note-Taking Model, p. T19
- 3-Minute Warm-Up, p. T20

Labs and Activities

- EXPLORE Energy, p. 71
- INVESTIGATE Potential Energy, p. 75
- Think Science, p. 77

UNIT RESOURCE BOOK
Datasheet, Potential Energy, p. 153

SECTION

3.2 Energy can change forms but is never lost.
pp. 78–85

Time: 3 periods (1.5 blocks)

Lesson Plan, pp. 155–156

Integrated Technology

UNIT TRANSPARENCY BOOK
- Daily Vocabulary Scaffolding, p. T18
- 3-Minute Warm-Up, p. T20
- "Converting Energy" Visual, p. T22

Labs and Activities

CHAPTER INVESTIGATION, Energy Conversions, pp. 84–85

UNIT RESOURCE BOOK
- Additional INVESTIGATION, Build a Roller Coaster, A, B, & C, pp. 192–200
- CHAPTER INVESTIGATION, Energy Conversions, A, B, & C, pp. 183–191

SECTION

3.3 Technology improves the ways people use energy.
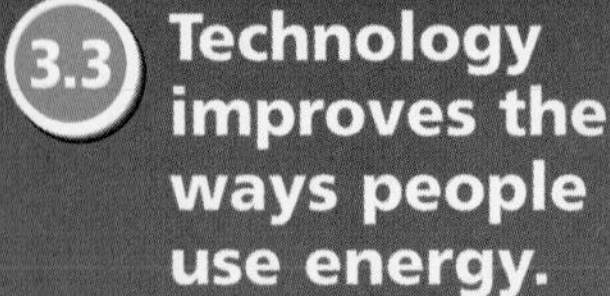
pp. 86–91

Time: 3 periods (1.5 blocks)

Lesson Plan, pp. 165–166

Integrated Technology

- **VISUALIZATION,** Solar Cells
- **RESOURCE CENTER,** Alternative Energy Sources
- **MATH TUTORIAL**

UNIT TRANSPARENCY BOOK
- Big Idea Flow Chart, p. T17
- Daily Vocabulary Scaffolding, p. T18
- 3-Minute Warm-Up, p. T21
- Chapter Outline, pp. T23–T24

Labs and Activities

- EXPLORE Solar Cells, p. 86
- INVESTIGATE Solar Energy, p. 89
- Math in Science, p. 91

UNIT RESOURCE BOOK
- Datasheet, Solar Energy, p. 174
- Math Support, p. 181
- Math Practice, p. 182

KEY TO ICONS  CD/CD-ROM  Teacher Edition  UNIT TRANSPARENCY BOOK  SPANISH ASSESSMENT BOOK

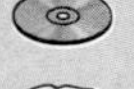 INTERNET  Pupil Edition  UNIT RESOURCE BOOK  UNIT ASSESSMENT BOOK  SCIENCE TOOLKIT

READING AND REINFORCEMENT

ASSESSMENT

STANDARDS

Reading and Reinforcement

- Frame Game, B26–27
- Mind Map, C40–41
- Daily Vocabulary Scaffolding, H1–8

 **UNIT RESOURCE BOOK**

- Vocabulary Practice, pp. 178–179
- Decoding Support, p. 180
- Summarizing the Chapter, pp. 201–202

Audio Readings CD
Listen to Pupil Edition.

Audio Readings in Spanish CD
Listen to Pupil Edition in Spanish.

Assessment

- Chapter Review, pp. 93–94
- Standardized Test Practice, p. 95

 **UNIT ASSESSMENT BOOK**

- Diagnostic Test, pp. 40–41
- Chapter Test, A, B, & C, pp. 45–56
- Alternative Assessment, pp. 57–58

 Spanish Chapter Test, pp. 221–224

 **Test Generator CD-ROM**
Generate customized tests.

 **Lab Generator CD-ROM**
Rubrics for Labs

Standards

National Standards
A.2–8, A.9.a–f, B.3.a, E.2–5, E.6.c–e, F.5.a–c

See p. 68 for the standards.

Reading and Reinforcement

UNIT RESOURCE BOOK

- Reading Study Guide, A & B, pp. 146–149
- Spanish Reading Study Guide, pp. 150–151
- Challenge and Extension, p. 152
- Reinforcing Key Concepts, p. 154

Assessment

Ongoing Assessment, pp. 72–76

Section 3.1 Review, p. 76

 **UNIT ASSESSMENT BOOK**
Section 3.1 Quiz, p. 42

Standards

National Standards
A.2–7, A.9.a–b, A.9.e–f, B.3.a, E.2–5, E.6.d–e

Reading and Reinforcement

UNIT RESOURCE BOOK

- Reading Study Guide, A & B, pp. 157–160
- Spanish Reading Study Guide, pp. 161–162
- Challenge and Extension, p. 163
- Reinforcing Key Concepts, p. 164

Assessment

Ongoing Assessment, pp. 78–83

 Section 3.2 Review, p. 83

 **UNIT ASSESSMENT BOOK**
Section 3.2 Quiz, p. 43

Standards

National Standards
A.2–8, A.9.a–f, B.3.a, E.6.c–e, F.5.a–c

Reading and Reinforcement

 **UNIT RESOURCE BOOK**

- Reading Study Guide, A & B, pp. 167–170
- Spanish Reading Study Guide, pp. 171–172
- Challenge and Extension, p. 173
- Reinforcing Key Concepts, p. 175
- Challenge Reading, pp. 176–177

Assessment

 Ongoing Assessment, pp. 87, 89–90

 Section 3.3 Review, p. 90

 **UNIT ASSESSMENT BOOK**
Section 3.3 Quiz, p. 44

Standards

National Standards
A.2–8, A.9.a–f, B.3.a, E.6.c–e, F.5.a–c

Previewing Resources for Differentiated Instruction

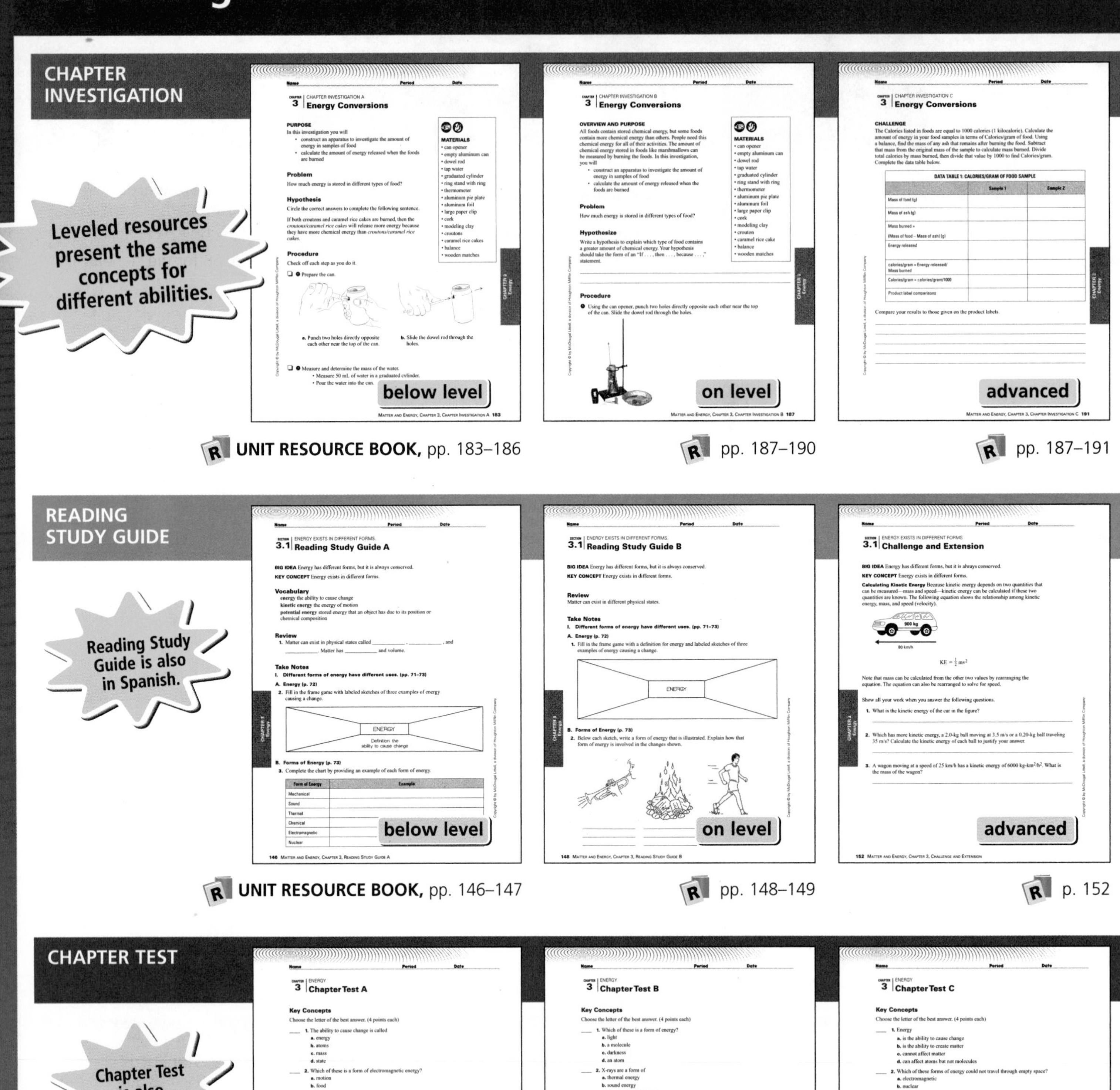

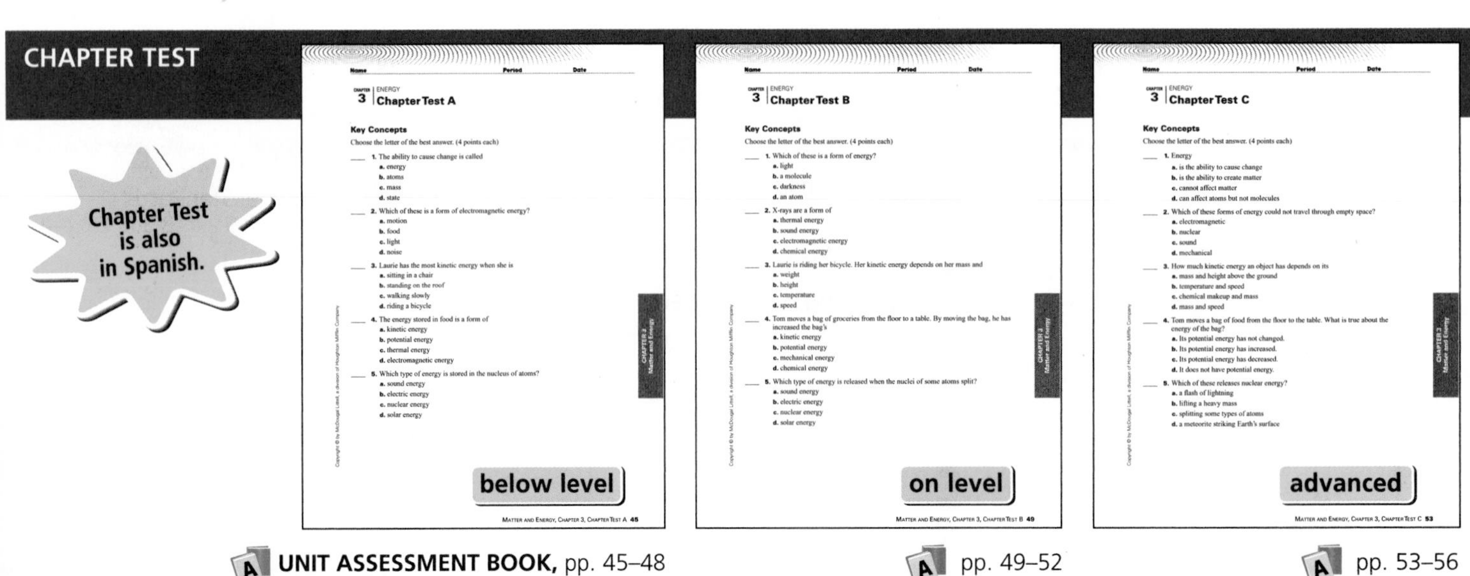

TECHNOLOGY

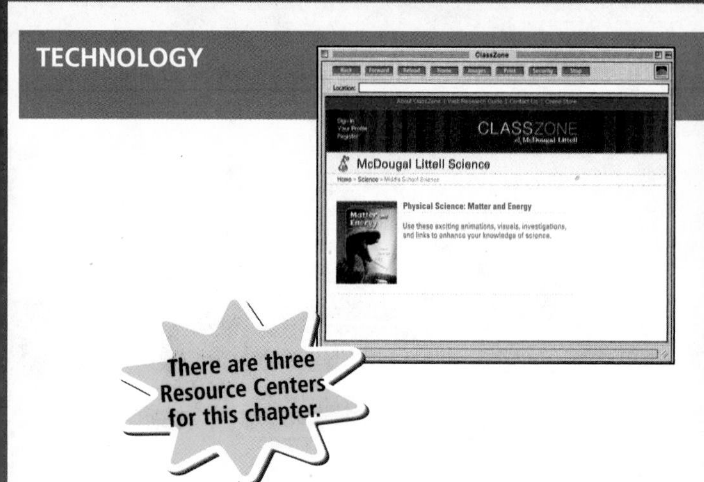

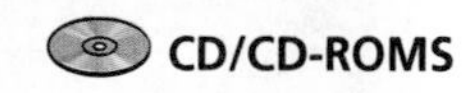

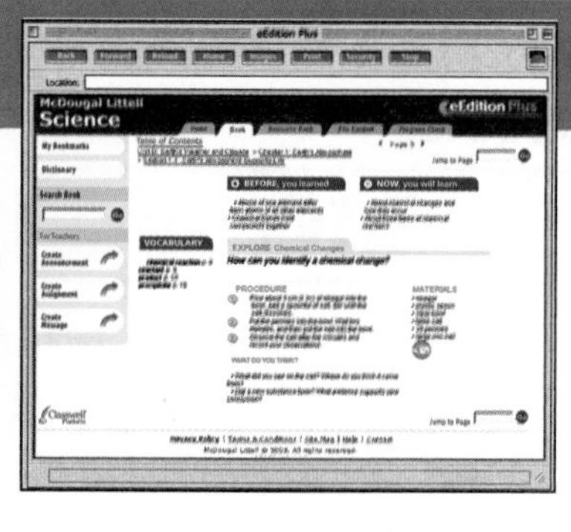

VISUAL CONTENT

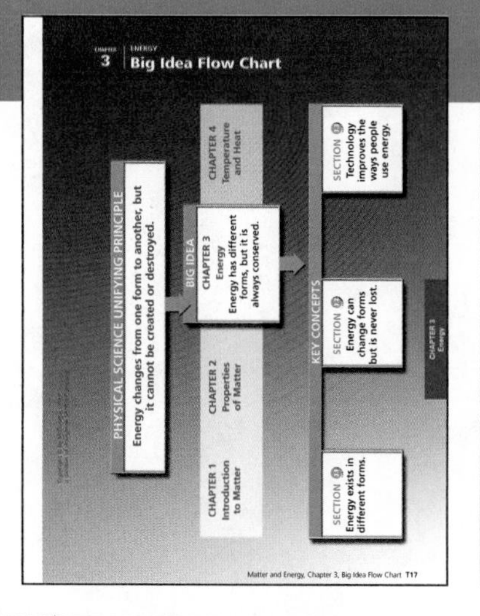

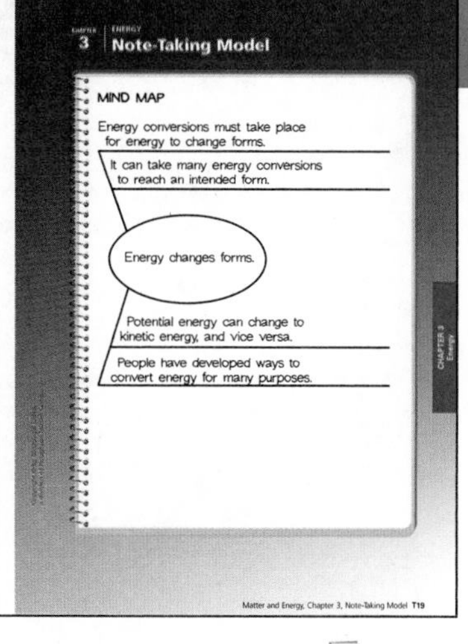

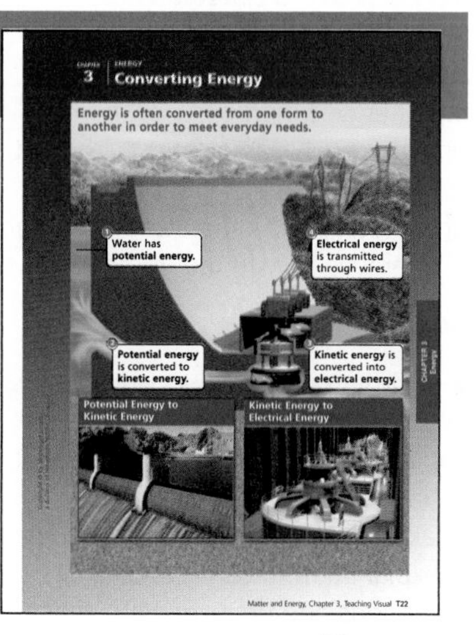

UNIT TRANSPARENCY BOOK, p. T17

p. T19

p. T22

MORE SUPPORT

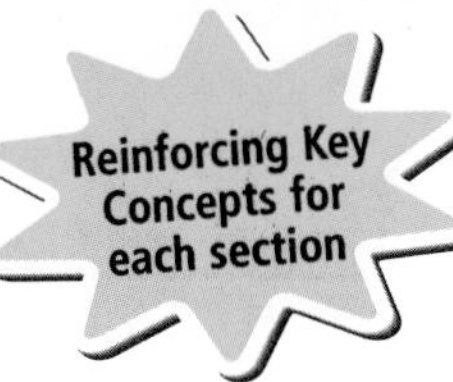

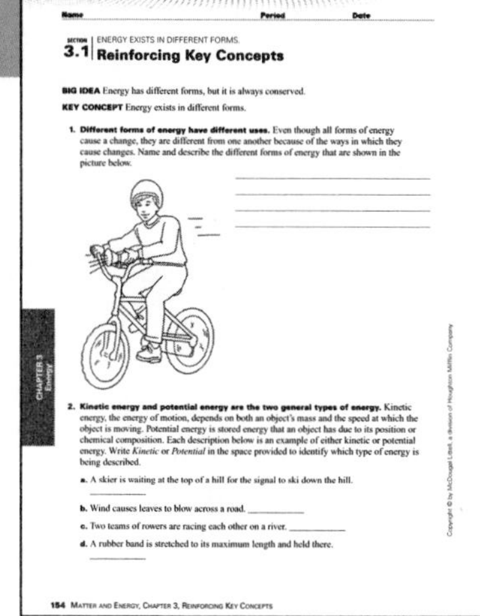

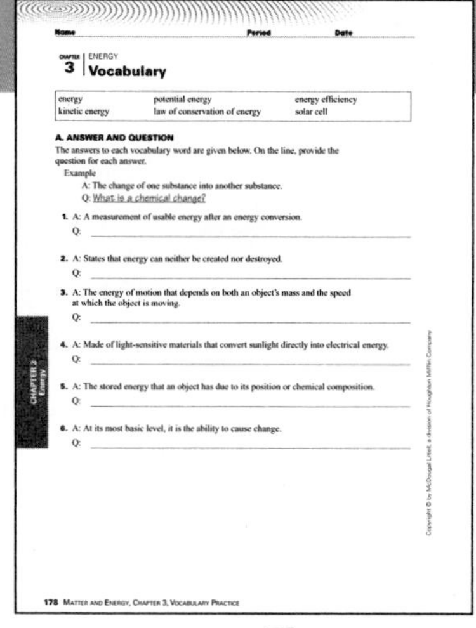

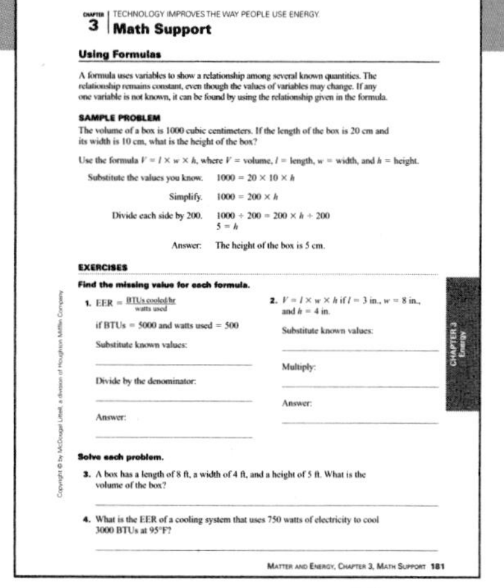

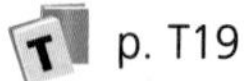

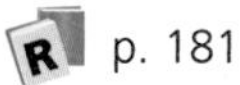

CHAPTER

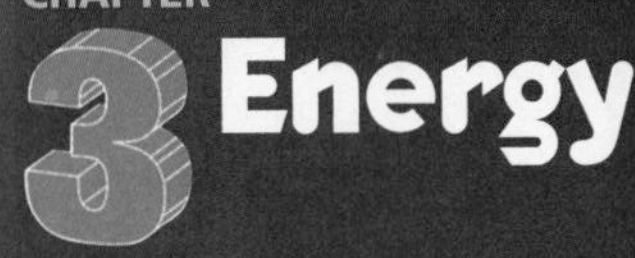

INTRODUCE

the BIG idea

Have students look at the photograph of cyclists and discuss how the question in the box links to the Big Idea:

- Where in the photograph can energy be observed?
- Where in the photograph can energy be inferred?

National Science Education Standards

Content

B.3.a Energy is a property of many substances and is associated with heat, light, electricity, mechanical motion, sound, nuclei, and the nature of a chemical. Energy is transferred in many ways.

Process

A.2–8 Design and conduct an investigation; use tools to gather and interpret data; use evidence to describe, predict, explain, model; think critically to make relationships between evidence and explanation; recognize different explanations and predictions; communicate scientific procedures and explanations; use mathematics.

A.9.a–f Understand scientific inquiry by using different investigations, methods, mathematics, technology, explanations based on logic, evidence, and skeptisicm.

E.2–5 Design, implement, and evaluate a solution or product; communicate technological design.

E.6.c–e Science and technology are reciprocal; technological designs have constraints.

F.5.a–c Science and technology in society.

G.1.b Science requires different abilities.

CHAPTER

Energy

the BIG idea

Energy has different forms, but it is always conserved.

What different forms of energy are shown in this photograph?

Key Concepts

SECTION

Energy exists in different forms.
Learn about several different forms of energy.

SECTION

Energy can change forms but is never lost.
Learn about the law of conservation of energy.

SECTION

Technology improves the ways people use energy.
Learn how technology can be used to make energy conversions more efficient.

Internet Preview

CLASSZONE.COM

Chapter 3 online resources: Content Review, Simulation, Visualization, three Resource Centers, Math Tutorial, Test Practice

INTERNET PREVIEW

CLASSZONE.COM For student use with the following pages:

Review and Practice

- Content Review, pp. 70, 92
- Math Tutorial: Rates, p. 91
- Test Practice, p. 95

Activities and Resources

- Internet Activity: p. 69
- Resource Centers: Kinetic and Potential Energy, p. 74; Electric Cars, p. 77; Alternative Energy Sources, p. 90. Visualization: Solar Cells, p. 88

Forms of Energy
Code: MDL063

EXPLORE the BIG idea

A Penny for Your Energy

Chill an empty glass bottle. Immediately complete the following steps: Rub a drop of cooking oil around the rim of the bottle. Place a coin on the rim so the oil forms a seal between the coin and the bottle. Wrap your hands around the bottle.

Observe and Think What happened to the coin? What do you think caused this to happen?

Hot Dog!

Cover a piece of cardboard with aluminum foil, and bend it into the shape of a U. Poke a wooden skewer through a hot dog, and through each side of the cardboard. Push corks over both ends of the skewer so the cardboard does not flatten out. Place your setup in direct sunlight for 30 minutes.

Observe and Think What happened to the hot dog? Were there any changes you had to make while the hot dog was in sunlight?

Internet Activity: Energy

Go to **ClassZone.com** to investigate the relationship between potential energy and kinetic energy.

Observe and Think How did you change potential energy? How do these changes affect kinetic energy?

NSTA scilinks.org SCILINKS
Forms of Energy **Code: MDL063**

EXPLORE the BIG idea

These inquiry-based activities are appropriate for use at home or as a supplement to classroom instruction.

A Penny for Your Energy

PURPOSE To explore the transfer of energy from a warm object to a cold object. Students make a coin vibrate by causing air to warm and expand.

TIP *10 min.* Make sure the coin will completely cover the opening of the bottle but not extend too far over the rim.

Answer: The coin vibrated and jumped on the top of the bottle. The air sealed inside the bottle warmed and expanded as a result of energy transferred from the student's hands to the trapped air.

REVISIT after p. 80.

Hot Dog!

PURPOSE To observe and use a solar-energy collector. Students cook a hot dog with solar energy.

TIPS *40 min.* Students should not eat the hot dog. Students should try this at home with an adult. The experiment should be done in direct sunlight during the middle of the day.

Answer: The hot dog cooked. The solar collector needed adjustment to keep sunlight reflected on the hot dog.

REVISIT after p. 89.

Internet Activity: Energy

PURPOSE To investigate the relationship between potential and kinetic energy.

TIP *20 min.* Students should understand that potential energy is related to position.

Answer: As the mass and height of an object was increased, the potential energy increased. The object will thus have more kinetic energy.

REVISIT after p. 76.

TEACHING WITH TECHNOLOGY

Video Camera You may wish to film students as they design their experiments for "Investigate Potential Energy" on p. 75. As they watch the video, encourage them to write down ideas for how to improve the design of their experiments.

CBL and Probeware If CBL equipment and probeware are available, have students substitute a temperature probe for the thermometer in the Chapter Investigation on pp. 84–85 and "Investigate Solar Energy" on p. 89.

PREPARE

CONCEPT REVIEW

Activate Prior Knowledge

- To demonstrate that matter has mass and volume, have students measure a small object such as a domino and calculate its volume. Have them find the mass of the object on a balance.
- Ask: Does all matter, even a gas that cannot be seen, have mass and volume? *Yes; the atoms or molecules that make up a gas have mass and take up space.*

TAKING NOTES

Mind Map

A mind map allows students to include as much information and detail about a concept as they choose. Encourage students to use the mind map to take notes in a way that will help them to remember relationships between concepts, definitions, and examples.

Vocabulary Strategy

A frame game diagram organizes characteristics of a vocabulary term into a coherent pattern. By filling in their own words, examples, and descriptions around the frame, students personalize their understanding of the term and can connect personal experience to the term's meaning.

Vocabulary and Note-Taking Resources

- Vocabulary Practice, pp. 178–179
- Decoding Support, p. 180

- Daily Vocabulary Scaffolding, p. T18
- Note-Taking Model, p. T19

- Frame Game, B26–27
- Mind Map, C40–41
- Daily Vocabulary Scaffolding, H1–8

CHAPTER 3

Getting Ready to Learn

CONCEPT REVIEW

- Matter has mass and is made of tiny particles.
- Matter can be changed physically or chemically.
- A change in the state of matter is a physical change.

VOCABULARY REVIEW

matter p. 9
mass p. 10
atom p. 16
physical change p. 44
chemical change p. 46

CONTENT REVIEW
CLASSZONE.COM
Review concepts and vocabulary.

TAKING NOTES

MIND MAP

Write each main idea, or blue heading, in an oval; then write details that relate to each other and to the main idea. Organize the details so that each spoke of the web has notes about one part of the main idea.

VOCABULARY STRATEGY

Write each new vocabulary term in the center of a **frame game** diagram. Decide what information to frame it with. Use examples, descriptions, parts, sentences that use the term in context, or pictures. You can change the frame to fit each term.

See the Note-Taking Handbook on pages R45–R51.

SCIENCE NOTEBOOK

ability to cause a change
different changes from different forms
DIFFERENT FORMS OF ENERGY HAVE DIFFERENT USES.
sunlight
motion
food
electromagnetic energy
mechanical energy
chemical energy

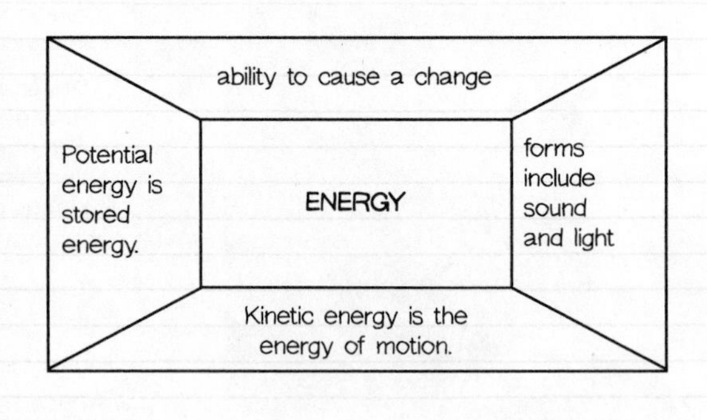
ability to cause a change
Potential energy is stored energy.
ENERGY
forms include sound and light
Kinetic energy is the energy of motion.

CHECK READINESS

Administer the Diagnostic Test to determine students' readiness for new science content and their mastery of requisite math skills.

 Diagnostic Test, pp. 40–41

Technology Resources

Students needing content and math skills should visit **ClassZone.com**.

- **CONTENT REVIEW**
- **MATH TUTORIAL**

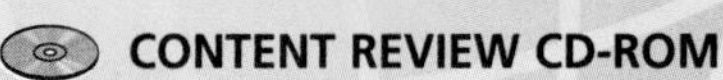
CONTENT REVIEW CD-ROM

KEY CONCEPT

Energy exists in different forms.

BEFORE, you learned

- All substances are made of matter
- Matter has both physical and chemical properties
- Matter can exist in different physical states

NOW, you will learn

- How energy causes change
- About common forms of energy
- About kinetic energy and potential energy

VOCABULARY

energy p. 72
kinetic energy p. 74
potential energy p. 75

EXPLORE Energy

How can you demonstrate energy?

PROCEDURE

1. Fill the bowl halfway with sand. Place the bowl on the floor as shown. Make sure the sand is level.
2. Place a pebble and a rock near the edge of a table above the bowl of sand.
3. Gently push the pebble off the table into the sand. Record your observations.
4. Remove the pebble, and make sure the sand is level. Gently push the rock off the table into the sand. Record your observations.

MATERIALS

- large plastic bowl
- sand
- pebble
- rock

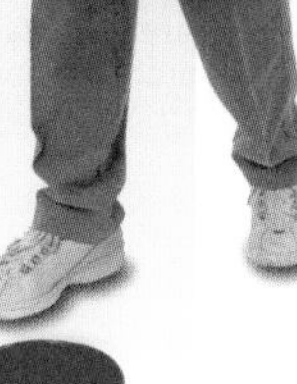

WHAT DO YOU THINK?

- What happened to the sand when you dropped the pebble? when you dropped the rock?
- How can you explain any differences you observed?

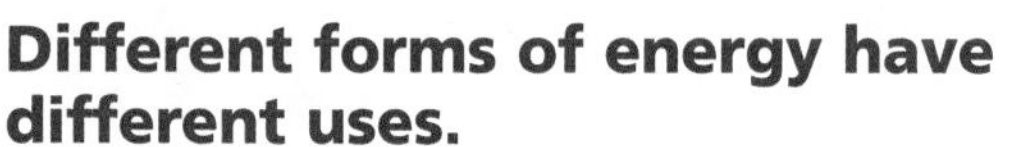

Different forms of energy have different uses.

Energy takes many different forms and has many different effects. Just about everything you see happening around you involves energy. Lamps and other appliances in your home operate on electrical energy. Plants use energy from the Sun to grow. You use energy provided by the food you eat to carry out all of your everyday activities—eating, exercising, reading, and even sitting and thinking. In this chapter, you will learn what these and other forms of energy have in common.

3.1 FOCUS

Set Learning Goals

Students will

- Recognize how energy causes change.
- Describe common forms of energy.
- Illustrate that the two general types of energy are kinetic energy and potential energy.
- Design an experiment to investigate and change potential energy.

3-Minute Warm-Up

Display Transparency 20 or copy this exercise on the board:

Draw diagrams that show the motion of particles in a solid, such as ice, and the motion of particles in a liquid, such as water. What is the process that changes the solid into the liquid? *melting* What must be added to cause the solid to change into the liquid? *energy*

3-Minute Warm-Up, p. T20

3.1 MOTIVATE

EXPLORE Energy

PURPOSE To observe that all objects have energy, even when they are stationary

TIPS *10 min.*

- Bowls with wide openings work best.
- Packing peanuts may be used instead of sand.

WHAT DO YOU THINK? *The pebble made a small dent in the sand; the rock made a much larger dent in the sand. The rock contains more (potential) energy due to the force of gravity.*

RESOURCES FOR DIFFERENTIATED INSTRUCTION

Below Level

UNIT RESOURCE BOOK

- Reading Study Guide A, pp. 146–147
- Decoding Support, p. 180

Advanced

UNIT RESOURCE BOOK

Challenge and Extension, p. 152

English Learners

UNIT RESOURCE BOOK

Spanish Reading Study Guide, pp. 150–151

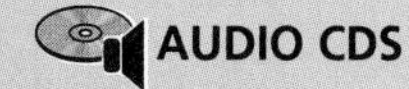

- Audio Readings in Spanish
- Audio Readings (English)

3.1 INSTRUCT

Address Misconceptions

IDENTIFY Ask: Is light a substance? If students say yes, they may hold the misconception that energy is a form of matter, that is, an object.

CORRECT Have students list the properties that matter must have. Discuss whether light and other forms of energy have mass and volume.

REASSESS Ask students to write a short paragraph differentiating between an object and a property of an object. Ask: Which is energy? *a property of an object*

Technology Resources

Visit **ClassZone.com** for background on common student misconceptions.

MISCONCEPTION DATABASE

Teach Difficult Concepts

Students may think that energy is associated only with living things. In fact, all things, both living and nonliving, have energy.

Place a domino at the base of a ramp. Have students roll a marble down the ramp so that it knocks over the domino. Ask: What knocked the domino over? *the marble's energy* Ask students to design another way to demonstrate that inanimate objects have energy that can cause a change.

Teach from Visuals

To help students identify energy and its effects, ask: What other changes might be occurring in the picture? *people moving in cars; trees moving; sounds*

Ongoing Assessment

Recognize how energy causes change.

Ask: What is the most fundamental quality of energy? *Its ability to cause change.*

CHECK YOUR READING *Sample answer: Hitting a baseball changes its direction and speed.*

Energy

A All forms of energy have one important point in common—they cause changes to occur. The flow of electrical energy through a wire causes a cool, dark bulb to get hot and glow. The energy of the wind causes a flag to flutter.

You are a source of energy that makes changes in your environment. For example, when you pick up a tennis racquet or a paintbrush, you change the position of that object. When you hit a tennis ball or smooth paint on a canvas, you cause further changes. Energy is involved in every one of these actions. At its most basic level, **energy** is the ability to cause change.

VOCABULARY
Remember to use a frame game diagram for *energy* and other vocabulary terms.

CHECK YOUR READING Provide your own example of energy and how it causes a change.

The photograph below shows a city street. All of the activities that take place on every street in any city require energy, so there are many changes taking place in the picture. Consider one of the cars. A person's energy is used to turn the key that starts the car. The key's movement starts the car's engine and gasoline begins burning. Gasoline provides the energy for the car to move. The person's hand, the turning key, and the burning gasoline all contain energy that causes change.

The motion of the cars and the shining of the street lights are changes produced by energy.

DIFFERENTIATE INSTRUCTION

More Reading Support

A What do all forms of energy have in common? *They cause changes.*

English Learners Help English learners understand complex sentences. Give students several difficult sentences and ask them to circle the subject and verb. Be sure students circle the entire subject—often it is more than one word. For example, in the sentence "Just about everything you see happening around you involves energy" (p. 71), students should circle "Just about everything you see happening around you" as the subject and "involves" as the verb.

Forms of Energy

Scientists classify energy into many forms, each of which causes change in a different way. Some of these forms are described below.

Mechanical Energy The energy that moves objects is mechanical energy. The energy that you use to put a book on a shelf is mechanical energy, as is energy that a person uses to turn a car key.

Sound Energy Sound results from the vibration of particles in a solid, liquid, or gas. People and other animals are able to detect these tiny vibrations with structures in their ears that vibrate due to the sound. So, when you hear a car drive past, you are detecting vibrations in the air produced by sound energy. Sound cannot travel through empty space. If there were no air or other substance between you and the car, you would not hear sounds from the car.

Chemical Energy Energy that is stored in the chemical composition of matter is chemical energy. The amount of chemical energy in a substance depends on the types and arrangement of atoms in the substance. When wood or gasoline burns, chemical energy produces heat. The energy used by the cells in your body comes from chemical energy stored in the foods you eat.

Thermal Energy The total amount of energy from the movement of particles in matter is thermal energy. Recall that matter is made of atoms, and atoms combined in molecules. The atoms and molecules in matter are always moving. The energy of this motion in an object is the object's thermal energy. You will learn more about thermal energy in the next chapter.

Electromagnetic Energy Electromagnetic (ih-LEHK-troh-mag-NEHT-ihk) energy is transmitted through space in the form of electromagnetic waves. Unlike sound, electromagnetic waves can travel through empty space. These waves include visible light, x-rays, and microwaves. X-rays are high energy waves used by doctors and dentists to look at your bones and teeth. Microwaves can be used to cook food or to transmit cellular telephone calls but contain far less energy than x-rays. The Sun releases a large amount of electromagnetic energy, some of which is absorbed by Earth.

Nuclear Energy The center of an atom—its nucleus—is the source of nuclear energy. A large amount of energy in the nucleus holds the nuclear particles together. When a heavy atom's nucleus breaks apart, or when the nuclei (NOO-klee-EYE) of two small atoms join together, energy is released. Nuclear energy released from the fusing of small nuclei to form larger nuclei keeps the Sun burning.

CHECK YOUR READING How does chemical energy cause a change? What about electromagnetic energy?

APPLY Where in this photograph can you find chemical, sound, and mechanical energy?

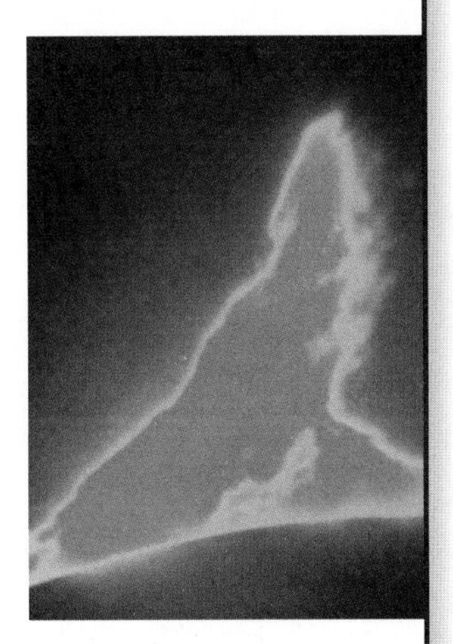

This solar flare releases electromagnetic energy and thermal energy produced by nuclear energy in the Sun.

DIFFERENTIATE INSTRUCTION

More Reading Support

B What is the energy that moves objects? *mechanical energy*

C What is the energy stored in the chemical composition of matter? *chemical energy*

Inclusion If you demonstrate sound energy to your class, students with hearing impairments will be able to feel the energy through its conversion to mechanical waves, that is, vibrations in the floor, the desk, and other objects.

Integrate the Sciences

The chemical energy stored in food is not used directly for energy by cells. A complex series of reactions known as cellular respiration changes the molecules in food into a usable source of energy.

Teach Difficult Concepts

Thermal energy is not heat, but the two are related. Heat is a process that transfers energy from warmer objects to cooler objects. The result of this transfer of energy is an increase in the cooler object's thermal energy and a decrease in the warmer object's thermal energy.

Ongoing Assessment

Describe common forms of energy.

Ask: What are three forms of electromagnetic energy. *Sample answer: visible light, x-rays, microwaves*

PHOTO CAPTION Answer: Sample answer: chemical energy in the drummers' bodies, sound energy from the drums, mechanical energy in the moving drumsticks

CHECK YOUR READING *Sample answers: Chemical energy can cause a change through its release when a substance is burned. Electromagnetic energy can cause a change when food is cooked in a microwave oven.*

Teacher Demo

To help students understand that decreasing an object's speed decreases its kinetic energy, do the following demonstration. Set up a ramp. Position a small object at the end of the ramp. Roll a small ball down the ramp so that it knocks over the object. Decrease the angle of the ramp to decrease the speed of the ball, then roll the ball down again. Continue to decrease the speed in this way until the ball has insufficient kinetic energy to knock over the object. To demonstrate the relationship between mass and kinetic energy, repeat the above procedure, keeping the ramp at the same angle but using balls with different mass. Find a ball with so little mass that the ball cannot knock over the object. Use a balance (or kitchen scale to measure objects in increments of 10 grams) to measure the mass of the balls to confirm that they have different masses.

Teach Difficult Concepts

Mechanical energy is a combination of potential energy and kinetic energy. Mechanical energy results from the position of objects (potential energy), the movement of objects (kinetic energy), or both.

Teach from Visuals

To help students interpret the photos of the speed skater, ask: Where did the skater's kinetic energy come from? *from chemical energy stored in the skater's muscles*

Ongoing Assessment

READING VISUALS *Answer: If the skater with less mass is moving faster, he or she could have more kinetic energy.*

Kinetic energy and potential energy are the two general types of energy.

RESOURCE CENTER CLASSZONE.COM

Learn more about kinetic energy and potential energy.

All of the forms of energy can be described in terms of two general types of energy—kinetic energy and potential energy. Anything that is moving, such as a car that is being driven or an atom in the air, has kinetic energy. All matter also has potential energy, or energy that is stored and can be released at a later time.

Kinetic Energy

READING TIP

Kinetic means "related to motion."

D

The energy of motion is called **kinetic energy.** It depends on both an object's mass and the speed at which the object is moving.

All objects are made of matter, and matter has mass. The more matter an object contains, the greater its mass. If you held a bowling ball in one hand and a soccer ball in the other, you could feel that the bowling ball has more mass than the soccer ball.

- **Kinetic energy increases as mass increases.** If the bowling ball and the soccer ball were moving at the same speed, the bowling ball would have more kinetic energy because of its greater mass.
- **Kinetic energy increases as speed increases.** If two identical bowling balls were rolling along at different speeds, the faster one would have more kinetic energy because of its greater speed. The speed skater in the photographs below has more kinetic energy when he is racing than he does when he is moving slowly.

E

High Speed

This skater has a large amount of kinetic energy when moving at a high speed.

Low Speed

When the same skater is moving more slowly, he has less kinetic energy.

READING VISUALS **APPLY** How could a skater with less mass than another skater have more kinetic energy?

DIFFERENTIATE INSTRUCTION

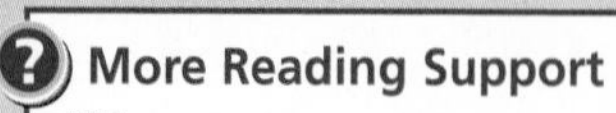

More Reading Support

D What is kinetic energy? *the energy of motion*

E What is the relationship between speed and kinetic energy? *As speed increases, kinetic energy increases.*

Advanced Introduce students to the formula for kinetic energy, $KE = 1/2\ mv^2$. Use examples to show that the kinetic energy of an object varies directly with its mass and with the square of the velocity (speed). Ask students to explain why changing the velocity of an object has a much greater impact on its kinetic energy than changing its mass.

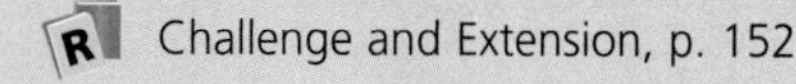

R Challenge and Extension, p. 152

Potential Energy

Suppose you are holding a soccer ball in your hands. Even if the ball is not moving, it has energy because it has the potential to fall. **Potential energy** is the stored energy that an object has due to its position or chemical composition. The ball's position above the ground gives it potential energy.

The most obvious form of potential energy is potential energy that results from gravity. Gravity is the force that pulls objects toward Earth's surface. The giant boulder on the right has potential energy because of its position above the ground. The mass of the boulder and its height above the ground determine how much potential energy it has due to gravity.

Because the boulder could fall, it has potential energy from gravity.

It is easy to know whether an object has kinetic energy because the object is moving. It is not so easy to know how much and what form of potential energy an object has, because objects can have potential energy from several sources. For example, in addition to potential energy from gravity, substances contain potential energy due to their chemical composition—the atoms they contain.

CHECK YOUR READING How can you tell kinetic energy and potential energy apart?

INVESTIGATE Potential Energy

How can you change the amount of potential energy?

DESIGN — YOUR OWN — EXPERIMENT

Use what you know about potential energy to design an experiment that shows how potential energy can be increased or decreased.

SKILL FOCUS
Designing experiments

MATERIALS
- model car
- meter stick
- weights
- balance
- tape
- cardboard
- books

TIME
30 minutes

PROCEDURE

1. Using the materials in the list, design an experiment to investigate the potential energy of the model car. Use the cardboard as a ramp.
2. Write up your hypothesis and your procedure. Remember to include the variables and constants in the experiment.
3. Conduct your experiment and record your results.

WHAT DO YOU THINK?

- What variables did you change? Why?
- How do your results demonstrate a change in potential energy?

DIFFERENTIATE INSTRUCTION

More Reading Support

F What is potential energy? *stored energy*

G What causes an object to have potential energy due to its position above the ground? *gravity*

Address Misconceptions

IDENTIFY Ask: What happens to energy when an object is at rest? If students reply the object has no energy, they may hold the misconception that objects at rest do not have any energy.

CORRECT Explain how potential energy relates to various forms of energy. Even though an object may be at rest, it has stored energy. Different substances have potential energy because of the types of atoms in them. Have students make a list of objects with potential energy.

REASSESS Ask: What type of energy is found in food? *chemical energy*

Technology Resources

Visit **ClassZone.com** for background on common student misconceptions.

MISCONCEPTION DATABASE

INVESTIGATE Potential Energy

PURPOSE To design an experiment that shows how potential energy can change

TIPS *30 min.* Pennies or washers can be used as the weights. Ask students to bring a toy car from home.

WHAT DO YOU THINK? *mass of car, height of ramp. Both alter the potential energy the car has. The farther the car travels, the greater its kinetic energy; thus the greater was its potential energy.*

Datasheet, Potential Energy, p. 153

Technology Resources

Customize this student lab as needed or look for an alternative. Print rubrics to assess student lab reports.

Lab Generator CD-ROM

Ongoing Assessment

CHECK YOUR READING *Answer: Kinetic energy—an object is in motion; potential energy—stored energy due to position or chemical composition.*

Ongoing Assessment

Illustrate that the two general types of energy are kinetic energy and potential energy.

Ask: How might riding down a waterslide involve both potential and kinetic energy? *At the top of the slide, you have potential energy and no kinetic energy. As you move down the slide, potential energy changes to kinetic energy.*

Answer: It is energy that is stored in molecules.

EXPLORE the BIG idea

Revisit "Internet Activity: Energy" on p. 69. Have students explain the relationship between potential and kinetic energy.

Reinforce the BIG idea

Have students relate the section to the Big Idea.

 Reinforcing Key Concepts, p. 154

3.1 ASSESS & RETEACH

Assess

 Section 3.1 Quiz, p. 42

Reteach

Ask students how water flowing over a waterfall involves kinetic and potential energy. *Flowing water has kinetic energy; the water has potential energy due to gravity.* Ask: What forms of energy can be observed? *mechanical in the motion of the water; sound from the water*

Technology Resources

Have students visit **ClassZone.com** for reteaching of Key Concepts.

 **CONTENT REVIEW**

 **CONTENT REVIEW CD-ROM**

Pulling the string, which bends the bow, gives the bow potential energy.

Another form of potential energy related to an object's position comes from stretching or compressing an object. Think about the spring that is pushed down in a jack-in-the-box. The spring's potential energy increases when the spring is compressed and decreases when it is released. Look at the bow that is being bent in the photograph on the left. When the bowstring is pulled, the bow bends and stores energy. When the string is released, both the string and the bow return to their normal shape. Stored energy is released as the bow and the string straighten out and the arrow is pushed forward.

When a rock falls or a bow straightens, potential energy is released. In fact, in these examples, the potential energy produced either by gravity or by bending is changed into kinetic energy.

Chemical energy, such as the energy stored in food, is less visible, but it is also a form of potential energy. This form of potential energy depends on chemical composition rather than position. It is the result of the atoms, and the bonds between atoms, that make up the molecules in food. When these molecules are broken apart, and their atoms rearranged through a series of chemical changes, energy is released.

Chemical energy in the fuel of a model rocket engine is potential energy.

The fuel in a model rocket engine also contains chemical energy. Like the molecules that provide energy in your body, the molecules in the fuel store potential energy. When the fuel ignites in the rocket engine, the arrangement of atoms in the chemical fuel changes and its potential energy is released.

Why is chemical energy a form of potential energy?

3.1 Review

KEY CONCEPTS

1. List three ways you use energy. How does each example involve a change?
2. What are some changes that can be caused by sound energy? by electromagnetic energy?
3. What two factors determine an object's kinetic energy?

CRITICAL THINKING

4. **Synthesize** How do the different forms of potential energy depend on an object's position or chemical composition?
5. **Infer** What forms of potential energy would be found in an apple on the branch of a tree? Explain.

CHALLENGE

6. **Synthesize** Describe a stone falling off a tabletop in terms of both kinetic energy and potential energy.

ANSWERS

1. Answers could include any activity, but should indicate what change results.

2. sound—vibrations in a solid, liquid, or gas; electromagnetic—lighting a dark room

3. mass and speed

4. an object's position above the ground; an object's compressed or stretched position; chemical potential depends on the atoms and bonds in an object.

5. gravitational, because of the apple's position; chemical, because the apple can be eaten for chemical energy

6. The stone has potential energy due to gravity. The amount of energy depends on its mass and height. The greater its potential energy, the greater its kinetic energy when it falls.

SKILL: FINDING SOLUTIONS

Gasoline or Electric?

Cars use a significant amount of the world's energy. Most cars get their energy from the chemical energy of gasoline, a fossil fuel. Cars can also get their energy from sources other than gasoline. For many years, engineers have been working to design cars that run only on electricity. The goals of developing these new cars include reducing air pollution and decreasing the use of fossil fuels. So why have electric cars not replaced gasoline-powered cars?

Advantages of Electric Cars

- Electric motors are more simple than gasoline engines.
- Electric cars use energy more efficiently than gasoline-powered cars, so they are cheaper to operate.
- Controlling pollution at power plants that produce electricity is easier than controlling pollution from cars.
- Electric motors are quieter than gasoline engines.
- Electric cars do not produce smog, which is a major health concern in large cities.

Disadvantages of Electric Cars

- At this time, electric cars can travel only about 100 miles on a single battery charge.
- It takes several hours to recharge the batteries of an electric car using today's charging systems.
- The batteries of an electric car need to be replaced after being recharged about 600 times.
- An electric car's range is decreased by heating or cooling the inside of the car because, unlike batteries in gasoline-powered cars, its batteries are not recharged during driving.

Finding Solutions

As a Group

What technology would need to be improved for electric cars to replace gasoline-powered cars? What facilities that do not exist today would be needed to serve electric cars?

As a Class

Compare your group's solutions to those of other groups. Use the Internet to research hybrid vehicles. How would these vehicles solve some of the problems that you identified?

Find out more about electric cars.

ANSWERS

AS A GROUP Battery technology needs to improve and facilities for recharging car batteries have to be built.

AS A CLASS Hybrid electric vehicles solve some of the problems of electric cars because they combine electric power with the power of the internal combustion engine. They are more efficient and less polluting.

THINK SCIENCE
Scientific Methods of Thinking

Set Learning Goal

To learn about electric cars and weigh their advantages and disadvantages

Present the Science

Batteries are the weak link in the development of economical and practical electric cars. The batteries are heavy, bulky, and expensive. They must be recharged (a slow process) and have to be replaced regularly. For this reason, research has centered on hybrid electric vehicles (HEVs) and on fuel cells. HEVs are an intermediate step between purely electric cars and gasoline-powered cars, and were meant to be a temporary solution until better batteries were developed.

Guide the Activity

DIFFERENTIATION TIP Have slower learners discuss the general characteristics of the advantages and disadvantages of electric cars and organize them into a chart. Be sure students identify health and the environment as advantages, and convenience as a disadvantage.

Point out that most energy sources have advantages and disadvantages. Discuss a major energy source for your area and how it benefits people and what problems it causes.

COOPERATIVE LEARNING STRATEGY Divide the class into groups of four. Have each group discuss what would have to be done to make electric cars convenient and practical.

Close

Ask: What might be the advantages if all future cars were electric powered? *Smog and pollution would be decreased.*

Technology Resources

Have students visit **ClassZone.com** to find out more about electric cars

RESOURCE CENTER

3.2 FOCUS

Set Learning Goals

Students will

- Explain how energy can be converted from one form to another.
- Restate the law of conservation of energy.
- Understand that energy conversions may be inefficient.

3-Minute Warm-Up

Display Transparency 20 or copy this exercise on the board:

Decide if these statements are true. If not true, correct them.

1. Chemical energy is based on the movement of particles within matter. *thermal energy*
2. Nuclear energy holds an atom's nucleus together. *true*
3. Electromagnetic energy is the energy used to move objects. *mechanical energy*
4. Kinetic energy is energy of motion, and potential energy is stored energy. *true*

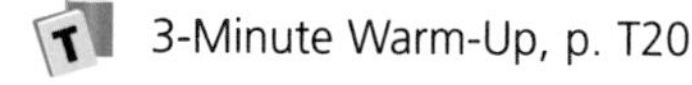

3.2 MOTIVATE

THINK ABOUT

PURPOSE To identify how energy changes forms several times when a match is lit

DISCUSS Have students suggest energy changes in other everyday events. *Using a stove, turning on a light, and using an electric appliance all involve energy changing forms.*

Answer: The energy to strike the match initially comes from the person who uses the match.

Ongoing Assessment

CHECK YOUR READING *Answer: Chemical energy becomes mechanical energy in the match. The match's mechanical energy changes into heat and light released by the burning match.*

KEY CONCEPT

3.2 Energy can change forms but is never lost.

BEFORE, you learned

- Energy causes change
- Energy has different forms
- Kinetic energy and potential energy are the two general types of energy

NOW, you will learn

- How energy can be converted from one form to another
- About the law of conservation of energy
- How energy conversions may be inefficient

VOCABULARY

law of conservation of energy p. 82
energy efficiency p. 83

THINK ABOUT

How does energy change form?

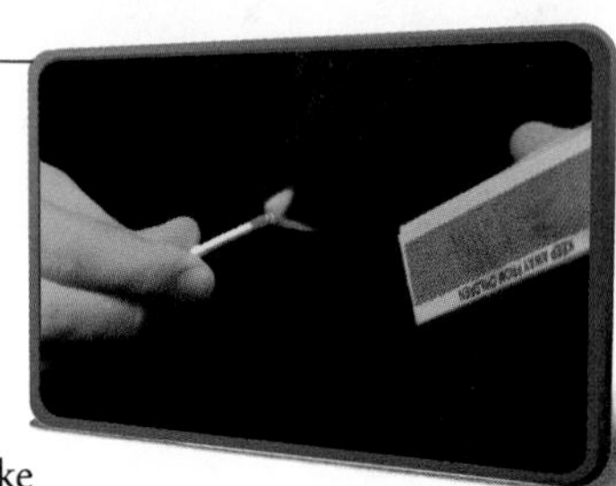

Potential energy is stored in the chemicals on the head of a match. The flame of a burning match releases that energy as light and heat. Where does the energy to strike the match come from in the first place?

MIND MAP
Use a mind map to take notes about how energy changes forms.

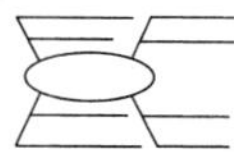

Energy changes forms.

A match may not appear to have any energy by itself, but it does contain potential energy that can be released. The chemical energy stored in a match can be changed into light and heat. Before the chemical energy in the match changes forms, however, other energy conversions must take place.

Plants convert energy from the Sun into chemical energy, which is stored in the form of sugars in their cells. When a person eats food that comes from plants—or from animals that have eaten plants—the person's cells can release this chemical energy. Some of this chemical energy is converted into the kinetic energy that a person uses to rub the match over a rough surface to strike it. The friction between the match and the striking surface produces heat. The heat provides the energy needed to start the chemical changes that produce the flame. From the Sun to the flame, at least five energy conversions have taken place.

How is a person's chemical energy changed into another form of energy in the lighting of a match?

RESOURCES FOR DIFFERENTIATED INSTRUCTION

Below Level

UNIT RESOURCE BOOK
- Reading Study Guide A, pp. 157–158
- Decoding Support, p. 180

AUDIO CDS

Advanced

UNIT RESOURCE BOOK
Challenge and Extension, p. 163

English Learners

UNIT RESOURCE BOOK
Spanish Reading Study Guide, pp. 161–162

- Audio Readings in Spanish
- Audio Readings (English)

Additional INVESTIGATION,
Build a Roller Coaster, A, B, & C, pp. 192–200;
Teacher Instructions, pp. 262–263

Conversions Between Potential Energy and Kinetic Energy

The results of some energy conversions are obvious, such as when electrical energy in a light bulb is changed into light and heat. Other energy conversions are not so obvious. The examples below and on page 80 explore, step by step, some ways in which energy conversions occur in the world around you.

Potential energy can be changed into kinetic energy and back into potential energy. Look at the illustrations and photograph of the ski jumper shown below.

1. At first, the ski jumper is at the top of the hill. This position gives him potential energy (PE) due to gravity.
2. As the ski jumper starts moving downhill, some of his potential energy changes into kinetic energy (KE). Kinetic energy moves him down the slope to the ramp.
3. When the ski jumper takes off from the ramp, some of his kinetic energy is changed back into potential energy as he rises in the air.

When the ski jumper descends to the ground, his potential energy once again changes into kinetic energy. After the ski jumper lands and stops moving, how might he regain the potential energy that he had at the top of the hill? The kinetic energy of a ski lift can move the ski jumper back up the mountain and give him potential energy again.

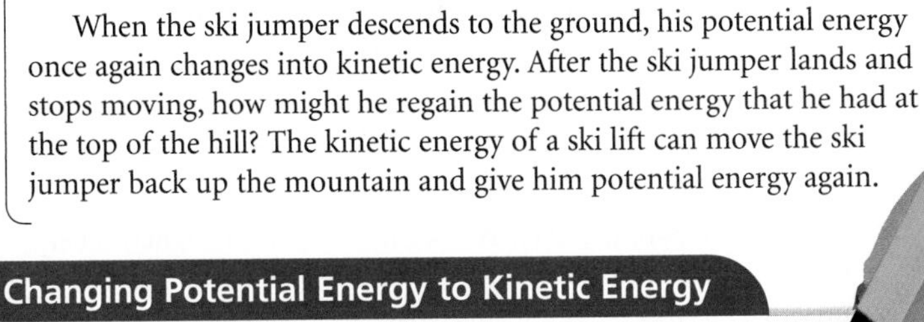

Changing Potential Energy to Kinetic Energy

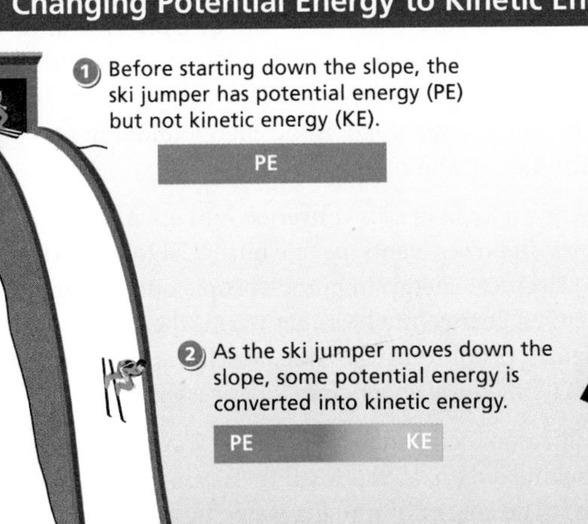

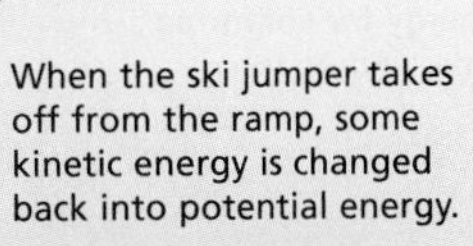

1. Before starting down the slope, the ski jumper has potential energy (PE) but not kinetic energy (KE).
2. As the ski jumper moves down the slope, some potential energy is converted into kinetic energy.
3. When the ski jumper takes off from the ramp, some kinetic energy is changed back into potential energy.

READING VISUALS What would the colored bar look like just before the ski jumper lands on the ground?

3.2 INSTRUCT

Teach from Visuals

To help students interpret the graphic of energy changes in ski jumping, ask:

- Why does the ski jumper have potential energy? *At the top, he has potential energy due to gravity.*
- Why is some potential energy converted to kinetic energy in step 2? *The skier has descended and moves faster as a result. Kinetic energy is the energy of motion.*
- When does the ski jumper have the most potential energy after he takes off from the jump? *at the skier's point of greatest height*
- Why does the ski jumper's potential energy increase for a short time after taking off from the jump? *because his height above the ground increases*

Develop Critical Thinking

INFER Have students infer what energy changes are involved if a different ski jumper slides down the slope. Ask:

- How could another ski jumper have a greater amount of potential energy than the one shown in the illustration? *start from a greater height, have a greater mass*
- How would the energies involved change if the ski jumper was much heavier? *His mass would be larger, and therefore his potential and kinetic energy would also be larger; he would exert a greater force on the ground when he landed.*
- When the ski jumper has landed and stopped moving, does he still possess a form of potential energy? Explain. *yes; chemical potential energy from molecules obtained from food.*

DIFFERENTIATE INSTRUCTION

More Reading Support

A What type of energy moves a skier down a hill? *kinetic energy*

Additional Investigation To reinforce Section 3.2 learning goals, use the following full-period investigation:

R **Additional INVESTIGATION,** Build a Roller Coaster, A, B, & C, pp. 192–200, 262–263 (Advanced students should complete Levels B and C.)

English Learners Students may need background knowledge of a ski jumper and ski lifts (p. 79). For the section review (p. 83), tell students what a trampoline is and make sure English learners understand the direction *Suppose*.

Ongoing Assessment

READING VISUALS *Answer: It would be all yellow.*

EXPLORE the BIG idea

Revisit "A Penny for Your Energy" on p. 69. Have students describe the energy transfer that took place.

Teach Difficult Concepts

Students may confuse the transfer of energy with transformation of energy. In a transfer of energy, energy moves from one object to another. This occurs in the activity "A Penny for Your Energy," where thermal energy is transferred from warm hands to the cold bottle. Transformation of energy is the conversion of one energy form to another, such as when generators convert the mechanical energy of moving water to electrical energy. Ask students to describe examples of energy transfer and transformation. Make a table listing student responses on the board.

Teacher Demo

To demonstrate the conversion of sound energy to mechanical energy, position a fully inflated balloon in front of a stereo speaker. Students will be able to feel the vibrations from a loud sound by placing their hands lightly on the balloon.

Ongoing Assessment

Explain how energy can be converted from one form to another.

Ask: What energy conversions take place when fireworks explode? *Chemical energy in the firework chemicals is converted into light, heat, and sound energy.*

CHECK YOUR READING *Sample answer: The potential energy of water behind a dam can be changed into electrical energy.*

Using Energy Conversions

People have developed ways to convert energy from one form to another for many purposes. Read about the energy conversion process below, and follow that process in the illustrations on page 81 to see how energy in water that is stored behind a dam is changed into electrical energy.

READING TIP
As you read about the process for producing electrical energy, follow the steps on the next page.

1. The water held behind the dam has potential energy because of its position.
2. Some of the water is allowed to flow through a tunnel within the dam. The potential energy in the stored water changes into kinetic energy when the water moves through the tunnel.
3. The kinetic energy of the moving water turns turbines within the dam. The water's kinetic energy becomes kinetic energy in the turbines. The kinetic energy of the turning turbines is converted into electrical energy by electrical generators.
4. Electrical energy is transported away from the dam through wires. The electrical energy is converted into many different forms of energy and is used in many different ways. For example, at a concert or a play, electrical energy is converted into light and heat by lighting systems and into sound energy by sound systems.

As you can see, several energy conversions occur in order to produce a usable form of energy—potential energy becomes kinetic energy, and kinetic energy becomes electrical energy.

The Hoover Dam produces a large amount of electrical energy for California, Nevada, and Arizona.

Other sources of useful energy begin with electromagnetic energy from the Sun. In fact, almost all of the energy on Earth began as electromagnetic energy from the Sun. This energy can be converted into many other forms of energy. Plants convert the electromagnetic energy of sunlight into chemical energy as they grow. This energy, stored by plants hundreds of millions of years ago, is the energy found in fossil fuels, such as petroleum, coal, and natural gas.

The chemical energy in fossil fuels is converted into other forms of energy for specific uses. In power plants, people burn coal to convert its chemical energy into electrical energy. In homes, people burn natural gas to convert its chemical energy into heat that warms them and cooks their food. In car engines, people burn gasoline, which is made from petroleum, to convert its chemical energy into kinetic energy.

One important difference between fossil fuels and sources of energy like the water held behind a dam, is that fossil fuels cannot be replaced once they are used up. The energy of moving water, by contrast, is renewable as long as the river behind the dam flows.

CHECK YOUR READING How can potential energy be changed into a usable form of energy?

DIFFERENTIATE INSTRUCTION

More Reading Support

B How did almost all of Earth's energy begin? *as electromagnetic energy from the Sun*

C What kind of energy do fossil fuels contain? *chemical energy*

Below Level Have students make a flow chart that follows the path of energy from the Sun to plants to fossil fuels to a coal-burning power plant to students' homes. Students should label each stage of energy conversion with the forms of energy.

Converting Energy
Energy is often converted from one form to another in order to meet everyday needs.
1 Water held behind the dam has potential energy.
2 Potential energy is converted to kinetic energy when the water moves through the tunnel.
3 Kinetic energy is used to turn turbines. This mechanical energy is converted into electrical energy by generators.
4 Electrical energy is transmitted through wires, and then converted into many other forms of energy.
Potential Energy to Kinetic Energy
Kinetic Energy to Electrical Energy

The potential energy of water behind the dam becomes the kinetic energy of moving water.

The kinetic energy of turning turbines becomes electrical energy in these generators.

READING VISUALS How many different energy conversions are described in this diagram?

Teach from Visuals

To help students interpret the diagrams of energy conversions in a hydroelectric power plant, ask:

- Why does the water behind the dam have potential energy? *Because of its position, it has potential energy due to gravity.*
- What is the function of the moving water? *It turns the blades of turbines, which power the generators.*

The visual "Converting Energy" is available as T22 in the Unit Transparency Book.

Metacognitive Strategy

Ask students to discuss whether or not they find it easier to understand the conversion of energy from one form to another if they study a large diagram. What changes would they make in the diagram on this page to make it more useful?

Real World Example

The Hoover Dam does more than provide electric power to a large portion of the southwestern United States. It also prevents annual spring flooding by the lower Colorado River, provides water for irrigation, and helps form an artificial lake (Lake Mead). During dam construction, the Colorado River was diverted through four concrete tunnels. At one time, Hoover Dam, which was completed in 1935, was the largest hydroelectric plant in the world.

Ongoing Assessment

READING VISUALS *Answer: three; potential to kinetic, kinetic (mechanical) to electrical, electrical to many other forms*

DIFFERENTIATE INSTRUCTION

Advanced Students could extend their knowledge of a hydroelectric dam by thinking about why a dam is needed. Ask: Why can't water simply be removed from a river and made to turn turbines? Ask them to design experiments using water and a pin wheel to observe that there must be a difference in elevation because the energy to turn the turbines comes from falling water.

R Challenge and Extension, p. 163

Address Misconceptions

IDENTIFY Ask: Is Earth running out of energy? If students answer yes, they may be confusing the idea of the "conservation of energy" with the meaning of conservation as preservation or protection.

CORRECT Explain that an energy source is a fuel that can provide energy. Energy sources are not conserved according to the law of conservation of energy and can be used up, but the energy itself is converted rather than lost. Have students make a list of energy sources and discuss those that are in danger of being used up.

REASSESS Ask: What happens to the energy produced by fuels? *It is converted into other forms of energy such as heat and light, and the total amount of energy remains the same.*

Technology Resources

Visit **ClassZone.com** for background on common student misconceptions.

MISCONCEPTION DATABASE

History of Science

Arguably the most famous equation of the 20th century is Albert Einstein's $E = mc^2$, which demonstrates that matter and energy are actually the same. Mass can be considered to be "solidified energy." Because the equation states that energy is equal to mass multiplied by the speed of light squared, a very small mass is equal to a very large amount of energy.

Ongoing Assessment

Restate the law of conservation of energy.

Ask students to restate the law of conservation of energy in their own words. *Sample answer: Energy is converted to other forms, not used up or created.*

Answer: Energy can neither be created nor destroyed.

Energy is always conserved.

When you observe energy conversions in your daily life, it may seem that energy constantly disappears. After all, if you give a soccer ball kinetic energy by kicking it along the ground, it will roll for a while but eventually stop. Consider what might have happened to the ball's kinetic energy.

As the ball rolls, it rubs against the ground. Some kinetic energy changes into heat as a result of friction. Some of the ball's energy also changes into sound energy that you can hear as the ball moves. Although the ball loses kinetic energy, the overall amount of energy in the universe does not decrease. The photograph below shows how the soccer ball's kinetic energy decreases.

The soccer ball's kinetic energy decreases as that energy is changed into sound energy and heat.

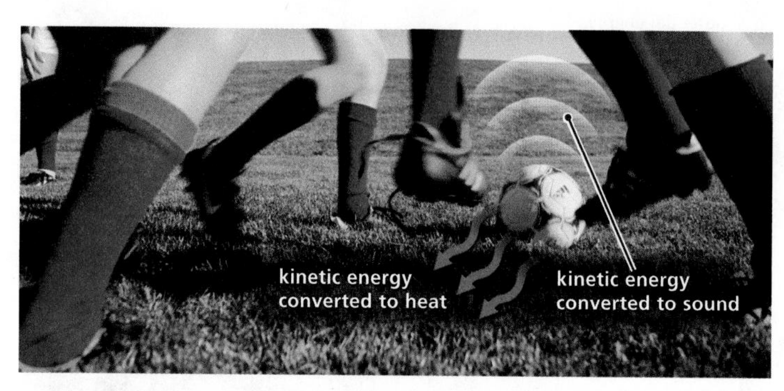

READING TIP

Conservation refers to a total that does not change.

In the soccer ball example, the ball loses energy, but this energy is transferred to other parts of the universe. Energy is conserved. The **law of conservation of energy** states that energy can neither be created nor destroyed. Conservation of energy is called a law because this rule is true in all known cases. Although in many instances it may appear that energy is gained or lost, it is really only changed in form.

CHECK YOUR READING Explain what is meant by the law of conservation of energy.

Conservation of energy is a balance of energy in the universe. When a soccer ball is kicked, a certain amount of energy is transferred by the kick. The ball gains an equal amount of energy, mostly in the form of kinetic energy. However, the ball's kinetic energy decreases as some of that energy is converted into sound energy and heat from the friction between the ball and the ground.

E

According to the law of conservation of energy, the amount of energy that a soccer player gives to the ball by kicking it is equal to the energy the ball gains. The energy the ball loses, in turn, is equal to the amount of energy that is transferred to the universe as sound energy and heat as the ball slows down.

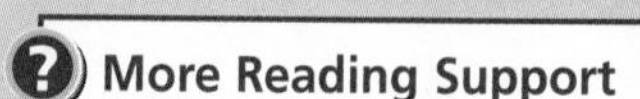

DIFFERENTIATE INSTRUCTION

More Reading Support

D Why is conservation of energy a law? *It is true in all known cases.*

E How does a ball's kinetic energy decrease? *It changes to heat and sound.*

Alternative Assessment Have students think of examples to explain the law of conservation of energy. For example, when food is warmed in a microwave oven, the thermal energy of the food increases through energy conversions. Electrical energy enters the microwave oven, in which it is converted into other forms of energy including electromagnetic energy (microwaves). The microwaves transfer energy to the food, but the total amount of energy is not changed.

Energy conversions may produce unwanted forms of energy.

When energy changes forms, the total amount of energy is conserved. However, the amount of useful energy is almost always less than the total amount of energy. For example, consider the energy used by an electric fan. The amount of electrical energy used is greater than the kinetic energy of the moving fan blades. Because energy is always conserved, some of the electrical energy flowing into the fan's motor is obviously changed into unusable or unwanted forms.

Some electrical energy is converted into unwanted sound energy.

Some electrical energy is converted into kinetic energy of the fan blades.

Some electrical energy is converted into unwanted heat.

The fan converts a significant portion of the electrical energy into the kinetic energy of the fan blades. At the same time, some electrical energy changes into heat in the fan's motor. If the fan shakes, some of the electrical energy is being turned into unwanted kinetic energy. The more efficiently the fan uses electrical energy, though, the more energy will be transformed into kinetic energy that moves the air.

Energy efficiency is a measurement of usable energy after an energy conversion. You may be familiar with energy-efficient household appliances. These appliances convert a greater percentage of energy into the desired form than inefficient ones. The more energy-efficient a fan is, the more electrical energy it turns into kinetic energy in the moving blades. Less electrical energy is needed to operate appliances that are energy efficient.

What does it mean when an energy conversion is efficient?

KEY CONCEPTS

1. Describe an energy conversion you have observed in your own life.
2. Explain the law of conservation of energy in your own words.
3. Give an example of an energy conversion that produces unwanted forms of energy.

CRITICAL THINKING

4. **Synthesize** Suppose you are jumping on a trampoline. Describe the conversions that occur between kinetic energy and potential energy.
5. **Infer** Look at the ski jumper on page 79. Has all of his potential energy likely been changed into kinetic energy at the moment he lands? Explain.

CHALLENGE

6. **Communicate** Draw and label a diagram that shows at least three different energy conversions that might occur when a light bulb is turned on.

ANSWERS

1. Sample answer: a flashlight, chemical energy becomes electrical energy, which becomes visible light.

2. Energy can change forms, but the overall amount of energy is constant.

3. Sample answer: Incandescent bulbs release heat in addition to light.

4. Potential energy is greatest at the highest point. Potential energy converts into kinetic energy until hitting the trampoline. Kinetic energy converts to potential energy in the trampoline to the lowest point, and is converted to kinetic energy back into the air.

5. No; some energy changed into heat and sound.

6. chemical energy to mechanical energy when turning a switch; mechanical energy to electrical energy; electrical energy to light and heat in the bulb

Ongoing Assessment

Understand that energy conversions may be inefficient.

Ask: Which energy conversion is more efficient—a soccer ball kicked along a smooth sidewalk or a soccer ball kicked in a field of tall weeds? Why? *the sidewalk, because less energy will be changed into heat by friction*

CHECK YOUR READING *Answer: Most of the energy that enters an energy conversion changes into the desired form of energy.*

Reinforce (the BIG idea)

Have students relate the section to the Big Idea.

Reinforcing Key Concepts, p. 164

Assess

Section 3.2 Quiz, p. 43

Reteach

Remind students about the law of conservation of energy and the concept of conversions between potential and kinetic energy. A roller coaster is a good example of these concepts. Draw a simple roller coaster on the board. Make sure each hill is lower than the previous hill. Ask students to identify points of substantial potential energy (at the top of hills) and kinetic energy (at the bottom of hills). Ask: Why must each hill be lower than the previous hill? *Kinetic energy decreases due to friction. If two adjacent hills were the same height, the coaster would not be able to make it to the top of the second hill because it would not have enough kinetic energy.*

Technology Resources

Have students visit **ClassZone.com** for reteaching of Key Concepts.

CONTENT REVIEW

CONTENT REVIEW CD-ROM

CHAPTER INVESTIGATION

Focus

PURPOSE To investigate the amount of energy stored in different types of food

OVERVIEW Students will construct an apparatus to trap the energy released from different food samples when they burn. Students will collect data and calculate the amount of energy each food contained. They will find the following:

- The water in the can traps the energy that is released by the burning food.
- The temperature increase in the water is greater when burning foods high in fats than when burning foods high in carbohydrates.
- Fats contain more energy than carbohydrates.

Lab Preparation

- Students can bring many of the materials, such as aluminum cans, from home.
- Punch holes in the cans before class to save time.
- Prior to the investigation, have students read through the investigation and prepare their data tables. You may wish to copy and distribute datasheets and rubrics.

UNIT RESOURCE BOOK, pp. 183–191

SCIENCE TOOLKIT, F14

Lab Management

- Warn students not to eat the food samples.

SAFETY Advise students with long hair to tie it back. Students wearing long, loose sleeves should roll them up. Desks should be cleared of all non-essential and flammable materials.

Teaching with Technology

Have students use a temperature probe to record temperature changes.

CHAPTER INVESTIGATION

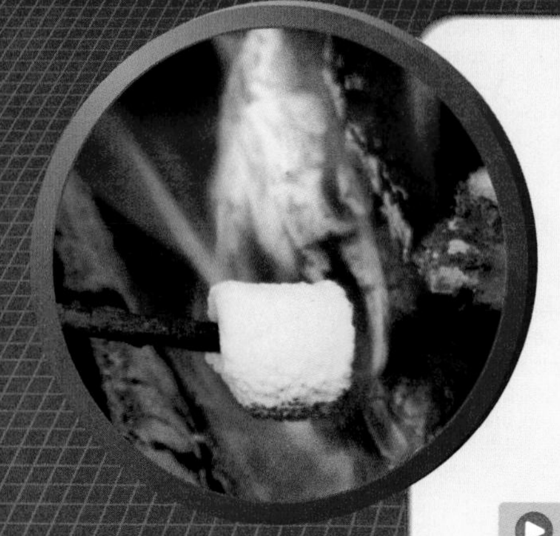

Energy Conversions

OVERVIEW AND PURPOSE All foods contain stored chemical energy, but some foods contain more chemical energy than others. People need this chemical energy for all of their activities. The amount of chemical energy stored in foods like marshmallows can be measured by burning the foods. In this investigation, you will

- construct an apparatus to investigate the amount of energy in samples of food
- calculate the amount of energy released when the foods are burned

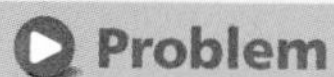

Problem

How much energy is stored in different types of food?

Hypothesize

Write a hypothesis to explain which type of food contains a greater amount of chemical energy. Your hypothesis should take the form of an "If . . . , then . . . , because . . ." statement.

MATERIALS
- can opener
- empty aluminum can
- dowel rod
- tap water
- graduated cylinder
- ring stand with ring
- thermometer
- aluminum pie plate
- aluminum foil
- large paper clip
- cork
- modeling clay
- crouton
- caramel rice cake
- balance
- wooden matches

Procedure

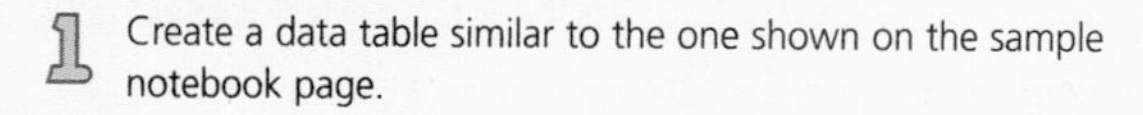

1. Create a data table similar to the one shown on the sample notebook page.

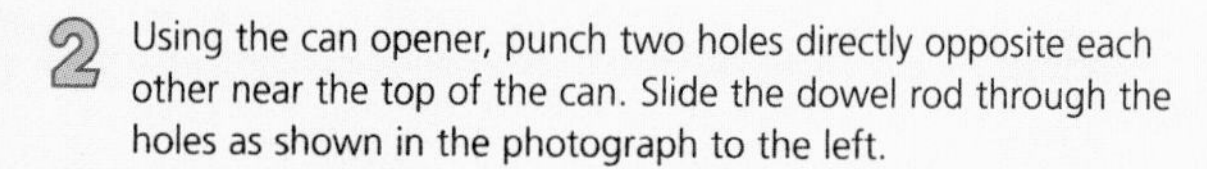

2. Using the can opener, punch two holes directly opposite each other near the top of the can. Slide the dowel rod through the holes as shown in the photograph to the left.
3. Measure 50 mL of water with a graduated cylinder, and pour the water into the can. Record the mass of the water. (**Hint:** 1 mL of water = 1 gram)
4. Rest the ends of the dowel rod on the ring in the ring stand to hold the can in the air. Carefully place the thermometer in the can. Measure and record the initial temperature (T1) of the water in the can.
5. Make a collar of aluminum foil around the bottom of the can as shown. Leave enough room to insert the burner platform and food sample.

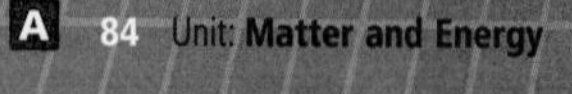

INVESTIGATION RESOURCES

CHAPTER INVESTIGATION, Energy Conversions
- Level A, pp. 183–186
- Level B, pp. 187–190
- Level C, p. 191

Advanced students should complete Levels B & C.

Writing a Lab Report, D12–13

Technology Resources

Customize this student lab as needed or look for an alternative. Print rubrics to assess student lab reports.

Lab Generator CD-ROM

6. Construct the burner platform as follows: Open up the paper clip. Push the straightened end into a cork, and push the bottom of the cork into the clay. Push the burner onto the pie plate so it will not move. Put the pie plate under the ring.

step 6

7. Find and record the mass of the crouton. Place the crouton on the flattened end of the burner platform. Adjust the height of the ring so the bottom of the can is about 4 cm above the crouton.

8. Use a match to ignite the crouton. Allow the crouton to burn completely. Measure and record the final temperature (T2) of the water.

9. Empty the water from the can and repeat steps 3–8 with a caramel rice cake. The mass of the rice cake should equal the mass of the crouton.

Observe and Analyze

Write It Up

1. **RECORD OBSERVATIONS** Make sure to record all measurements in the data table.
2. **CALCULATE** Find the energy released from the food samples by following the next two steps.

   Calculate and record the change in temperature.
   change in temperature = T2 – T1

   Calculate and record the energy released in calories. One calorie is the energy needed to raise the temperature of 1 g of water by 1°C.
   energy released = (mass of water • change in temperature • 1 cal/g°C)
3. **GRAPH** Make a bar graph showing the number of calories in each food sample. Which type of food contains a greater amount of chemical energy?

Conclude

1. **INTERPRET** Answer the question posed in the problem.
2. **INFER** Did your results support your hypothesis? Explain.
3. **EVALUATE** What happens to any energy released by the burning food that is not captured by the water? How could you change the setup for a more accurate measurement?
4. **APPLY** Find out how much fat and carbohydrate the different foods contain. Explain the relationship between this information and the number of calories in the foods.

INVESTIGATE Further

CHALLENGE The Calories listed in foods are equal to 1000 calories (1 kilocalorie). Calculate the amount of energy in your food samples in terms of Calories per gram of food (Calories/g). Using a balance, find the mass of any ash that remains after burning the food. Subtract that mass from the original mass of the sample to calculate mass burned. Divide total calories by mass burned, then divide that value by 1000 to find Calories/g. Compare your results to those given on the product labels.

Energy Conversions
Problem How much energy is stored in different types of food?
Hypothesize
Observe and Analyze
Table 1. Energy in Food

	Sample 1	Sample 2
Mass of water (g)		
Initial water temp. (T1) (°C)		
Final water temp. (T2) (°C)		
Mass of food (g)		
Change in temp. (T2 - T1) (°C)		
Energy released (mass•change in temp.•cal/g°C)		

Conclude

Observe and Analyze

1. *Sample data: Rice snack (Sample 1): mass of water 50 g; T1 20°C; T2 24°C; Crouton (Sample 2): mass of water 50 g; T1 20°C; T2 32°C. Mass of food will depend on the size of the sample.*
2. *For the sample data, the rice snack produced 200 calories and the crouton 600 calories.*
3. *See students' graphs; croutons*

Conclude

1. *The amount of energy stored in food depends on the nutrients in it. Fats store more energy than carbohydrates.*
2. *Answers will vary depending on the initial hypothesis.*
3. *It went into the air. A container that is completely closed and insulated to prevent the loss of heat would improve accuracy.*
4. *Fats contain more calories than carbohydrates, so the food that contains more fat should release more energy when it is burned.*

INVESTIGATE Further

CHALLENGE Students will need to keep the ashes from their food samples to find how much the mass changed. Students' results will probably vary from the values given on product labels because the equipment used is crude, contributing to a high percent of error.

Post-Lab Discussion

- Ask: Why was it important that the same amount of water be used in each trial? *A greater mass of water would heat up more slowly with a given amount of energy.*
- Ask: What energy conversions occurred in this lab? Where was energy transferred to? *Chemical energy from the food molecules changed to thermal energy (heat), electromagnetic energy (light), and sound energy when the food was burned. Some of the thermal energy was captured by the water. Additional thermal energy was lost to the air and to the equipment.*

3.3 FOCUS

Set Learning Goals

Students will

- Summarize how technology can improve energy conversions.
- Evaluate advantages and disadvantages of different types of energy conversions.
- Recognize how technology can improve the use of natural resources.
- Experiment to observe how the collection of solar energy is affected by the color of a solar collector.

3-Minute Warm-Up

Display Transparency 21 or copy this exercise on the board:

Match each definition with the correct term.

Definitions

1. energy you have when you are running *b*
2. energy you have while standing still on a diving board *d*
3. measure of useable energy after an energy conversion *e*

Terms

a. conservation of energy
b. kinetic energy
c. sound energy
d. potential energy
e. energy efficiency

 3-Minute Warm-Up, p. T21

3.3 MOTIVATE

EXPLORE Solar Cells

PURPOSE To explore the size of a solar cell needed to provide electrical energy for a solar calculator

TIP *10 min.* The calculator must not have a battery backup.

WHAT DO YOU THINK? *Answers will depend on the calculator and light conditions. A large solar cell would keep the calculator working under relatively poor lighting conditions.*

KEY CONCEPT

Technology improves the ways people use energy.

BEFORE, you learned

- Energy can change forms
- When energy changes forms, the overall amount of energy remains the same
- Energy conversions usually produce unwanted forms of energy

NOW, you will learn

- How technology can improve energy conversions
- About advantages and disadvantages of different types of energy conversions
- How technology can improve the use of natural resources

VOCABULARY

solar cell p. 88

EXPLORE Solar Cells

Why does a solar calculator need a large solar cell?

PROCEDURE

1. Measure the area of the calculator's solar cell. (**Hint:** area = length • width)
2. Turn the calculator on. Make sure that there is enough light for the calculator to work.
3. Gradually cover the solar cell with the index card. Observe the calculator's display as you cover more of the cell.
4. Measure the uncovered area of the solar cell when the calculator no longer works.

MATERIALS

- solar calculator without backup battery
- ruler
- index card

WHAT DO YOU THINK?

- How much of the solar cell is needed to keep the calculator working?
- Why might a solar calculator have a solar cell that is larger than necessary?

MIND MAP
Use a mind map to take notes about technology that improves energy conversions.

Technology improves energy conversions.

In many common energy conversions, most of the wasted energy is released as heat. One example is the common incandescent light bulb. Amazingly, only about 5 percent of the electrical energy that enters an incandescent light bulb is converted into light. That means that 95 percent of the electrical energy turns into unwanted forms of energy. Most is released as heat and ends up in the form of thermal energy in the surrounding air. To decrease this amount of wasted energy, scientists have investigated several more efficient types of lights.

RESOURCES FOR DIFFERENTIATED INSTRUCTION

Below Level

UNIT RESOURCE BOOK
- Reading Study Guide A, pp. 167–168
- Decoding Support, p. 180

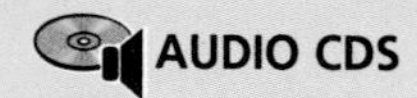 **AUDIO CDS**

Advanced

UNIT RESOURCE BOOK
- Challenge and Extension, p. 173
- Challenge Reading, pp. 176–177

English Learners

UNIT RESOURCE BOOK
Spanish Reading Study Guide, pp. 171–172

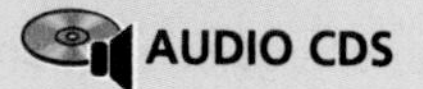 **AUDIO CDS**
- Audio Readings in Spanish
- Audio Readings (English)

Efficient Lights

Research to replace light bulbs with a more energy-efficient source of light has resulted in the light-emitting diode, or LED. LEDs have the advantage of converting almost all of the electrical energy they use into light.

The first LEDs were not nearly as bright as typical light bulbs, but over time scientists and engineers have been able to produce brighter LEDs. LEDs have many uses, including television remote controls, computer displays, outdoor signs, giant video boards in stadiums, and traffic signals. LEDs are also used to transmit information through fiber optic cables that connect home audio and visual systems.

CHECK YOUR READING How are LEDs more efficient than incandescent lights?

LEDs that produce infrared light are used in remote controls.

Efficient Cars

Another common but inefficient energy conversion is the burning of gasoline in cars. A large percentage of gasoline's chemical energy is not converted into the car's kinetic energy. Some of the kinetic energy is then wasted as heat from the car's engine, tires, and brakes. Here, too, efficiency can be improved through advances in technology.

Fuel injectors, common in cars since the 1980s, have improved the efficiency of engines. These devices carefully monitor and control the amount of gasoline that is fed into a car's engine. This precise control of fuel provides a significant increase in the distance a car can travel on a tank of gasoline. More recently, hybrid cars have been developed. These cars use both gasoline and electrical energy from batteries. These cars are very fuel efficient. Even better, some of the kinetic energy lost during braking in hybrid cars is used to generate electrical energy to recharge the car's batteries.

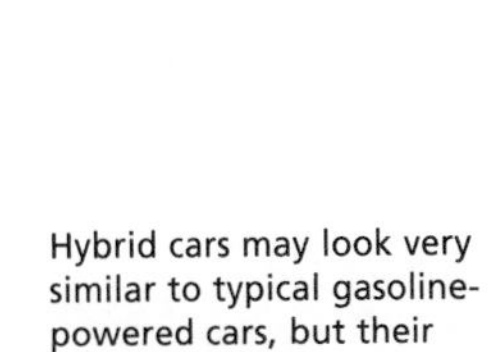

Hybrid cars may look very similar to typical gasoline-powered cars, but their engines are different.

3.3 INSTRUCT

Teach Difficult Concepts

Students may think that some energy conversions are 100% efficient. Have students hammer a nail into a board. They should immediately feel the nail head, which should be warm. Ask students to write a short description of the energy conversions that took place and an explanation of why the nail head became warm. *Chemical energy in muscles converted to mechanical energy. Some mechanical energy was converted into heat due to friction.*

History of Science

The first visible LED, a red light, was developed in the late 1960s. LEDs that produce yellow, orange, green, blue, and white light were later developed. LED technology is known as electroluminescence and uses semiconductors such as gallium arsenide. Different semiconductors have different properties, so they can be used to produce LEDs with different characteristics.

Ongoing Assessment

Summarize how technology can improve energy conversions.

Ask: What is one aspect of an energy conversion that technology tries to improve? *efficiency*

CHECK YOUR READING *Answer: They convert almost all of the electrical energy that enters them into light.*

DIFFERENTIATE INSTRUCTION

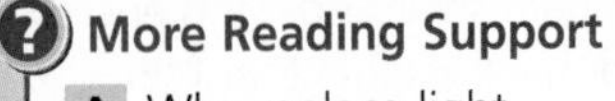

More Reading Support

A Why replace light bulbs with LEDs? *LEDs are more efficient*

B How is a large amount of gasoline's energy wasted? *as heat*

English Learners The English language uses many words and phrases in ways that make little sense when taken literally. Examples from this section include "ends up" (p. 86) and "over time" (p. 90 in the Check Your Reading question). Help students by pointing out such phrases and tell them not to read such prepositions literally. Encourage them to use the context of the sentence to decipher the phrases' meanings.

Real World Example

In July 2003, teams of university students from the United States and Canada raced the solar-powered cars they had spent months designing and building. The race, called the American Solar Challenge, is sponsored by the U.S. Department of Energy. The route follows Route 66 across the Great Plains, the Rocky Mountains, and the southwestern desert, from Chicago to southern California. In the 2003 race, speeds of 75 miles per hour were recorded.

Teacher Demo

Connect a set of solar cells to a small electric motor that drives a propeller. Have students predict what will happen when the apparatus is placed under a light source. This will work outdoors on a bright, sunny day or in the classroom with a bright lamp. Ask: where does the energy come from and where does it go? *Electromagnetic energy is converted into mechanical energy in the propeller.*

Teach from Visuals

To help students interpret the visual of the solar car, ask:

- Where would the solar cells have to be on the car? *on the top exterior*
- Why might such a large area of the car be covered with solar cells? *to capture as much light as possible because solar cells are inefficient*
- What would happen to the car on a rainy day? *The car would not run unless it had a backup energy supply.*

Develop Critical Thinking

COMPARE Have students compare the conversion of solar energy in a solar car and in the leaves of a plant. *In a solar car, solar energy is converted to electrical energy. In a plant, solar energy is converted to chemical energy during photosynthesis.*

Ongoing Assessment

unlimited supplies, no harmful waste products

Technology improves the use of energy resources.

Much of the energy used on Earth comes from fossil fuels such as coal, petroleum, and natural gas. However, the supply of fossil fuels is limited. So, scientists and engineers are exploring the use of several alternative energy sources. Today, for example, both solar energy and wind energy are used on a small scale to generate electrical energy.

Solar energy and wind energy have several advantages compared to fossil fuels. Their supply is not limited, and they do not produce the same harmful waste products that fossil fuels do. However, there are also many obstacles that must be overcome before solar energy and wind energy, among other alternative energy sources, are as widely used as fossil fuels.

What are the advantages of solar energy and wind energy as compared to fossil fuels?

Solar Energy

Observe how solar cells produce electricity.

Solar cells are important in today's solar energy technology. Modern **solar cells** are made of several layers of light-sensitive materials, which convert sunlight directly into electrical energy. Solar cells provide the electrical energy for such things as satellites in orbit around Earth, hand-held calculators, and, as shown below, experimental cars.

Solar Energy for Electricity

Solar cells can produce electricity to run a car.

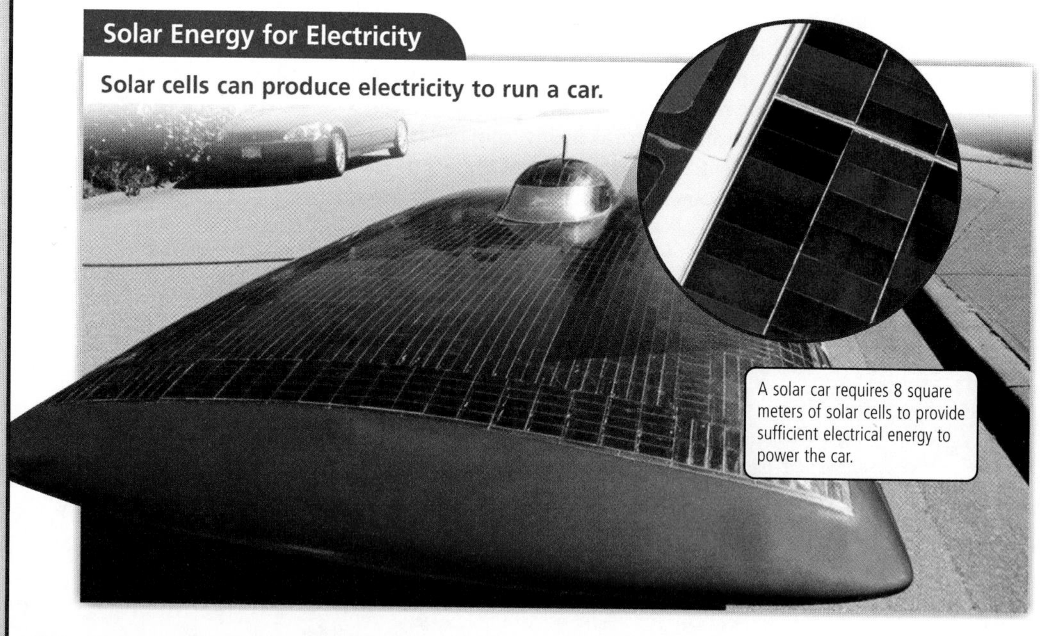
A solar car requires 8 square meters of solar cells to provide sufficient electrical energy to power the car.

DIFFERENTIATE INSTRUCTION

More Reading Support

C What energy conversion occurs in solar cells? *Sunlight is converted to electrical energy.*

Below Level Have students compare solar and wind energy to fossil fuel energy. Students should make a table that lists each type of energy source and some of the characteristics of each type. Have students mark an "X" in the box on the table to indicate that the energy has that characteristic. Discuss with students how the energy sources are alike and how they are different.

Solar cells produce electrical energy quietly and cleanly. However, they are not yet commonly used because the materials used to make them are very expensive. What's more, solar cells are not very efficient in producing electrical energy. Large numbers of solar cells produce only a relatively small amount of electrical energy. Typical solar cells convert only about 12 to 15 percent of the sunlight that reaches them into electrical energy. However, solar cells presently being developed could have efficiencies close to 40 percent.

Solar energy can be used in homes to provide heat and electrical energy.

In addition to converting the Sun's light into electrical energy, people have used the Sun's heat. In ancient Rome, glass was used to trap solar energy indoors so that plants could be grown in the winter. Today heat from the Sun is still used to grow plants in greenhouses and to warm buildings. The photograph above shows a house that uses both solar cells and the Sun's heat. The solar cells on the roof provide electrical energy, and the large windows help to trap the Sun's warmth. In fact, some solar power systems also use that warmth to produce additional electrical energy.

CHECK YOUR READING How can energy from the Sun be used by people?

INVESTIGATE Solar Energy

What improves the collection of solar energy?

PROCEDURE

1. Cover the top of one cup with white plastic, and cover the top of the other cup with black plastic. Secure the plastic with a rubber band.
2. Use the scissors to make a small hole in the center of each cup's plastic lid. Insert a thermometer through each opening.
3. Place the cups in direct sunlight, and record their temperatures every minute for 10 minutes.

WHAT DO YOU THINK?

- Which cup showed a greater temperature change? Why do you think this happened?
- Make a line graph of your results to show the change in temperature in each cup.

CHALLENGE Try the experiment again, using aluminum foil instead of white plastic. How do the results differ with the aluminum foil? Why might this be the case?

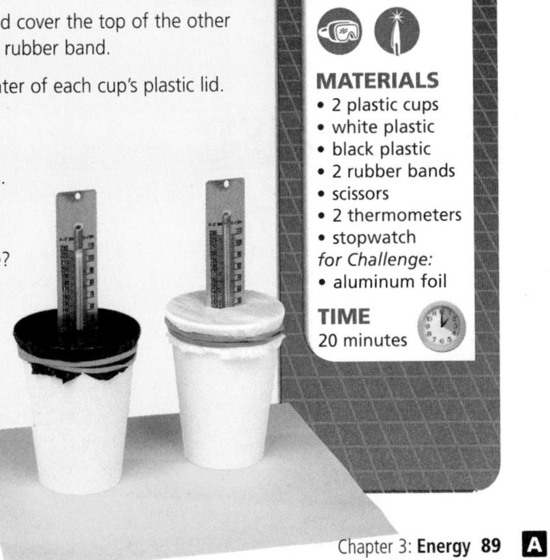

SKILL FOCUS
Observing

MATERIALS
- 2 plastic cups
- white plastic
- black plastic
- 2 rubber bands
- scissors
- 2 thermometers
- stopwatch

for Challenge:
- aluminum foil

TIME
20 minutes

DIFFERENTIATE INSTRUCTION

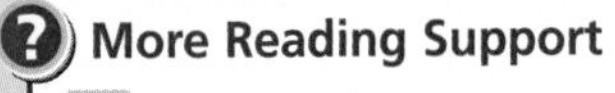

More Reading Support

D Why is solar power not commonly used? *It is inefficient and expensive.*

E How can the Sun's heat be used? *It can be trapped to provide heat.*

Alternative Assessment Have small groups orally present their data and graphs for peer review. Then lead a discussion to reach a consensus for each question.

Advanced Have interested students conduct research and make a poster presenting limitations and real sites or locations of one of the alternative energy sources discussed on pp. 88–90.

- Challenge and Extension, p. 173
- Challenge Reading, pp. 176–177

INVESTIGATE Solar Energy

PURPOSE To observe how the collection of solar energy is affected by the color of the collector

TIPS *20 min.*

- The plastics should be the same thickness; garbage bags can be used.
- Be careful not to rip the plastic and aluminum; a faulty seal may lead to inaccurate data.

WHAT DO YOU THINK? *The cup with black plastic; the black plastic absorbs more sunlight. Check students' graphs.*

CHALLENGE *Aluminum foil will reflect sunlight but absorb more energy than the white plastic; a cup covered with aluminum foil should show a temperature increase between black plastic and white plastic.*

Datasheet, Solar Energy, p. 174

Technology Resources

Customize this student lab as needed or look for an alternative. Print rubrics to assess student lab reports.

Lab Generator CD-ROM

Teaching with Technology

If probeware is available, students can use a temperature probe in place of a thermometer. Graphing calculators can be used to graph students' data.

EXPLORE the BIG idea

Revisit "Hot Dog!" on p. 69. Have students explain why the hot dog cooked.

Ongoing Assessment

Evaluate advantages and disadvantages in energy conversions.

Ask: What is an advantage to heating a home with solar power? What is a disadvantage? *advantage: nonpolluting; disadvantage: materials are expensive*

Sample answer: to produce electrical energy; to warm buildings

Ongoing Assessment

Recognize how technology can improve the use of natural resources.

Ask: How can technology help conserve natural resources? *by making energy conversion more efficient*

CHECK YOUR READING *Answer: Wind has been used to propel ships, pump water, and grind grains. Now it is also used to produce electricity at windfarms.*

PHOTO CAPTION Answer: to capture as much energy from the wind as possible

Integrate the Sciences

A new branch of meteorology involves offshore wind power. Wind and wave characteristics are studied at sea for the purpose of developing offshore wind farms for electric power. Denmark is a global leader in this field because of strong political support and the mandatory purchase of wind power output by utilities.

Reinforce the BIG idea

Have students relate the section to the Big Idea.

Reinforcing Key Concepts, p. 175

3.3 ASSESS & RETEACH

Assess

Section 3.3 Quiz, p. 44

Reteach

Have students brainstorm the best locations (in terms of weather conditions) for using solar and wind energy. Use a weather map for reference. Then ask students to list the limitations of these alternative energy sources and brainstorm how technology might help.

Technology Resources

Have students visit **ClassZone.com** for reteaching of Key Concepts.

CONTENT REVIEW

CONTENT REVIEW CD-ROM

INFER Why might so many windmills be needed at a windfarm?

Wind Energy

For many centuries, people have used the kinetic energy of wind to sail ships, and, by using windmills, to grind grain and pump water. More recently, windmills have been used to generate electrical energy. In the early 1900s, for example, windmills were already being used to produce electrical energy in rural areas of the United States.

Like the technological advances in the use of solar energy, advances in capturing and using wind energy have helped to improve its efficiency and usefulness. One way to better capture the wind's energy has been to build huge windmill farms in areas that receive a consistent amount of wind. Windmill farms are found in several states, including Kansas, California, and New York. Other methods of more efficiently capturing wind energy include the use of specially shaped windmill blades that are made of new, more flexible materials.

RESOURCE CENTER CLASSZONE.COM

Find out more about alternative energy sources.

CHECK YOUR READING How has the use of wind energy changed over time?

3.3 Review

KEY CONCEPTS

1. Provide an example of a common technology that does not efficiently convert energy. Explain.
2. Describe two ways in which hybrid cars are more energy-efficient than gasoline-powered cars.
3. List two advantages and two disadvantages of solar power.

CRITICAL THINKING

4. **Compare and Contrast** How are LEDs similar to incandescent light bulbs? How are they different?
5. **Synthesize** What are two ways in which the Sun's energy can be captured and used? How can both be used in a home?

CHALLENGE

6. **Draw Conclusions** Satellites orbiting Earth use solar cells as their source of electrical energy. Why are solar cells ideal energy sources for satellites?

ANSWERS

1. Sample answer: incandescent lights—heat instead of light

2. They use less fuel and convert heat from brakes into electrical energy.

3. advantages: clean, unlimited supplies; disadvantages: inefficient, expensive materials

4. Both produce light. LEDs are more efficient and have several applications. Incandescent lights are brighter than most LEDs.

5. solar cells for electricity; trapping heat to warm the building

6. They receive more constant sunlight.

MATH in SCIENCE
MATH TUTORIAL
CLASSZONE.COM
Click on Math Tutorial for more help with rates.

SKILL: USING FORMULAS

Cool Efficiency

Energy efficiency is important because energy supplies are limited. The energy used by appliances such as air conditioners is measured in British thermal units, or BTUs. One BTU warms one pound of water by 1°F. The cooling ability of an air conditioner is measured by the number of BTUs it can cool. Consider the amount of BTUs that an air conditioning system must cool in an ice rink.

An air conditioner typically has an energy efficiency ratio (EER) rating. The EER measures how efficiently a cooling system operates when the outdoor temperature is 95°F. The EER is the ratio of cooling per hour to the amount of electricity used, which is measured in watts. The higher the EER, the more energy efficient the air conditioner is.

$$\text{EER} = \frac{\text{BTUs cooled/hr}}{\text{watts used}}$$

Example

Suppose an air conditioner uses 750 watts of electricity to cool 6000 BTUs per hour at 95°F. Calculate the air conditioner's EER.

(1) Use the formula above to calculate the EER.

$$\text{EER} = \frac{\text{BTUs cooled/hr}}{\text{watts used}}$$

(2) Enter the known values into the formula.

$$\text{EER} = \frac{\text{6000 BTUs cooled/hr}}{\text{750 watts used}}$$

(3) Solve the formula for the unknown value.

$$\text{EER} = \frac{\text{6000 BTUs cooled/hr}}{\text{750 watts used}} = 8$$

ANSWER EER = 8 BTUs cooled/hr per watt used

Answer the following questions.

1. What is the EER of a cooling system that uses 500 watts of electricity to cool 6000 BTUs per hour at 95°F?
2. What is the EER of a cooling system that uses 1500 watts of electricity to cool 12,000 BTUs per hour at 95°F?
3. Which air conditioner in the two questions above is more efficient?

CHALLENGE How many BTUs per hour would an air conditioner cool at 95°F if it had an EER of 10 and used 1200 watts of electricity?

Indoor ice rinks require cooling systems that can keep ice frozen even when the outdoor temperature is 95°F.

ANSWERS

1. 6000 BTUs/500 watts = 12 BTUs cooled/hr per watt used

2. 12,000 BTUs/1500 watts = 8 BTUs cooled/hr per watt used

3. the one in item 1

CHALLENGE EER = 10 = x BTUs/1200 watts
x = 10 (1200) = 12,000

MATH IN SCIENCE

Math Skills Practice for Science

Set Learning Goal

To use a mathematical formula to calculate the energy efficiency of cooling systems

Present the Science

A cooling system with a high EER rating is more efficient in that it uses less electricity to deliver a given amount of cooling power. Similarly, heating systems with high EER ratings more efficiently use energy to produce a given amount of warming power. An EER of 14 is close to the upper limit of current technology. Efficient window air conditioners have EER ratings of about 10. Central air conditioning systems have EER ratings of around 12.

Develop Algebra Skills

- Remind students that the numerator of a fraction is above the line and the denominator is below.
- To help students use the formula for calculating EER, remind them that the line in the formula means "divided by." Thus, the numerator, BTUs, per hour, should be divided by the denominator, watts.
- The units of an EER rating are BTUs per hour per watt. The rating measures output of the air conditioner divided by input.

Close

Ask: What other types of appliances could have an EER rating? *refrigerators, freezers, furnaces, hot water heaters*

- Math Support, p. 181
- Math Practice, p. 182

Technology Resources

Students can visit **ClassZone.com** for practice with rates.

MATH TUTORIAL

BACK TO

the BIG idea

Make two sets of index cards with each card naming a form of energy. One set should be labeled "Original energy form" and the other "Converted to." Call on students to choose one card from each set and give an example of that energy conversion.

KEY CONCEPTS SUMMARY

SECTION 3.1

Ask: What do all forms of energy have in common? *They all have the ability to produce a change.*

Ask: How does changing a skater's mass and speed affect his or her kinetic energy? *Increasing mass or speed increases kinetic energy. Decreasing either decreases kinetic energy.*

SECTION 3.2

Ask: When will kinetic energy decrease and potential energy increase? *after taking off from the jump as the skier is rising in the air*

Ask: Is kinetic energy destroyed when that occurs? *No, energy is never created nor destroyed*

SECTION 3.3

Ask: Why is it important to use technology to better use natural resources? *Many sources of energy are limited in supply, and those that aren't limited are used inefficiently.*

Review Concepts

- Big Idea Flow Chart, p. T17
- Chapter Outline, pp. T23–T24

3 Chapter Review

the BIG idea

Energy has different forms, but it is always conserved.

KEY CONCEPTS SUMMARY

3.1 Energy exists in different forms.

- Energy is the ability to cause a change.
- Different forms of energy produce changes in different ways.
- Kinetic energy depends on mass and speed.

Potential energy depends on position and chemical composition.

VOCABULARY
energy p. 72
kinetic energy p. 74
potential energy p. 75

3.2 Energy can change forms but is never lost.

- Energy often needs to be transformed in order to produce a useful form of energy.
- The law of conservation of energy states that energy is never created or destroyed.

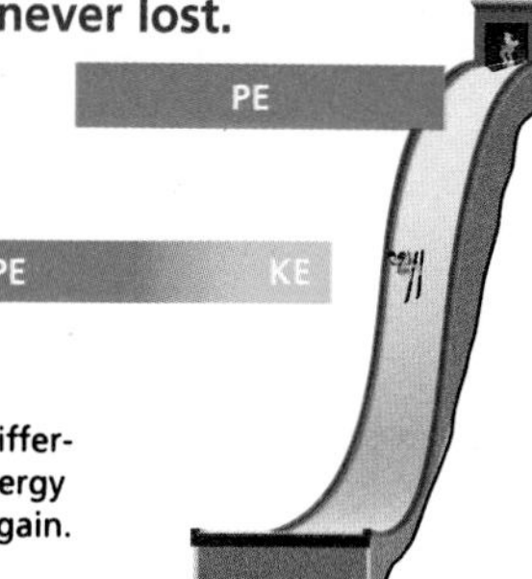

Energy can be transformed in many different ways, including from potential energy (PE) to kinetic energy (KE) and back again.

VOCABULARY
law of conservation of energy p. 82
energy efficiency p. 83

3.3 Technology improves the ways people use energy.

- Different forms of technology are being developed and used to improve the efficiency of energy conversions.
- Solar cells convert sunlight into electrical energy.

New solar cells convert light into electrical energy more efficiently than those in the past.

VOCABULARY
solar cell p. 88

Technology Resources

Have students visit **ClassZone.com** or use the CD-ROM for a cumulative review of concepts.

CONTENT REVIEW

CONTENT REVIEW CD-ROM

Engage students in a whole-class interactive review of Key Concepts. Edit content as you wish.

POWER PRESENTATIONS

Reviewing Vocabulary

Review vocabulary terms by making a four square diagram for each term as shown in the example below. Include a definition, characteristics, examples from real life, and, if possible, nonexamples of the term.

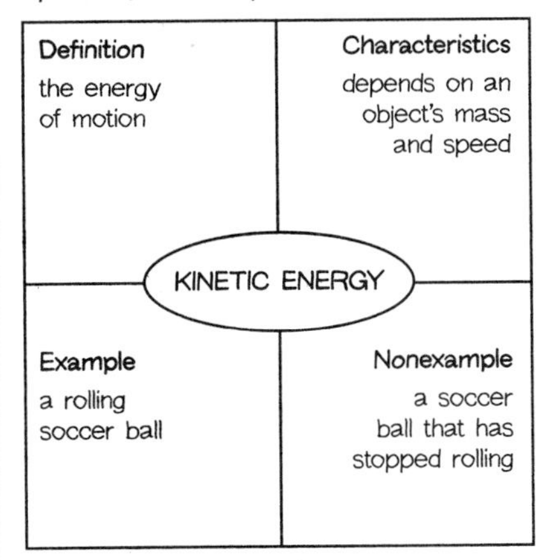

Definition
the energy of motion
Characteristics
depends on an object's mass and speed
KINETIC ENERGY
Example
a rolling soccer ball
Nonexample
a soccer ball that has stopped rolling

1. energy
2. potential energy
3. conservation of energy
4. energy efficiency

Reviewing Key Concepts

Multiple Choice *Choose the letter of the best answer.*

5. All forms of energy are a combination of
   a. mechanical energy and chemical energy
   b. chemical energy and kinetic energy
   c. potential energy and thermal energy
   d. potential energy and kinetic energy

6. Which type of energy is transmitted by vibrations of air?
   a. electromagnetic
   b. sound
   c. nuclear
   d. chemical

7. When energy is converted from one form to another, what is usually produced?
   a. chemical energy
   b. gravity
   c. heat
   d. potential energy

8. An object's kinetic energy is determined by its
   a. position and composition
   b. speed and position
   c. mass and speed
   d. height and width

9. Which of the following is a conversion from chemical energy to mechanical energy?
   a. a dark light bulb starting to glow
   b. food being heated in an oven
   c. a ball rolling down a hill
   d. a person lifting a weight

10. An energy-efficient electric fan converts a large portion of the electrical energy that enters it into
    a. an unwanted form of energy
    b. kinetic energy of the fan blades
    c. thermal energy in the fan's motor
    d. sound energy in the fan's motor

11. The energy in wind used to generate electricity is
    a. chemical energy
    b. sound energy
    c. potential energy
    d. kinetic energy

12. A skier on a hill has potential energy due to
    a. speed
    b. energy efficiency
    c. compression
    d. position

Short Answer *Write a short answer to each question.*

13. Explain how the law of conservation of energy might apply to an energy conversion that you observe in your daily life.

14. Describe a situation in which chemical energy is converted into mechanical energy. Explain each step of the energy conversion process.

ASSESSMENT RESOURCES

UNIT ASSESSMENT BOOK
- Chapter Test A, pp. 45–48
- Chapter Test B, pp. 49–52
- Chapter Test C, pp. 53–56
- Alternative Assessment, pp. 57–58

SPANISH ASSESSMENT BOOK
Spanish Chapter Test, pp. 221–224

Technology Resources

Edit test items and answer choices.

Test Generator CD-ROM

Visit **ClassZone.com** to extend test practice.

Test Practice

Reviewing Vocabulary

1. *Sample answer—Definition: the ability to cause a change; Characteristics: is conserved; Examples: sound, chemical, mechanical; Nonexample: matter*
2. *Sample answer—Definition: stored energy; Characteristics: depends on position or chemical composition; Example: skier at top of ramp; Nonexample: skier at bottom of hill due to position*
3. *Sample answer—Definition: energy cannot be created or destroyed; Characteristics: energy may seem to disappear when it changes form, but it is transferred or changed; Example: air gains thermal energy (heat) transferred during a conversion*
4. *Sample answer—Definition: useful energy after an energy conversion; Characteristics: high efficiency—less energy is needed to run the appliance; Example: Inefficient light bulbs convert electrical energy to light and a large amount of heat.*

Reviewing Key Concepts

5. *d*
6. *b*
7. *c*
8. *c*
9. *d*
10. *b*
11. *d*
12. *d*
13. *Answers should suggest that the amount of energy entering a process may appear to decrease during an energy conversion but actually changes into a different energy form.*
14. *Answers could involve a process that uses chemical energy in a person's muscles to move or lift an object.*

Thinking Critically

15. potential energy: 5 (or 1) because height is greatest; kinetic energy: 4 (or 2), because speed is greatest

16. 2; the skater's height above the ground begins to increase.

17. No, because the conversions between potential and kinetic energy are not 100% efficient, so some energy is being converted into unwanted forms.

18. Energy will be converted into heat and sound due to the friction of the skates against the ramp.

19. 5—all potential; 4—mostly kinetic, some potential; 3—all kinetic; 2—mostly kinetic, some potential; 1—all potential

20. Both convert sunlight into another energy form. Plants turn light into chemical energy; solar cells convert light into electrical energy.

21. Both are relatively inefficient and require a steady input of light or wind. New materials are being developed, and large groups of solar cells or windmills are used to capture as much energy as possible.

22. the machine that does not get hot, because less energy is being transformed through unwanted heat and more is being turned into the desired form of energy

23. 40; Energy must be conserved.

24. Answers might include striking a match; check students' diagrams.

the BIG idea

25. Answers will vary; check students' answers.

26. Answers will vary; check students' answers.

UNIT PROJECTS

Students should have begun designing their models or presentations by this time. Remind them to continue researching as needed. Encourage them to try different solutions to the problems they encounter.

R Unit Projects, pp. 5–10

Thinking Critically

The illustrations below show an in-line skater on a ramp. Use the illustrations to answer the next five questions.

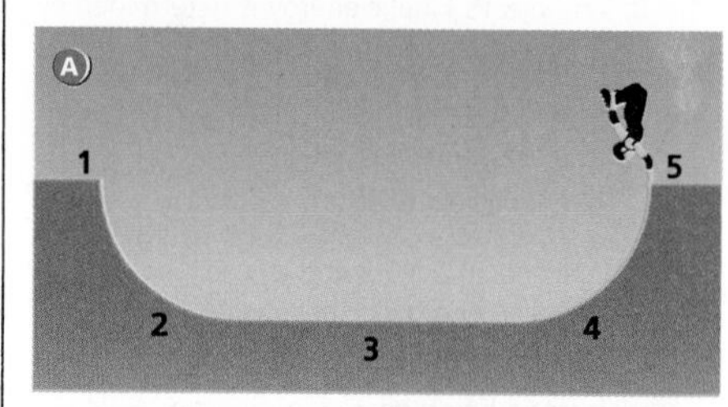

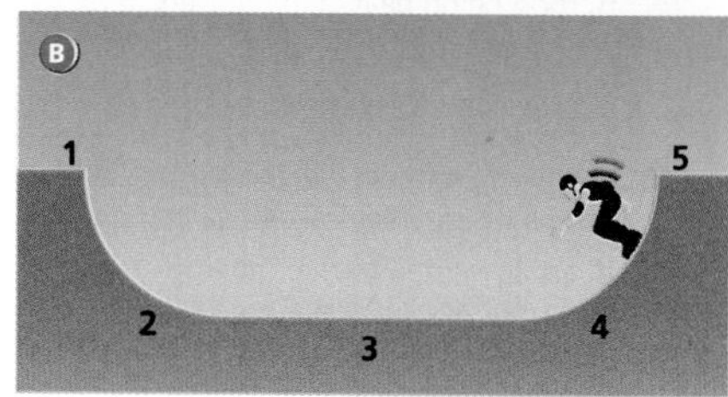

15. OBSERVE At what point in the illustrations would the skater have the most potential energy? the most kinetic energy? Explain.

16. SYNTHESIZE At what point in illustration B will the skater's kinetic energy begin to be changed back into potential energy? Explain.

17. INFER When the skater's kinetic energy is changed back into potential energy, will this amount of potential energy likely be equal to the skater's potential energy in illustration A? Why or why not?

18. PREDICT Describe how energy may appear to decrease in the example shown above. What energy conversions that produce unwanted forms of energy are occurring? Explain.

19. SYNTHESIZE Draw colored bars that might represent the potential energy and kinetic energy of the skater at each of the five labeled points on illustration A. Explain why you drew the bars the way you did. (**Hint:** See the illustration on p. 79)

20. SYNTHESIZE How are plants and solar cells similar? How are the ways in which they capture sunlight and convert it into other forms of energy different? Explain.

21. COMPARE Explain how energy sources such as solar energy and wind energy have similar problems that must be overcome. How have scientists tried to address these problems?

22. INFER Suppose that one air conditioner becomes very hot when it is working but another air conditioner does not. Which air conditioner is more energy efficient? How can you tell?

23. DRAW CONCLUSIONS Suppose a vacuum cleaner uses 100 units of electrical energy. All of this energy is converted into heat and sound (from the motor), and into the kinetic energy of air being pulled into the vacuum cleaner. 60 units of electrical energy are converted into heat and sound. How much electrical energy is converted into the desired form of energy? How do you know?

24. COMMUNICATE Describe a process in which energy changes forms at least twice. Draw and label a diagram that shows these energy conversions.

the BIG idea

25. APPLY Look again at the photograph on pages 68 and 69 and consider the opening question. How might your answer have changed after reading the chapter?

26. COMMUNICATE How have your ideas about energy and its different forms changed after reading the chapter? Provide an example from your life to describe how you would have thought of energy compared to how you might think about it now.

UNIT PROJECTS

If you need to do an experiment for your unit project, gather the materials. Be sure to allow enough time to observe results before the project is due.

MONITOR AND RETEACH

If students have trouble applying the concept of conservation of energy in items 13, 17, 18, and 23, explain that when one object gains energy, another object must have lost the same amount of energy. Divide a group of 25 paper clips into several smaller groups that represent different forms of energy. Explain that the paper clips can be moved from group to group, but that the total number will still equal 25.

Students may benefit from summarizing one or more sections of the chapter.

 Summarizing the Chapter, pp. 201–202

Standardized Test Practice

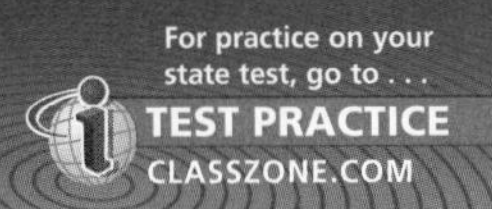

Interpreting Graphs

Study the graph below. Then answer the first five questions.

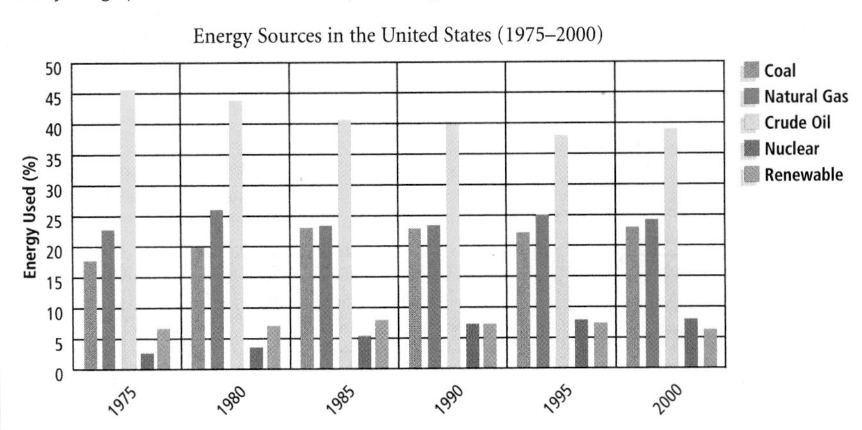

Source: U.S. Energy Information Administration, Monthly Energy Review (June 2003)

1. In which year did the greatest percentage of energy used in the United States come from crude oil?

a. 1975 **b.** 1980 **c.** 1995 **d.** 2000

2. What three sources of energy account for about 80 percent of all energy used in each year shown?

a. coal, crude oil, nuclear
b. natural gas, crude oil, renewable
c. coal, natural gas, crude oil
d. crude oil, nuclear, renewable

3. Which sources of energy show a greater percentage in 2000 as compared to 1980?

a. crude oil, renewable **b.** natural gas, crude oil **c.** coal, nuclear **d.** coal, crude oil

4. The use of which energy source tended to decrease between 1975 and 2000?

a. coal **b.** natural gas **c.** crude oil **d.** nuclear

5. The use of which source of energy has steadily increased between 1975 and 2000?

a. coal **b.** crude oil **c.** nuclear **d.** renewable

Extended Response

Answer the questions in detail. Include some of the terms from the word box on the right. Underline each term you use in your answers.

chemical energy	potential energy
electrical energy	sound energy
mechanical energy	thermal energy

6. When gasoline is burned in a moving car's engine, which forms of energy are being used? Which forms of energy are produced? Explain.

7. Name two appliances in your home that you believe are inefficient. What about them indicates that they may be inefficient?

METACOGNITIVE ACTIVITY

Have students answer the following questions in their **Science Notebook:**

1. What misconceptions about energy did you have? Have these misconceptions been corrected?
2. Which topics in this chapter would you like to learn more about?
3. What have you learned from your research on your Unit Project?

Interpreting Graphs

1. a *2. c* *3. c* *4. c* *5. c*

Extended Response

6. RUBRIC

4 points for a response that correctly answers both questions, gives an accurate explanation, and uses the following terms correctly:

- chemical energy
- mechanical energy
- sound energy
- thermal energy
- potential energy

Sample: The chemical energy of gasoline is potential energy that is converted into several other forms of energy listed when a car is driven. For example, sound energy and thermal energy are produced in the car's engine and mechanical energy moves the car.

3 points correctly answers one of the questions, gives an accurate explanation, and uses the terms correctly
2 points correctly answers at least one question but fails to give an accurate explanation, and uses at least three terms correctly
1 point does not correctly answer either question and fails to give an accurate explanation, but uses at least three terms correctly

7. RUBRIC

4 points for a response that correctly answers both questions and gives two correct examples.

Sample: The refrigerator and air conditioner are inefficient. They produce unwanted heat and sound energy from electrical energy.

3 points correctly answers both questions and gives one correct example
2 points correctly answers both questions but does not give any correct examples
1 point correctly answers one question and gives one correct example

FOCUS

Set Learning Goals

Students will

- Observe how scientists created new theories of temperature and heat by building on earlier observations.
- Learn the characteristics of temperature and heat and how they are measured.
- Write a procedure for an experiment to test a specific method of calculating temperature.

National Science Education Standards

A.9.a–b, A.9.d–g, Understandings About Scientific Inquiry

E.6.a–c Understandings About Science and Technology

F.5.a–e, F.5.g Science and Technology in Society

G.1.a–b Science as a Human Endeavor

G.2.a Nature of Science

G.3.a–c History of Science

INSTRUCT

Point out to students that the top half of the timeline shows major events in the scientific study of temperature and heat and the years in which they occurred. The bottom half addresses the developments in technology based on the scientific discoveries in the top half. The gap between 320 B.C. and A.D. 1600 represents a block of time that has been omitted.

Teach from Visuals

350 B.C. To help students better understand the ancient Greek theory of matter, have them review the diagram of the basic qualities of matter. Ask students to create a table of the four basic substances and their characteristics to illustrate which substances were thought to have which two qualities.

TIMELINES in Science

ABOUT TEMPERATURE AND HEAT

Most likely, the first fires early people saw were caused by lightning. Eventually, people realized that fire provided warmth and light, and they learned how to make it themselves. During the Stone Age 25,000 years ago, people used firewood to cook food as well as to warm and light their shelters. Wood was the first fuel.

This timeline shows a few of the many steps on the path toward understanding temperature and heat. Notice how the observations and ideas of previous thinkers sparked new theories by later scientists. The boxes below the timeline show how technology has led to new insights and to applications related to temperature and heat.

445 B.C.

Four Basic Substances Named

Greek philosopher Empedocles says that everything on Earth is made of some combination of four basic substances: earth, air, fire, and water. Different types of matter have different qualities depending on how they combine these substances.

350 B.C.

Aristotle Expands Theory of Matter

Greek philosopher Aristotle names four basic qualities of matter: dryness, wetness, hotness, and coldness. Each of the four basic substances has two of these qualities.

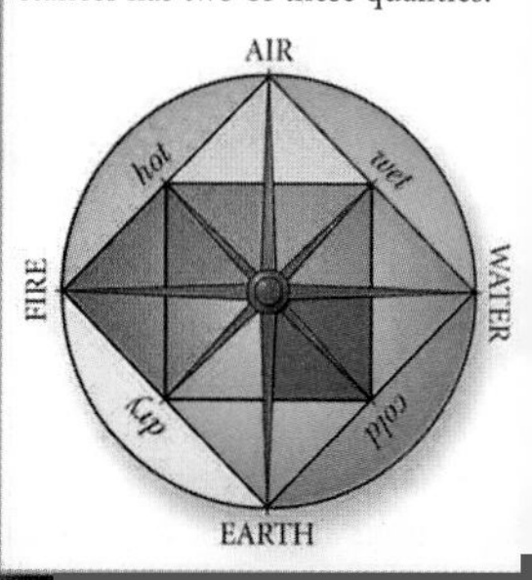

EVENTS

480 B.C. | 440 B.C. | 400 B.C. | 360 B.C. | 320 B.C.

APPLICATIONS AND TECHNOLOGY

People have been trying to understand and control heat since early times.

DIFFERENTIATE INSTRUCTION

Below Level For students who may have difficulty understanding how information is organized on a timeline, point out the dates on the center line. Explain how the dates become more recent when read from left to right. Show them the lines connecting the event boxes to the specific dates on the timeline. Discuss how timelines are a good way to show the order in which events happened.

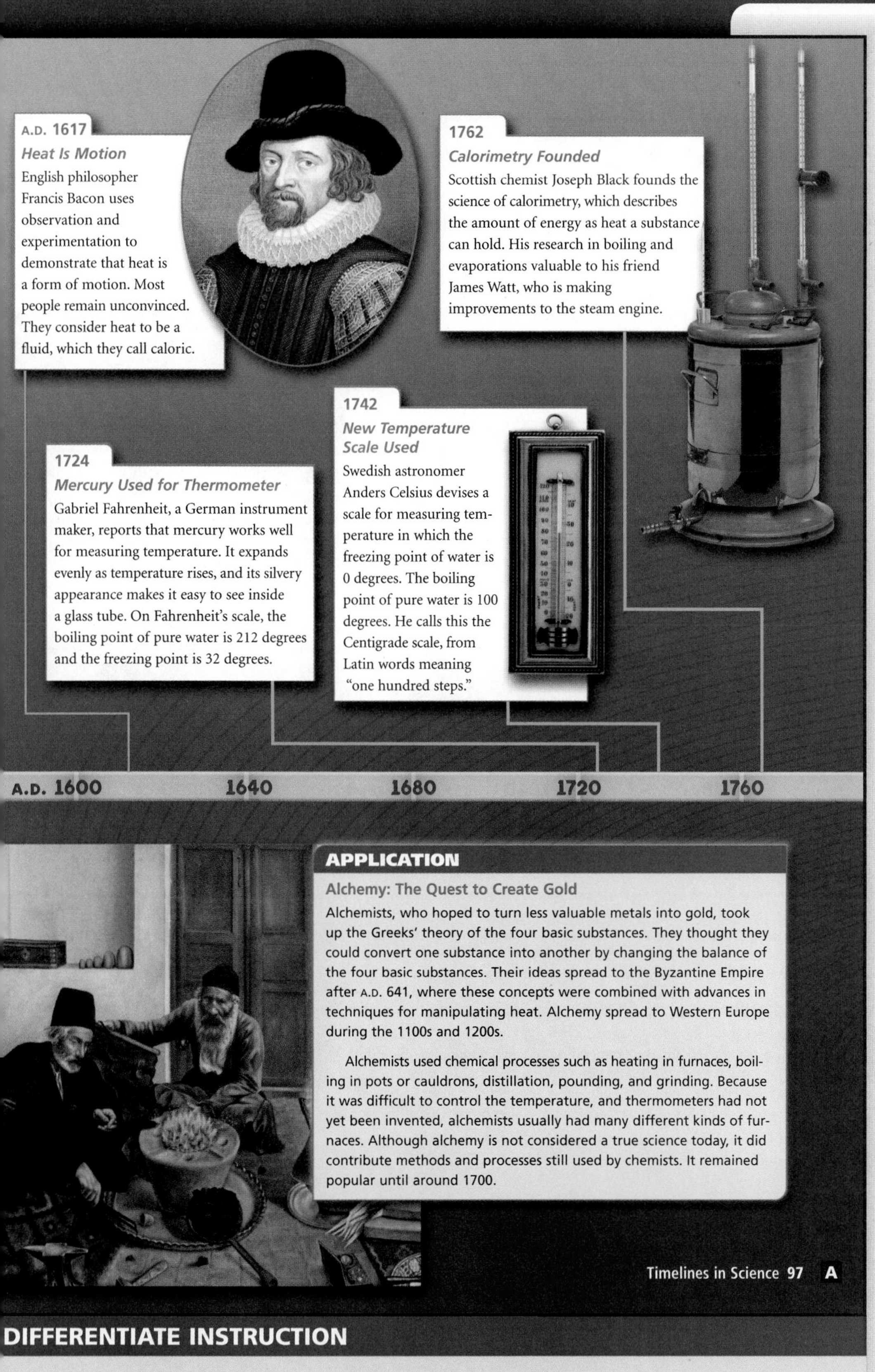

A.D. 1617
Heat Is Motion
English philosopher Francis Bacon uses observation and experimentation to demonstrate that heat is a form of motion. Most people remain unconvinced. They consider heat to be a fluid, which they call caloric.
1762
Calorimetry Founded
Scottish chemist Joseph Black founds the science of calorimetry, which describes the amount of energy as heat a substance can hold. His research in boiling and evaporations valuable to his friend James Watt, who is making improvements to the steam engine.
1742
New Temperature Scale Used
Swedish astronomer Anders Celsius devises a scale for measuring temperature in which the freezing point of water is 0 degrees. The boiling point of pure water is 100 degrees. He calls this the Centigrade scale, from Latin words meaning "one hundred steps."
1724
Mercury Used for Thermometer
Gabriel Fahrenheit, a German instrument maker, reports that mercury works well for measuring temperature. It expands evenly as temperature rises, and its silvery appearance makes it easy to see inside a glass tube. On Fahrenheit's scale, the boiling point of pure water is 212 degrees and the freezing point is 32 degrees.
A.D. 1600
1640
1680
1720
1760
APPLICATION
Alchemy: The Quest to Create Gold
Alchemists, who hoped to turn less valuable metals into gold, took up the Greeks' theory of the four basic substances. They thought they could convert one substance into another by changing the balance of the four basic substances. Their ideas spread to the Byzantine Empire after A.D. 641, where these concepts were combined with advances in techniques for manipulating heat. Alchemy spread to Western Europe during the 1100s and 1200s.
Alchemists used chemical processes such as heating in furnaces, boiling in pots or cauldrons, distillation, pounding, and grinding. Because it was difficult to control the temperature, and thermometers had not yet been invented, alchemists usually had many different kinds of furnaces. Although alchemy is not considered a true science today, it did contribute methods and processes still used by chemists. It remained popular until around 1700.
Timelines in Science 97 A

DIFFERENTIATE INSTRUCTION

Advanced Have students use the temperature conversion formula shown in the Mathematics Connection on this page to convert Fahrenheit to Celsius. Ask students what the formula for the conversion is. *°C = 5/9 (°F − 32)* Then ask students to convert 98.6°F to Celsius. *37°C*

Scientific Process

1724 Gabriel Fahrenheit improved on existing technology to measure temperature through his observations and investigations. He found that mercury was more accurate in measuring temperature than the alcohol thermometers of the time. Fahrenheit then developed the mercury thermometer, which we still use today.

Sharing Results

1762 Chemist Joseph Black experimented with the amount of heat a substance can hold. His results helped James Watt, who was working on improvements to the steam engine. Ask students what research Black was conducting that was critical to Watt's work. *boiling and evaporation*

Mathematics Connection

1742 The Celsius scale is the metric scale for measuring temperature. Most countries use the Celsius scale for everyday temperature measurement. Scientists also use this scale for their experiments. In the United States, the Fahrenheit scale is used to measure temperature. To convert a Celsius temperature to a Fahrenheit temperature, multiply the Celsius temperature by 9/5 and then add 32 to the result. Write the formula °F = 9/5(°C) + 32 on the board. Ask students to convert 35°C to Fahrenheit. *95°F*

Application

ALCHEMY While alchemy is no longer considered a true science, for some time it was a major source of chemical knowledge. Because alchemists experimented with turning metals such as lead into gold, they gained wide knowledge about chemical substances. The alchemist's workshop became the forerunner of the modern chemistry laboratory. Alchemists used tools such as funnels, beakers, and balances. Although alchemists failed to produce gold from other materials, their experimentation was successful in other ways. Ask students what impact alchemy had on science and technology. *Alchemy created many chemical processes and tools that are still used by chemists today.*

Scientific Process

Refer students to Francis Bacon's observations on page 97. Despite his experiments, people still believed that heat is fluid rather than a form of motion. Thompson's observations about friction and heat provided evidence that contradicted the leading hypothesis of the time. With evidence against the fluid theory of heat mounting, scientists began to consider seriously other ideas about the nature of heat.

Technology

VACUUM FLASK The reflective silver coating that a vacuum flask, or thermos, uses to keep fluids hot is the same type of technology that NASA used on the Mars rover. To keep heat from escaping out of the rover body and cold air from entering during landing, the outside of the rover's body was painted gold. This coating helps reduce energy that is spread outward from the rover's body. It also prevents the body from emitting heat energy into its cold surroundings.

Integrate the Sciences

One way in which clouds are formed is by convection, that is, when warm air rises. The Sun's heat causes Earth's water (from lakes, oceans, and rivers) to evaporate into the air. When that air is heated, it becomes less dense than the surrounding air. As a result, it rises. As the moist air continues to rise, it expands and becomes cooler. The water vapor in the air condenses and forms clouds.

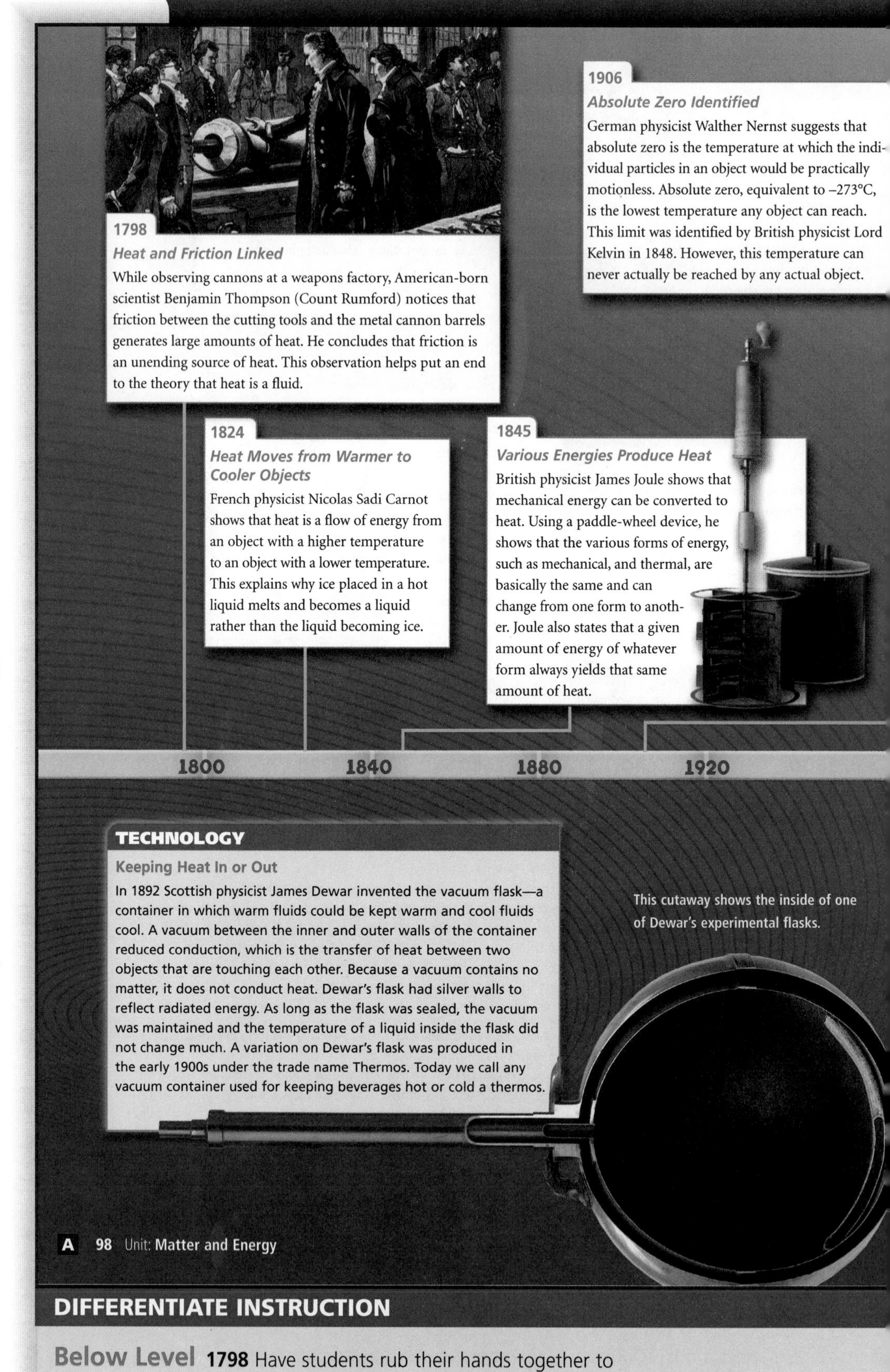

1798
Heat and Friction Linked
While observing cannons at a weapons factory, American-born scientist Benjamin Thompson (Count Rumford) notices that friction between the cutting tools and the metal cannon barrels generates large amounts of heat. He concludes that friction is an unending source of heat. This observation helps put an end to the theory that heat is a fluid.
1824
Heat Moves from Warmer to Cooler Objects
French physicist Nicolas Sadi Carnot shows that heat is a flow of energy from an object with a higher temperature to an object with a lower temperature. This explains why ice placed in a hot liquid melts and becomes a liquid rather than the liquid becoming ice.
1845
Various Energies Produce Heat
British physicist James Joule shows that mechanical energy can be converted to heat. Using a paddle-wheel device, he shows that the various forms of energy, such as mechanical, and thermal, are basically the same and can change from one form to another. Joule also states that a given amount of energy of whatever form always yields that same amount of heat.
1906
Absolute Zero Identified
German physicist Walther Nernst suggests that absolute zero is the temperature at which the individual particles in an object would be practically motionless. Absolute zero, equivalent to –273°C, is the lowest temperature any object can reach. This limit was identified by British physicist Lord Kelvin in 1848. However, this temperature can never actually be reached by any actual object.
1800
1840
1880
1920
TECHNOLOGY
Keeping Heat In or Out
In 1892 Scottish physicist James Dewar invented the vacuum flask—a container in which warm fluids could be kept warm and cool fluids cool. A vacuum between the inner and outer walls of the container reduced conduction, which is the transfer of heat between two objects that are touching each other. Because a vacuum contains no matter, it does not conduct heat. Dewar's flask had silver walls to reflect radiated energy. As long as the flask was sealed, the vacuum was maintained and the temperature of a liquid inside the flask did not change much. A variation on Dewar's flask was produced in the early 1900s under the trade name Thermos. Today we call any vacuum container used for keeping beverages hot or cold a thermos.
A 98 Unit: Matter and Energy

This cutaway shows the inside of one of Dewar's experimental flasks.

DIFFERENTIATE INSTRUCTION

Below Level **1798** Have students rub their hands together to demonstrate how friction produces heat.

2003

Wasps Stay Cool

Scientists in Israel have found evidence that some wasps have an internal air-conditioning system. Like a refrigerator, the wasp uses energy to stay cooler than the air around it. The energy may come from several sources, such as the energy generated by an electric current produced when the wasp's shell is exposed to sunlight. This ability to stay cool allows wasps to hunt for food even on very hot days.

RESOURCE CENTER
CLASSZONE.COM

Learn about current temperature and heat research.

1960 2000

APPLICATION

Using Thermal Energy from Ponds

Ponds can be used to store solar energy. The goal is to turn the solar energy into energy people can use. Salt must be added to the ponds, however, so that the water at the bottom is denser than the water at the top. This prevents thermal energy stored on the bottom from moving up to the surface, where it would be lost to the air through evaporation. A net on the surface helps prevent wind from mixing the water layers.

INTO THE FUTURE

As scientists are able to create colder and colder temperatures in the laboratory, they gain new insight into the scientific theories that explain temperature and heat. Advances in our knowledge of temperature and heat will lead to future applications.

- Scientists have developed a car that can run on hydrogen cooled into its liquid state. Before cars that run on this supercooled fuel become common, a system of refueling stations must be established.
- Understanding how some materials, such as silicon, conduct energy as heat may result in medical advances through better scanning and imaging technology.
- At temperatures approaching absolute zero (–273°C), a unique state of matter can be formed that is different from a solid, liquid, or gas. This rare state of matter could possibly be used to help produce extremely small circuits for use in miniature computers or other electronics.

ACTIVITIES

Design a Procedure

Many people claim that it is possible to determine the temperature by listening to the chirping of crickets. Crickets are sensitive to changes in air temperature and chirp more quickly when the temperature rises. To calculate the temperature in degrees Celsius, count the number of chirps in 7 seconds and add 5.

Write a procedure for an experiment that would test this claim. What factors would you consider testing? What range of temperatures would you test?

Writing About Science

Alchemy has fascinated people for centuries. Research its influence on both the technology and procedures of modern chemistry. Write a short report.

DIFFERENTIATE INSTRUCTION

Advanced Encourage students to trace the development of ideas from the ancient Greeks to the present. Students might create a visual or model that represents each new idea as building on the previous idea.

Application

USING THERMAL ENERGY FROM PONDS The idea of using ponds with a salt gradient to collect and store thermal energy was developed after natural examples of such ponds were discovered. The energy from these ponds can be used for applications such as purifying water and producing electricity. Ask students what some of the benefits of using solar-energy ponds might be. *provides clean, cost-effective electricity*

INTO THE FUTURE

Have students divide into small groups. Have each group come up with a list of possible inventions or ideas based on what they read about temperature and heat. An example might be a portable solar-powered DVD player. Then have each group prepare a presentation about their invention, describing what it is and how it works. Presentations might consist of bulletin-board displays, videos done as news segments, or oral reports with visual aids.

ACTIVITIES

Design a Procedure

Refer students to the steps of the scientific process as they write their procedure for the experiment. Remind them to include in their procedure ways to record findings clearly.

Writing About Science

Suggest that students look up the history of alchemy on the Internet or in the library. Some might focus on contributions of a specific culture, such as Egyptian, Chinese, Indian, or Islamic alchemists.

Technology Resources

Students can visit **ClassZone.com** for information about temperature and heat.

CHAPTER

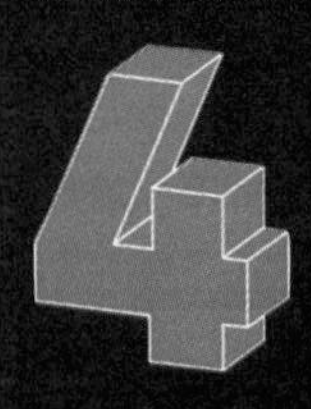

Temperature and Heat

Physical Science
UNIFYING PRINCIPLES

PRINCIPLE 1
Matter is made of particles too small to see.

PRINCIPLE 2
Matter changes form and moves from place to place.

PRINCIPLE 3
Energy changes from one form to another, but it cannot be created or destroyed.

PRINCIPLE 4
Physical forces affect the movement of all matter on Earth and throughout the universe.

Unit: Matter and Energy
BIG IDEAS

CHAPTER 1
Introduction to Matter
Everything that has mass and takes up space is matter.

CHAPTER 2
Properties of Matter
Matter has properties that can be changed by physical and chemical processes.

CHAPTER 3
Energy
Energy has different forms, but it is always conserved.

CHAPTER 4
Temperature and Heat
Heat is a flow of energy due to temperature differences.

CHAPTER 4
KEY CONCEPTS

SECTION 4.1

Temperature depends on particle movement.
1. All matter is made of moving particles.
2. Temperature can be measured.

SECTION 4.2

Energy flows from warmer to cooler objects.
1. Heat is different from temperature.
2. Some substances change temperature more easily than others.

SECTION 4.3

The transfer of energy as heat can be controlled.
1. Energy moves as heat in three ways.
2. Different materials are used to control the transfer of energy.

T The Big Idea Flow Chart is available on p. T25 in the **UNIT TRANSPARENCY BOOK.**

Previewing Content

SECTION

Temperature depends on particle movement. pp. 103–109

1. All matter is made of moving particles.

The **kinetic theory of matter** states that all the particles in matter are constantly in motion. As a result, all particles have kinetic energy. Particles in solids, liquids, and gases move differently.

- Particles in a solid vibrate in fixed positions but do not move past each other. Particles in a liquid are not tightly bound to each other, as in a solid, and slide past each other. Particles in a gas are separated by greater distances than those in a solid or liquid.
- All the particles in a substance are not moving at the same speed and can change speeds.
- **Temperature** is a measurement of the average kinetic energy of all particles in an object or location.

2. Temperature can be measured.

The two common temperature scales (Fahrenheit and Celsius) are based on the physical properties of pure water and are expressed in terms of degrees. A third temperature scale is the Kelvin scale, which is an absolute temperature scale. The zero point of the Kelvin scale is absolute zero, which is the complete absence of particle movement. Absolute zero is 0 K, which is equal to −273.15°C.

Thermometers measure temperature. Often, the physical property used to measure temperature is expansion or contraction. All gases, many liquids, and most solids expand when temperature increases.

- Thermometers filled with a liquid (alcohol or mercury) measure temperature through the uniform expansion or contraction of the liquid over a wide range of temperatures.
- Thermometers can also measure temperature through electrical resistance, infrared radiation, and the differential expansion of materials.

SECTION

Energy flows from warmer to cooler objects. pp. 110–115

1. Heat is different from temperature.

Heat, temperature, and thermal energy are closely related but not the same.

- Temperature is the average kinetic energy of particles in a substance or location.
- **Heat** is a flow of energy from an object or location at a higher temperature to an object or location at a lower temperature. The transfer of energy through heat continues as long as the temperature difference exists. When energy is transferred in this way, the thermal energy of both objects or locations changes.
- **Thermal energy** is the total kinetic energy of particles in a substance or location.

The most common units of heat measurement are the calorie and the joule.

- A **calorie** is the amount of energy needed to raise the temperature of 1 gram of water by 1° C.
- A Calorie with a capital *C*—the measure used with food and nutrition—is a kilocalorie, or 1000 calories.
- A **joule** is the standard scientific unit for measuring energy. One calorie is equal to 4.18 joules, so 4.18 joules of energy raises the temperature of 1 gram of water by 1°C.

2. Some substances change temperature more easily than others.

Each substance needs to absorb a different amount of energy in order for its temperature to increase. A substance's **specific heat** is the amount of energy that is required for 1 gram of that substance to increase in temperature by 1°C.

Any amount of a particular substance has the same specific heat. However, the more mass an object has, the more energy is required to produce an increase in its temperature and, conversely, the more energy must be released to produce a decrease in temperature.

Common Misconceptions

MISCONCEPTION DATABASE

CLASSZONE.COM Background on student misconceptions

CONSTANT MOTION OF PARTICLES Students may have difficulty understanding that all particles in matter are in constant motion. Particles in solids vibrate in place, particles in liquids slide past one another, and particles in gases move freely in all directions.

This misconception is addressed on p. 104.

DEFINITION OF TEMPERATURE Students may think that temperature is a measure of an object's heat. Temperature measures the average kinetic energy of the particles in an object.

This misconception is addressed in Teach Difficult Concepts on p. 105.

SECTION

The transfer of energy as heat can be controlled. pp. 116–123

1. Energy moves as heat in three ways.
Energy is transferred between objects or locations when there are temperature differences between them. Depending on both the medium and the objects themselves, energy can be transferred by conduction, convection, or radiation.

- **Conduction** is the process through which energy is transferred through physical contact. Particles of a warmer object collide with particles of a cooler object and transfer some of their energy to the cooler object. Materials that easily transfer energy are **conductors;** those that are poor conductors are called **insulators.**
- **Convection** is the process that transfers energy in gases and liquids. Differences in density between substances are produced by differences in temperature. A warmer region of gas or liquid is less dense than a cooler region, due to thermal expansion. The warmer, less dense gas or liquid is pushed up by cooler, denser gas or liquid that sinks in underneath. The cycle of convection accounts for currents in bodies of water and winds in the atmosphere.
- **Radiation** is energy that travels as electromagnetic waves, such as visible light, infrared light, and x-rays. All objects radiate at least a small amount of energy. Often, when radiation is absorbed by an object, the transfer of energy as heat occurs. Radiation differs from conduction and convection in that it can transfer energy through a vacuum.

2. Different materials are used to control the transfer of energy.
Materials are used for different purposes depending on whether they are good or poor conductors of energy. Many insulators contain or trap a layer of air, which is a poor conductor. Human-made insulators are similar to, and often based upon, insulators found in nature.

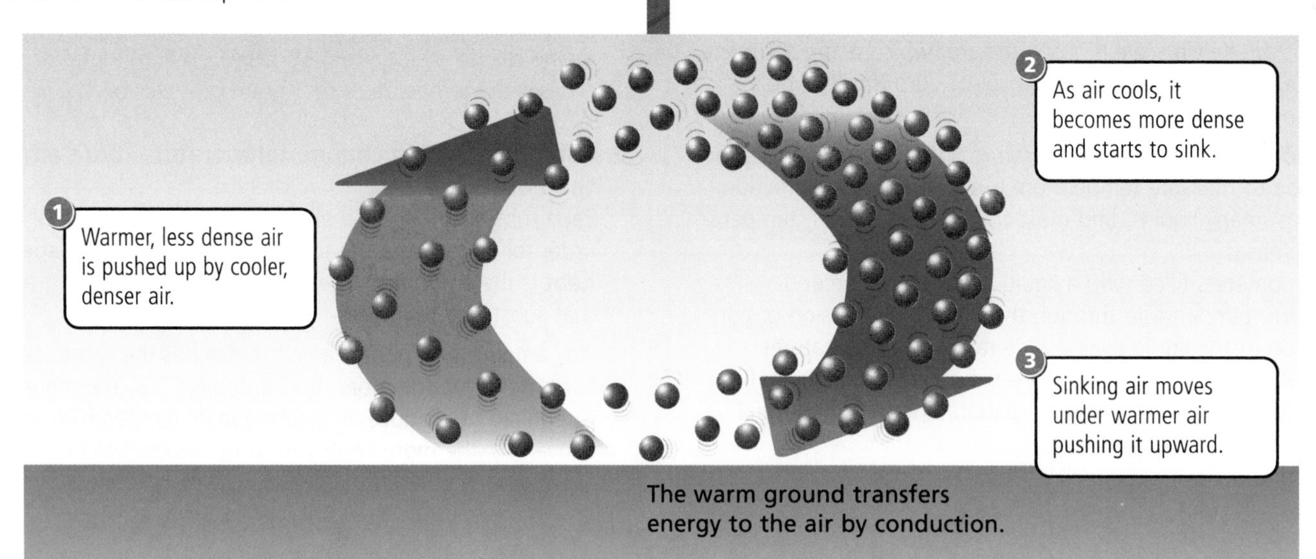

1
Warmer, less dense air is pushed up by cooler, denser air.
2
As air cools, it becomes more dense and starts to sink.
3
Sinking air moves under warmer air pushing it upward.
The warm ground transfers energy to the air by conduction.

Common Misconceptions

MISCONCEPTION DATABASE
CLASSZONE.COM Background on student misconceptions

HEAT Some students may think that heat is a fluid that literally flows from one object to another and that heat and cold are different, rather than being at opposite ends of an energy flow. Heat is the flow of energy from a warm substance to a cooler substance.

 This misconception is addressed on p. 111.

DIRECTION OF HEAT FLOW Because students have often heard that heat rises, they may think that heat only travels upward. Heat can transfer energy in all directions, depending upon the process that is involved.

 This misconception is addressed on p. 118.

Previewing Labs

Lab Generator CD-ROM
Edit these Pupil Edition labs and generate alternative labs.

EXPLORE the BIG idea

Moving Colors, p. 101 Students observe food coloring in hot and cold water to investigate motion and temperature.	**TIME** 10 minutes **MATERIALS** 2 plastic cups, hot and cold water, eyedropper, food coloring
Does It Chill? p. 101 Students investigate how soil acts as an insulator.	**TIME** 30 minutes **MATERIALS** outdoor thermometer, paper cup, freezer, soil, stopwatch
Internet Activity: Kinetic Theory, p. 101 Students observe the relationship between the kinetic energy of particles and temperature.	**TIME** 20 minutes **MATERIALS** computer with Internet access

SECTION 4.1

EXPLORE Temperature, p. 103 Students discover how a transfer of energy and increased motion produce an increase in temperature.	**TIME** 10 minutes **MATERIALS** large rubber band
INVESTIGATE Temperature Measurements, p. 107 Students make thermometers and observe how thermal expansion can be used to measure temperature.	**TIME** 30 minutes **MATERIALS** small plastic bottle, alcohol solution, food coloring, clear plastic straw, clay, bowl, ice water, hot tap water

SECTION 4.2

INVESTIGATE Heat Transfer, p. 112 Students investigate the specific heat of different materials by measuring a change in temperature.	**TIME** 30 minutes **MATERIALS** graduated cylinder, balance, room-temperature water, pennies, aluminum foil, 100 mL beaker, 3 plastic cups, hot tap water, thermometer, stopwatch

SECTION 4.3

EXPLORE Conduction, p. 116 Students observe the direction in which energy is transferred through direct contact between objects at different temperatures.	**TIME** 10 minutes **MATERIALS** 500 mL beaker, hot tap water, cold water, 2 thermometers, stopwatch, 200 mL beaker
CHAPTER INVESTIGATION Insulators, pp. 122–123 Students design and test an insulated bottle to slow a change in the temperature of water as compared to a noninsulated control bottle.	**TIME** 40 minutes **MATERIALS** 2 small plastic bottles, 2 thermometers, modeling clay, graduated cylinder, hot or cold tap water, foam packing peanuts, plastic wrap, aluminum foil, soil, sand, rubber bands, coffee can, beaker, stop-watch, graph paper

Additional INVESTIGATION, Observing Convection, A, B, & C, pp. 251–259; Teacher Instructions, pp. 262–263

Previewing Chapter Resources

	INTEGRATED TECHNOLOGY	LABS AND ACTIVITIES
CHAPTER 4 **Temperature and Heat**	 CLASSZONE.COM • eEdition Plus • EasyPlanner Plus • Misconception Database • Content Review • Test Practice • Simulations • Resource Centers • Internet Activity: Kinetic Theory • Math Tutorial  SCILINKS.ORG 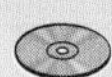 **CD-ROMS** • eEdition • EasyPlanner • Power Presentations • Content Review • Lab Generator • Test Generator 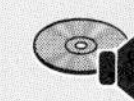 **AUDIO CDS** • Audio Readings • Audio Readings in Spanish	 EXPLORE the Big Idea, p. 101 • Moving Colors • Does It Chill? • Internet Activity: Kinetic Theory  **UNIT RESOURCE BOOK** Unit Projects, pp. 5–10  **Lab Generator CD-ROM** Generate customized labs.
SECTION  4.1 **Temperature depends on particle movement.** pp. 103–109 Time: 2 periods (1 block)  Lesson Plan, pp. 203–204	 • **RESOURCE CENTER,** Temperature and Temperature Scales • **MATH TUTORIAL**  **UNIT TRANSPARENCY BOOK** • Big Idea Flow Chart, p. T25 • Daily Vocabulary Scaffolding, p. T26 • Note-Taking Model, p. T27 • 3-Minute Warm-Up, p. T28	 • EXPLORE Temperature, p. 103 • INVESTIGATE Temperature Measurements, p. 107 • Math in Science, p. 109  **UNIT RESOURCE BOOK** • Datasheet, Temperature Measurements, p. 212 • Math Support, p. 240 • Math Practice, p. 241
SECTION   4.2 **Energy flows from warmer to cooler objects.** pp. 110–115 Time: 2 periods (1 block)  Lesson Plan, pp. 214–215	 **RESOURCE CENTER,** Thermal Energy  **UNIT TRANSPARENCY BOOK** • Daily Vocabulary Scaffolding, p. T26 • 3-Minute Warm-Up, p. T28 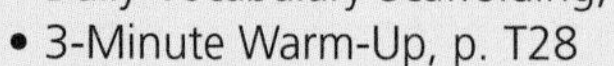	 • INVESTIGATE Heat Transfer, p. 112 • Science on the Job, p. 115  **UNIT RESOURCE BOOK** Datasheet, Heat Transfer, p. 223
SECTION  4.3 **The transfer of energy as heat can be controlled.** pp. 116–123 Time: 4 periods (2 blocks)  Lesson Plan, pp. 225–226	 **SIMULATION,** Conduction, Convection, or Radiation  **UNIT TRANSPARENCY BOOK** • Big Idea Flow Chart, p. T25 • Daily Vocabulary Scaffolding, p. T26 • 3-Minute Warm-Up, p. T29 • "Insulation" Visual, p. T30 • Chapter Outline, pp. T31–T32	 • EXPLORE Conduction, p. 116 • CHAPTER INVESTIGATION, Insulators, pp. 122–123  **UNIT RESOURCE BOOK** • CHAPTER INVESTIGATION, Insulators, A, B, & C, pp. 242–250 • Additional INVESTIGATION, Observing Convection, A, B, & C, pp. 251–259

KEY TO ICONS

 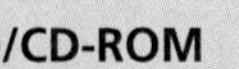 CD/CD-ROM

INTERNET

 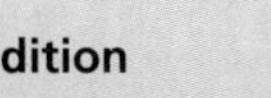 Pupil Edition

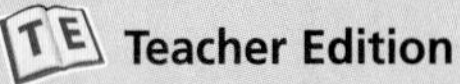 Teacher Edition

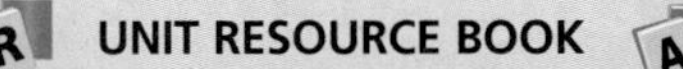 UNIT RESOURCE BOOK

 UNIT TRANSPARENCY BOOK

 UNIT ASSESSMENT BOOK

 SPANISH ASSESSMENT BOOK

SCIENCE TOOLKIT

READING AND REINFORCEMENT

- Description Wheel, B20–21
- Choose Your Own Strategy, C35–44
- Daily Vocabulary Scaffolding, H1–8

 **UNIT RESOURCE BOOK**
- Vocabulary Practice, pp. 237–238
- Decoding Support, p. 239
- Summarizing the Chapter, pp. 260–261

 **Audio Readings CD**
Listen to Pupil Edition.

 **Audio Readings in Spanish CD**
Listen to Pupil Edition in Spanish.

ASSESSMENT

- Chapter Review, pp. 125–126
- Standardized Test Practice, p. 127

UNIT ASSESSMENT BOOK
- Diagnostic Test, pp. 59–60
- Chapter Test, A, B, & C, pp. 64–75
- Alternative Assessment, pp. 76–77
- Unit Test, A, B, C, pp. 78–89

- Spanish Chapter Test, pp. 225–228
- Spanish Unit Test, pp. 229–232

 **Test Generator CD-ROM**
Generate customized tests.

 **Lab Generator CD-ROM**
Rubrics for Labs

STANDARDS

National Standards
A.1–8, A.9.a–g, B.3.a–b, E.2–5

See p. 100 for the standards.

READING AND REINFORCEMENT

 **UNIT RESOURCE BOOK**
- Reading Study Guide, A & B, pp. 205–208
- Spanish Reading Study Guide, pp. 209–210
- Challenge and Extension, p. 211
- Reinforcing Key Concepts, p. 213

ASSESSMENT

 Ongoing Assessment, pp. 105–108

 Section 4.1 Review, p. 108

 **UNIT ASSESSMENT BOOK**
Section 4.1 Quiz, p. 61

STANDARDS

National Standards
A.2–8, A.9.a–c, A.9.e–f, B.3.a

READING AND REINFORCEMENT

 **UNIT RESOURCE BOOK**
- Reading Study Guide, A & B, pp. 216–219
- Spanish Reading Study Guide, pp. 220–221
- Challenge and Extension, p. 222
- Reinforcing Key Concepts, p. 224

ASSESSMENT

 Ongoing Assessment, pp. 110–114

 Section 4.2 Review, p. 114

 **UNIT ASSESSMENT BOOK**
Section 4.2 Quiz, p. 62

STANDARDS

National Standards
A.2–7, A.9.a–b, A.9.d–f, B.3.a–b

READING AND REINFORCEMENT

 **UNIT RESOURCE BOOK**
- Reading Study Guide, A & B, pp. 227–230
- Spanish Reading Study Guide, pp. 231–232
- Challenge and Extension, p. 233
- Reinforcing Key Concepts, p. 234
- Challenge Reading, pp. 235–236

ASSESSMENT

 Ongoing Assessment, pp. 117, 119–121

 Section 4.3 Review, p. 121

 **UNIT ASSESSMENT BOOK**
Section 4.3 Quiz, p. 63

STANDARDS

National Standards
A.1–8, A.9.a–g, B.3.a, E.2–5

Previewing Resources for Differentiated Instruction

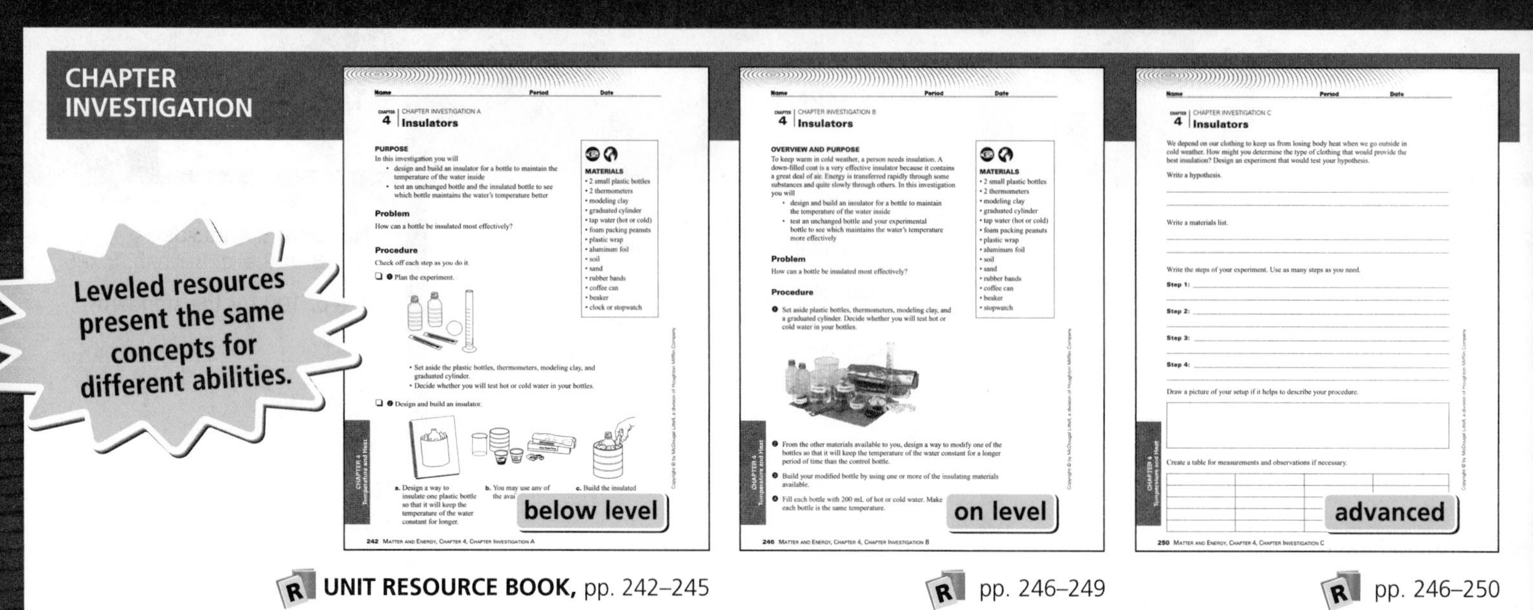

UNIT RESOURCE BOOK, pp. 242–245

pp. 246–249

pp. 246–250

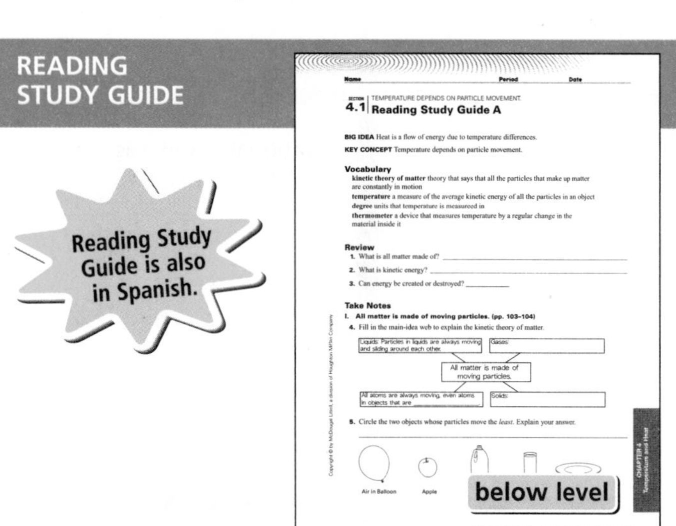

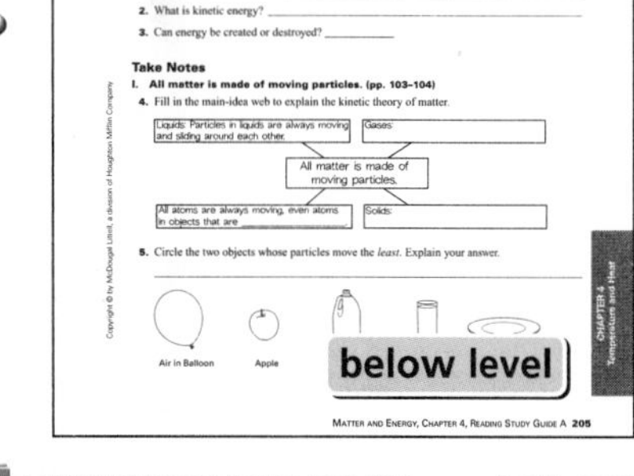

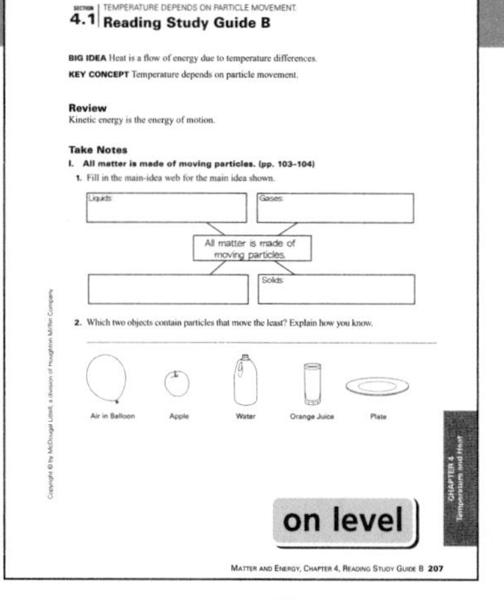

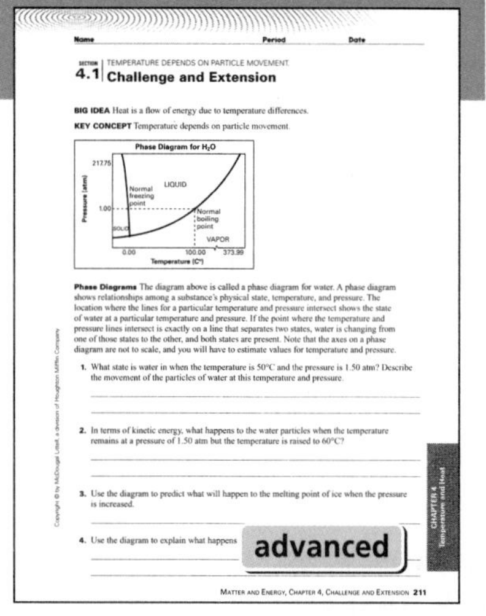

UNIT RESOURCE BOOK, pp. 205–206

pp. 207–208

p. 211

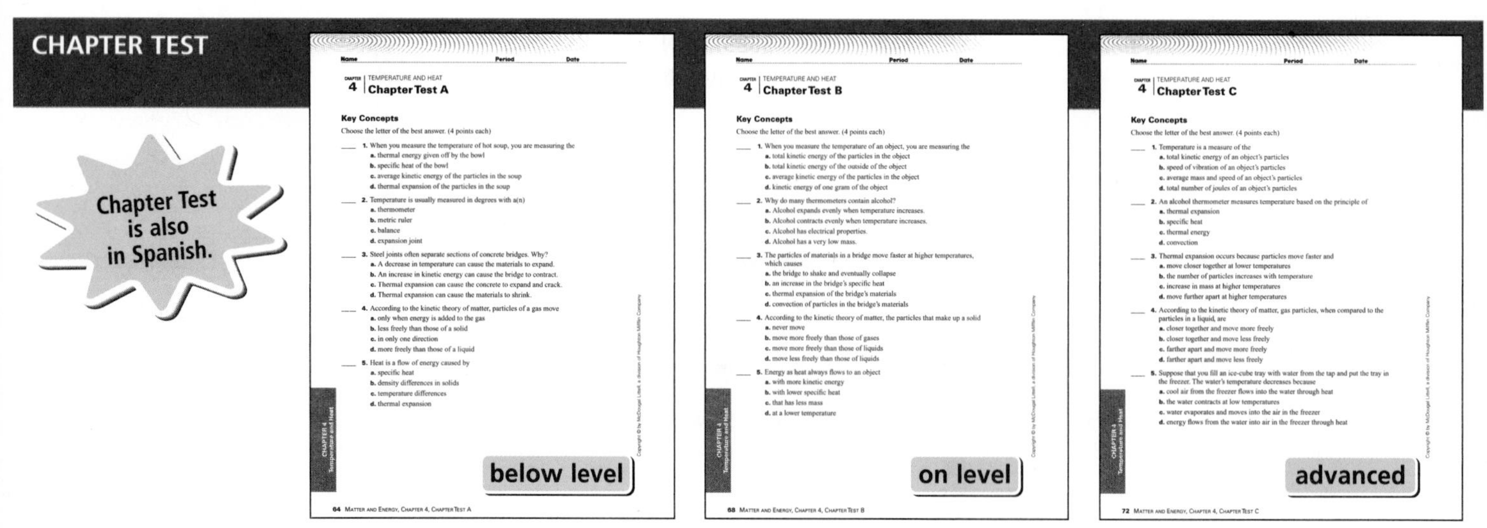

UNIT ASSESSMENT BOOK, pp. 64–67

pp. 68–71

pp. 72–75

TECHNOLOGY

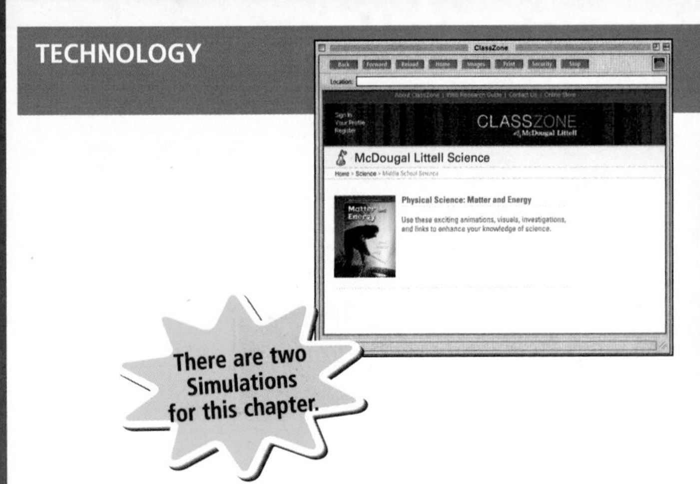

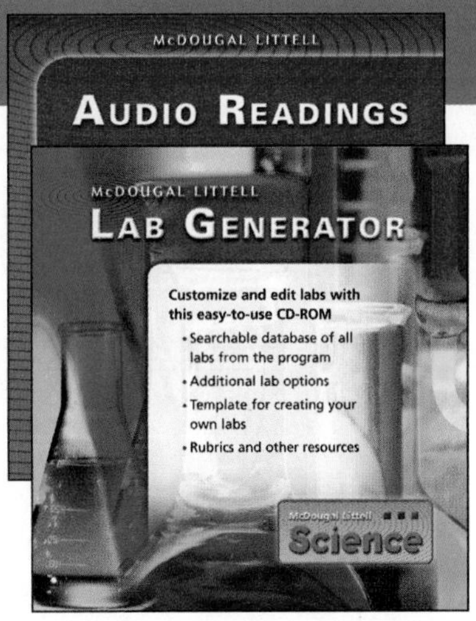

 CLASSZONE.COM

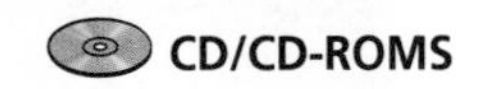 CD/CD-ROMS

 CLASSZONE.COM

VISUAL CONTENT

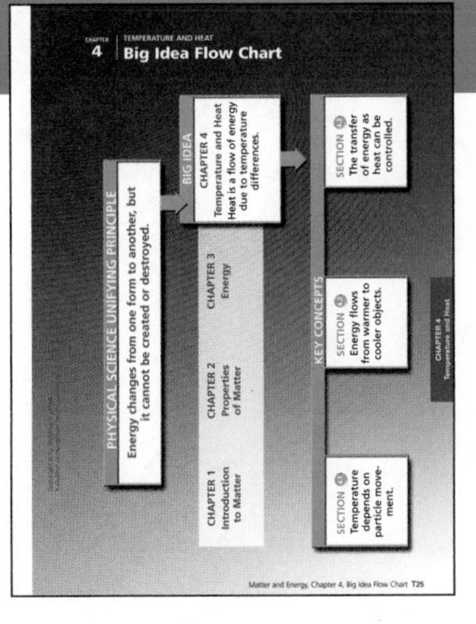

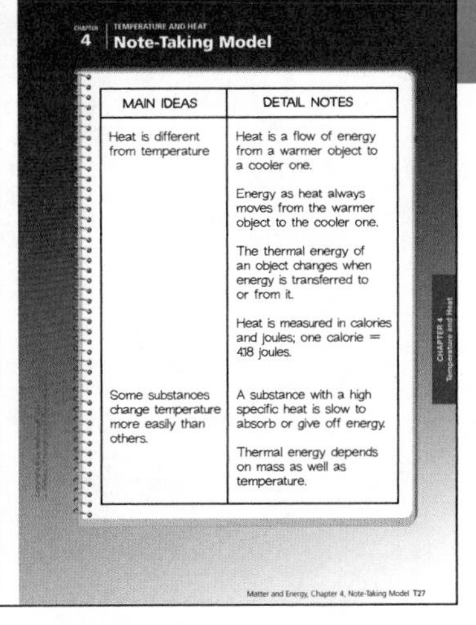

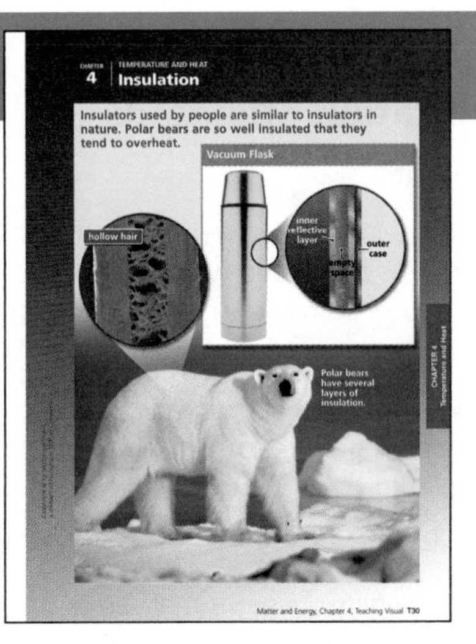

 UNIT TRANSPARENCY BOOK, p. T25

 p. T27

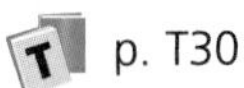 p. T30

MORE SUPPORT

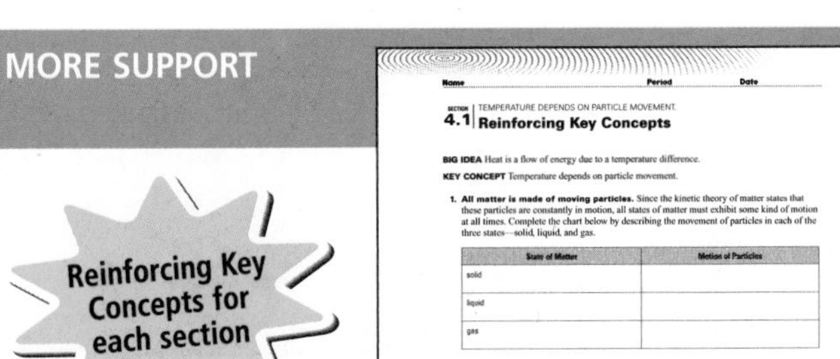

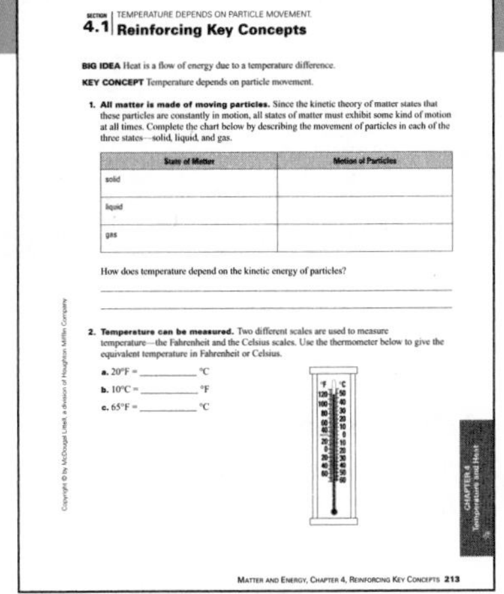

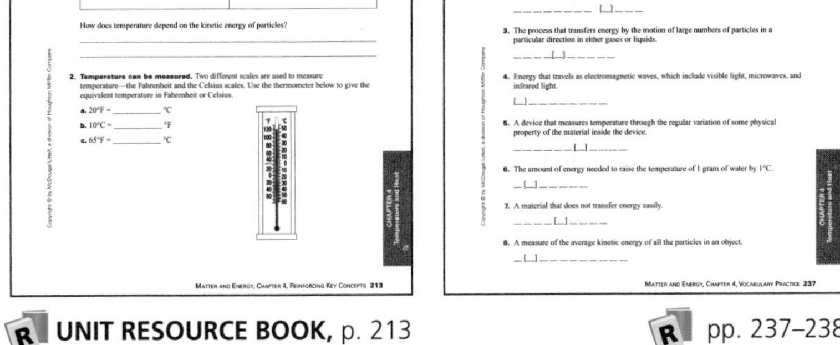

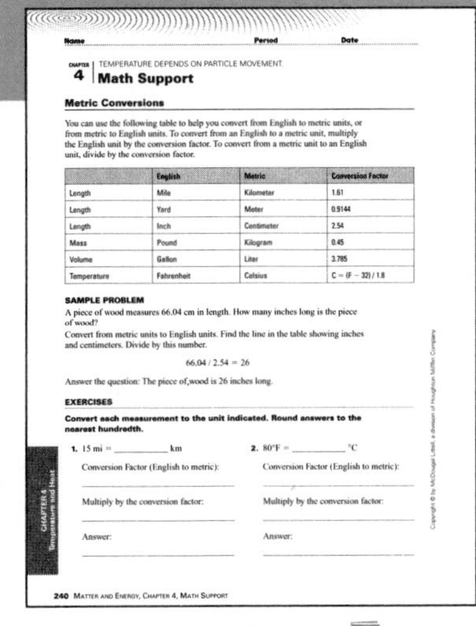

UNIT RESOURCE BOOK, p. 213

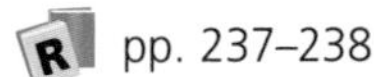 pp. 237–238

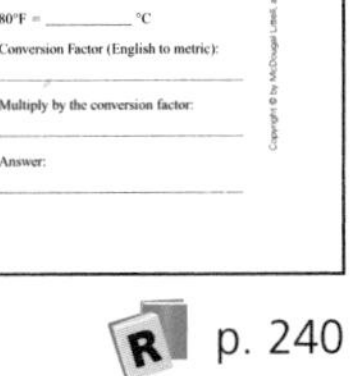 p. 240

INTRODUCE

the BIG idea

Have students look at the photograph of a giraffe in a sunny environment and discuss how the question in the box links to the Big Idea. Ask:

- How can you tell that energy travels from the Sun to Earth?
- What sources of energy are there after the Sun has set?

National Science Education Standards

Content

B.3.a Energy is a property of many substances, is associated with heat, and is transferred in many ways.

B.3.b Heat moves in predictable ways, flowing from warmer objects to cooler ones, until both reach the same temperature.

Process

A.1–8 Identify questions that can be answered through scientific investigations; design and conduct an investigation; use tools to gather and interpret data; use evidence to describe, predict, explain, model; think critically to make relationships between evidence and explanation; recognize different explanations and predictions; communicate scientific procedures and explanations; use mathematics.

A.9.a–g Understand scientific inquiry by using different investigations, methods, mathematics, technology, and explanations based on logic, evidence, and skepticism. Data often results in new investigations.

E.2–5 Design, implement, and evaluate a solution or product; communicate technological design.

CHAPTER

Temperature and Heat

the BIG idea

Heat is a flow of energy due to temperature differences.

Key Concepts

SECTION

4.1 Temperature depends on particle movement.
Learn how kinetic energy is the basis of temperature.

SECTION

4.2 Energy flows from warmer to cooler objects.
Learn about differences between temperature and heat, and how temperature changes in different substances.

SECTION

4.3 The transfer of energy as heat can be controlled.
Learn how energy is transferred through heat, and how that transfer can be controlled.

Internet Preview

CLASSZONE.COM

Chapter 4 online resources: Content Review, two Simulations, two Resource Centers, Math Tutorial, Test Practice

How does heat from the Sun increase this giraffe's temperature?

A 100 Unit: Matter and Energy

INTERNET PREVIEW

CLASSZONE.COM For student use with the following pages:

Review and Practice
- Content Review, pp. 102, 124
- Math Tutorial: Temperature Conversions, p. 109
- Test Practice, p. 127

Activities and Resources
- Internet Activity, p. 101
- Resource Centers: Temperature & Temperature Scales, p. 106; Thermal Energy, p. 111
- Simulations: Conduction, Convection, or Radiation, p. 119

Kinetic Theory
Code: MDL064

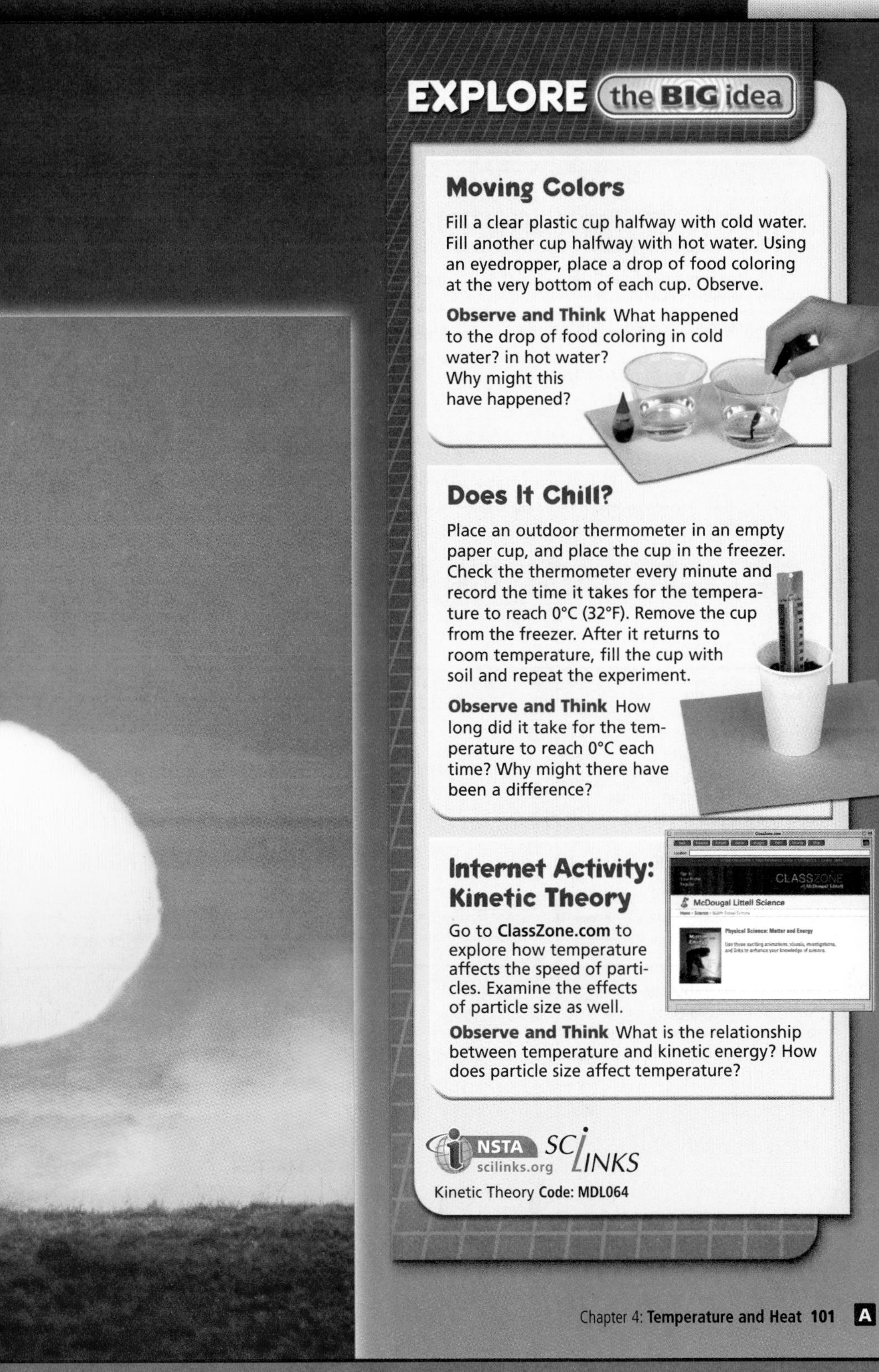

EXPLORE the BIG idea

Moving Colors

Fill a clear plastic cup halfway with cold water. Fill another cup halfway with hot water. Using an eyedropper, place a drop of food coloring at the very bottom of each cup. Observe.

Observe and Think What happened to the drop of food coloring in cold water? in hot water? Why might this have happened?

Does It Chill?

Place an outdoor thermometer in an empty paper cup, and place the cup in the freezer. Check the thermometer every minute and record the time it takes for the temperature to reach 0°C (32°F). Remove the cup from the freezer. After it returns to room temperature, fill the cup with soil and repeat the experiment.

Observe and Think How long did it take for the temperature to reach 0°C each time? Why might there have been a difference?

Internet Activity: Kinetic Theory

Go to **ClassZone.com** to explore how temperature affects the speed of particles. Examine the effects of particle size as well.

Observe and Think What is the relationship between temperature and kinetic energy? How does particle size affect temperature?

NSTA scilinks.org SCILINKS

Kinetic Theory **Code: MDL064**

TEACHING WITH TECHNOLOGY

CBL and Probeware If a probeware system is available, students can use a temperature probe to measure temperature changes for "Investigate Heat Transfer" on p. 112, "Explore Conduction" on p. 116, and to record and graph temperature changes in the Chapter Investigation, pp. 122–123.

EXPLORE the BIG idea

These inquiry-based activities are appropriate for use at home or as a supplement to classroom instruction.

Moving Colors

PURPOSE To observe that particles in matter move faster at higher temperatures than at lower temperatures. Students observe a drop of food coloring in cold water and in hot water.

TIP *10 min.* Have students use clear cups. Caution students not to get food coloring on their hands or clothing.

Answer: The food coloring spreads out faster in hot water than in cold water because the molecules in the hot water are moving faster.

REVISIT after p. 105.

Does It Chill?

PURPOSE To investigate how soil, as an insulator, slows changes in temperature. Students observe and record the temperature in a freezer.

TIP *30 min.* Have students read the thermometer and return it to the freezer as quickly as possible each time. Remind students to keep track of the time.

Answer: The thermometer in the soil should take longer to reach 0°C because soil acts as an insulator.

REVISIT after p. 117.

Internet Activity: Kinetic Theory

PURPOSE To observe that particle size and speed are related to temperature.

TIP *20 min.* Students should observe the kinetic energy of particles.

Answer: Temperature is the average kinetic energy of particles in an object; the greater the kinetic energy, the higher the temperature. The size of a particle affects its average kinetic energy.

REVISIT after p. 104.

PREPARE

CONCEPT REVIEW

Activate Prior Knowledge

Place an ice cube in a beaker, and place the beaker on a hot plate. Ask:

- How will the ice cube change? *It will melt into water, then evaporate and boil into water vapor.*
- What stays the same when the ice cube undergoes these changes? *The molecules remain as water.*
- How is energy related to these changes? *The addition of energy raises the temperature and produces the changes in physical state.*

TAKING NOTES

Choose Your Own Strategy

Choosing different strategies for taking notes can help students learn which strategies work best for them. Students can choose their own strategy or use the strategy suggested on the first page of each section. Encourage students to use their notes to test themselves.

Vocabulary Strategy

Description wheels can include as much information as students want to add. They are handy study devices when students look back through their notes.

Vocabulary and Note-Taking Resources

- Vocabulary Practice, pp. 237–238
- Decoding Support, p. 239

- Daily Vocabulary Scaffolding, p T26
- Note-Taking Model, p. T27

- Description Wheel, B20–21
- Choose Your Own Strategy, C35–44
- Daily Vocabulary Scaffolding, H1–8

CHAPTER 4

Getting Ready to Learn

CONCEPT REVIEW

- Matter is made of particles too small to see.
- Matter can be solid, liquid, or gas.
- Energy is the ability to cause a change.
- There are different forms of energy.

VOCABULARY REVIEW

matter p. 9
energy p. 72
kinetic energy p. 74

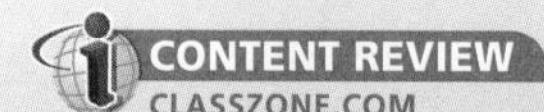

CONTENT REVIEW
CLASSZONE.COM
Review concepts and vocabulary.

TAKING NOTES

CHOOSE YOUR OWN STRATEGY

Take notes using one or more of the strategies from earlier chapters—**main idea and detail notes, main idea web,** or **mind map.** Feel free to mix and match the strategies, or use an entirely different note-taking strategy.

VOCABULARY STRATEGY

Place each vocabulary term at the center of a **description wheel** diagram. Write some words describing it on the spokes.

SCIENCE NOTEBOOK

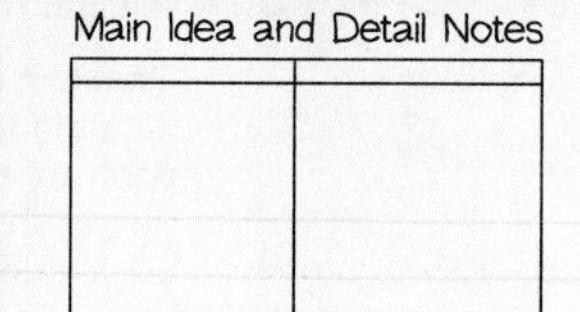

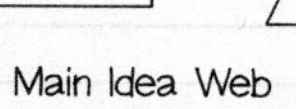

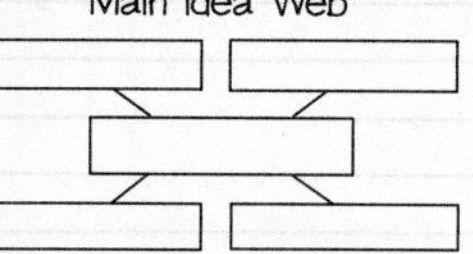

See the Note-Taking Handbook on pages R45–R51.

A 102 Unit: Matter and Energy

CHECK READINESS

Administer the Diagnostic Test to determine students' readiness for new science content and their mastery of requisite math skills.

 Diagnostic Test, pp. 59–60

Technology Resources

Students needing content and math skills should visit **ClassZone.com**.

- **CONTENT REVIEW**
- **MATH TUTORIAL**

 **CONTENT REVIEW CD-ROM**

KEY CONCEPT

Temperature depends on particle movement.

BEFORE, you learned
- All matter is made of particles
- Kinetic energy is the energy of motion
- Energy can be transferred or changed but is never created or destroyed

NOW, you will learn
- How temperature depends on kinetic energy
- How temperature is measured
- How changes in temperature can affect matter

VOCABULARY

kinetic theory of matter p. 104
temperature p. 105
degree p. 106
thermometer p. 107

EXPLORE Temperature

What can cause a change in temperature?

PROCEDURE

1. Work with a partner. Hold the rubber band with both hands. Without stretching it, hold it to the underside of your partner's wrist.
2. Move the rubber band away, then quickly stretch it once and keep it stretched. Hold it to the underside of your partner's wrist.
3. Move the rubber band away and quickly let it return to its normal size. Hold it to the underside of your partner's wrist.

MATERIALS

large rubber band

WHAT DO YOU THINK?
- What effect did stretching the rubber band have on the temperature of the rubber band?
- What may have caused this change to occur?

All matter is made of moving particles.

NOTE-TAKING STRATEGY
You could take notes on the movement of particles in matter by using a main idea web.

You have read that any object in motion has kinetic energy. All the moving objects you see around you—from cars to planes to butterflies—have kinetic energy. Even objects so small that you cannot see them, such as atoms, are in motion and have kinetic energy.

You might think that a large unmoving object, such as a house or a wooden chair, does not have any kinetic energy. However, all matter is made of atoms, and atoms are always in motion, even if the objects themselves do not change their position. The motion of these tiny particles gives the object energy. The chair you are sitting on has some amount of energy. You also have energy, even when you are not moving.

4.1 FOCUS

Set Learning Goals

Students will
- Explain how temperature depends on kinetic energy.
- Describe how temperature is measured.
- Describe how changes in temperature can affect matter.
- Observe experimentally how thermal expansion can be used to measure temperature.

3-Minute Warm-Up

Display Transparency 28 or copy this exercise on the board:

Decide if these statements are true. If not, correct them.

1. Solids and liquids are made of particles, but gases are made of air, which is not made of particles. *Gases are also made of particles.*
2. Kinetic energy is the energy of motion. *true*
3. Kinetic energy depends on position and chemical composition. *Kinetic energy depends on mass and speed. (Potential energy depends on position and chemical composition.)*

 3-Minute Warm-Up, p. T28

4.1 MOTIVATE

EXPLORE Temperature

PURPOSE To discover how a transfer of energy and increased motion produce an increase in temperature

TIP *10 min.* Use thick rubber bands that will not break easily. Students should wear safety goggles. Caution students to avoid stretching the rubber bands to their breaking point.

WHAT DO YOU THINK? *The temperature increased because stretching the rubber band added energy to it.*

RESOURCES FOR DIFFERENTIATED INSTRUCTION

Below Level

UNIT RESOURCE BOOK
- Reading Study Guide A, pp. 205–206
- Decoding Support, p. 239

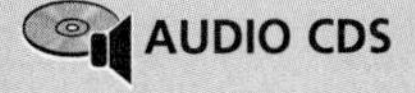

Advanced

UNIT RESOURCE BOOK
Challenge and Extension, p. 211

English Learners

UNIT RESOURCE BOOK
Spanish Reading Study Guide, pp. 209–210

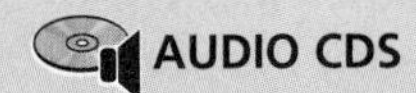

- Audio Readings in Spanish
- Audio Readings (English)

4.1 INSTRUCT

Address Misconceptions

IDENTIFY Ask: What states of matter have particles always in motion? If students do not mention all states of matter, they may think that some particles in matter are not in constant motion.

CORRECT Demonstrate the kinetic theory of matter in relation to solids, liquids, and gases using a clear plastic container holding beads or other small objects. Fill the container. Shake it to demonstrate a solid. If the container is completely full, the "particles" will move around slightly and will be in constant contact. Remove a portion of the objects, then shake the container to demonstrate a liquid. The "particles" will still collide but will move past each other and move more freely. Finally, remove all but a few objects, then shake the container to demonstrate a gas. The "particles" will move most freely and barely interact.

REASSESS Ask students to describe the motion of particles in their desks and chairs. *Particles are vibrating in place.* Reiterate that particles in all matter are always in motion.

Technology Resources

Visit **ClassZone.com** for background on common student misconceptions.

MISCONCEPTION DATABASE

Teach from Visuals

To help students interpret the diagrams of particles in solid, liquid, and gas, ask:

- What do the particles in the solid, liquid, and gas have in common? *They are all in constant motion.*
- How are the solid, liquid, and gas different? *The freedom with which the particles can move varies.*

EXPLORE the BIG idea

Revisit "Internet Activity: Kinetic Theory" on p. 101. Have students explain their observations.

Ongoing Assessment

CHECK YOUR READING *that particles in matter are in constant motion and have kinetic energy*

REMINDER

Kinetic energy is the energy of motion.

The Kinetic Theory of Matter

Physical properties and physical changes are the result of how particles of matter behave. The **kinetic theory of matter** states that all of the particles that make up matter are constantly in motion. As a result, all particles in matter have kinetic energy. The kinetic theory of matter helps explain the different states of matter—solid, liquid, and gas.

1. The particles in a solid, such as concrete, are not free to move around very much. They vibrate back and forth in the same position and are held tightly together by forces of attraction.
2. The particles in a liquid, such as water in a pool, move much more freely than particles in a solid. They are constantly sliding around and tumbling over each other as they move.
3. In a gas, such as the air around you or in a bubble in water, particles are far apart and move around at high speeds. Particles might collide with one another, but otherwise they do not interact much.

READING TIP

In illustrations of particle movement, more motion lines mean a greater speed.

Particles do not always move at the same speed. Within any group of particles, some are moving faster than others. A fast-moving particle might collide with another particle and lose some of its speed. A slow-moving particle might be struck by a faster one and start moving faster. Particles have a wide range of speeds and often change speeds.

CHECK YOUR READING What is the kinetic theory of matter?

Matter in Motion

All particles in this pool, from those in the concrete structure to those in air bubbles, are always moving.

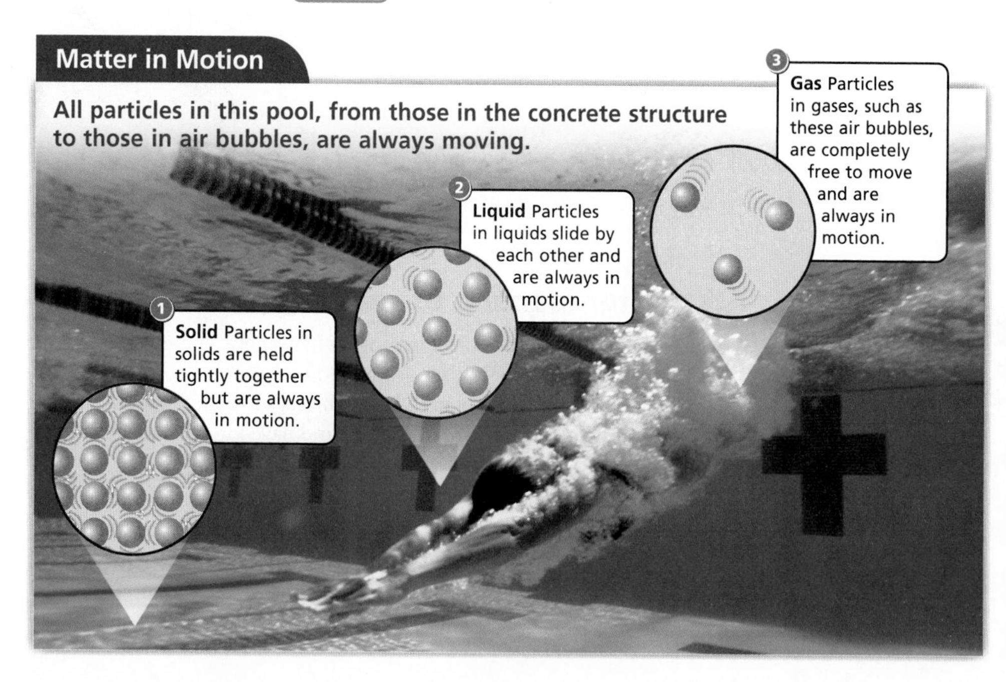

1
Solid Particles in solids are held tightly together but are always in motion.
2
Liquid Particles in liquids slide by each other and are always in motion.
3
Gas Particles in gases, such as these air bubbles, are completely free to move and are always in motion.

DIFFERENTIATE INSTRUCTION

More Reading Support

A What does the kinetic theory of matter say? *All the particles in matter are constantly in motion.*

English Learners Have students write the definitions of *temperature, degree,* and *thermometer* in their Science Word Dictionaries. English learners may require background knowledge of "smoothie" (p. 105), and the phrase "into account" (p. 108). Tell students to take something into account means to consider it.

Temperature and Kinetic Energy

Particles of matter moving at different speeds have different kinetic energies because kinetic energy depends on speed. It is not possible to know the kinetic energy of each particle in an object. However, the average kinetic energy of all the particles in an object can be determined.

Temperature is a measure of the average kinetic energy of all the particles in an object. If a liquid, such as hot cocoa, has a high temperature, the particles in the liquid are moving very fast and have a high average kinetic energy. The cocoa feels hot. If a drink, such as a fruit smoothie, has a low temperature, the particles in the liquid are moving more slowly and have a lower average kinetic energy. The smoothie feels cold.

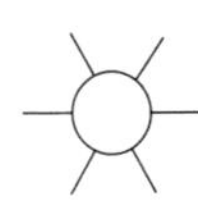

VOCABULARY
Remember to make a description wheel diagram for *temperature* and other vocabulary terms.

You experience the connection between temperature and the kinetic energy of particles every day. For example, to raise the temperature of your hands on a cold day—to warm your hands—you have to add energy, perhaps by putting your hands near a fire or a hot stove. The added energy makes the particles in your hands move faster. If you let a hot bowl sit on a table for a while, the particles in the bowl slow down due to collisions with particles in the air and in the table. The temperature of the bowl decreases, and it becomes cooler.

Temperature is the measurement of the average kinetic energy of particles, not just their speed. Recall that kinetic energy depends on mass as well as speed. Particles in a metal doorknob do not move as fast as particles in air. However, the particles in a doorknob have more mass and they can have the same amount of kinetic energy as particles in air. As a result, the doorknob and the air can have equal temperatures.

CHECK YOUR READING How does temperature change when kinetic energy increases?

DIFFERENTIATE INSTRUCTION

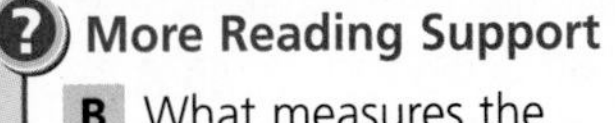

More Reading Support

B What measures the average kinetic energy of particles in an object? *temperature*

C What does kinetic energy depend on in addition to speed? *mass*

Below Level Students may have trouble with the idea that a large object with slow-moving particles can have a higher temperature than air with fast-moving particles. Have students feel a metal object and the air around it to compare their temperatures. Place the metal object under a lamp. After the metal is warm, have students compare how the metal and the air around it feel. Point out that even though particles of a gas move faster than particles of a solid, the particles in metal have more mass, and the average kinetic energy of its particles has increased more than that of the air particles.

Teach from Visuals

To help students interpret the diagrams of particles in hot and cold liquids, ask: How are the particles in the two liquids different? *The particles in the hot liquid are moving faster and have a higher average kinetic energy.*

Teach Difficult Concepts

Students might think that temperature is a measure of an object's heat. To help students understand what temperature actually measures, put out three large containers of water. One should have cold tap water, one should have hot tap water, and one should have room-temperature water. Have students put one hand in the cold and hot water for a few seconds, then have them put both hands in the room-temperature water. Ask students what temperature measures. *average kinetic energy of all the particles in an object* Point out that room-temperature water feels different to each hand because of temperature differences, not temperature measure itself.

Technology Resources

Visit **ClassZone.com** for background on common student misconceptions.

MISCONCEPTION DATABASE

EXPLORE the BIG idea

Revisit "Moving Colors" on p. 101. Have students explain their results.

Ongoing Assessment

Explain how temperature depends on kinetic energy.

Ask: If the average kinetic energy of an object's particles decreases, what happens to its temperature? *It decreases.*

CHECK YOUR READING *Answer: It increases.*

History of Science

In the 1800s, scientist William Thomson, Lord Kelvin, developed an absolute temperature scale with 0 representing absolute zero. At this theoretical temperature the particles in matter stop moving and have no kinetic energy. The Celsius scale uses the same magnitude of units as the Kelvin scale, and 0 K is equal to −273.15° Celsius. Therefore, to convert Celsius to Kelvin, add 273.15° to the Celsius temperature. This conversion is only necessary in the case of particular temperature values, but not for a change in temperature. That is, an increase of 10 K is equal to an increase of 10°C.

Note that the Kelvin scale employs no degree symbol, unlike the Fahrenheit and Celsius scales.

Teach from Visuals

To help students interpret the photograph of the temperature scales, ask:

- What is the temperature of Death Valley in the photograph? *49°C or 120°F*
- Why is it important to include the scale when giving a temperature? *The number is meaningless if the scale is unknown.*

Ongoing Assessment

CHECK YOUR READING *Answer: They both measure temperature in terms of degrees. They have different numbers of degrees between the freezing point and boiling point of water, and the zero point of the Celsius scale is fixed at the freezing point of water, whereas the zero point of the Fahrenheit scale is not.*

Find out more about temperature and temperature scales.

Temperature can be measured.

You have read that a warmer temperature means a greater average kinetic energy. How is temperature measured and what does that measurement mean? Suppose you hear on the radio that the temperature outside is 30 degrees. Do you need to wear a warm coat to spend the day outside? The answer depends on the temperature scale being used. There are two common temperature scales, both of which measure the average kinetic energy of particles. However, 30 degrees on one scale is quite different from 30 degrees on the other scale.

Temperature Scales

To establish a temperature scale, two known values and the number of units between the values are needed. The freezing and boiling points of pure water are often used as the standard values. These points are always the same under the same conditions and they are easy to reproduce. In the two common scales, temperature is measured in units called **degrees** (°), which are equally spaced units between two points.

D

The scale used most commonly in the United States for measuring temperature—in uses ranging from cooking directions to weather reports—is the Fahrenheit (FAR-uhn-HYT) scale (°F). It was developed in the early 1700s by Gabriel Fahrenheit. On the Fahrenheit scale, pure water freezes at 32°F and boils at 212°F. Thus, there are 180 degrees—180 equal units—between the freezing point and the boiling point of water.

During a summer day in Death Valley, California, the temperature can reach 49°C (120°F).

The temperature scale most commonly used in the rest of the world, and also used more often in science, is the Celsius (SEHL-see-uhs) scale (°C). This scale was developed in the 1740s by Anders Celsius. On the Celsius scale, pure water freezes at 0°C and boils at 100°C, so there are 100 degrees—100 equal units—between these two temperatures.

Recall the question asked in the first paragraph of this page. If the outside temperature is 30 degrees, do you need to wear a warm coat? If the temperature is 30°F, the answer is yes, because that temperature is colder than the freezing point of water. If the temperature is 30°C, the answer is no—it is a nice warm day (86°F).

CHECK YOUR READING How are the Fahrenheit and Celsius temperature scales different? How are they similar?

DIFFERENTIATE INSTRUCTION

More Reading Support

D What are the equally spaced units between two points? *degrees*

E What temperature scale is often used by scientists? *Celsius*

Advanced Have students calibrate the thermometers they make in the investigation on p. 107. Students should make a mark on the straw to indicate the alcohol level at a low temperature, and another mark for the level at a high temperature. The actual temperature should be measured with a real thermometer at the same time. Have students test a temperature midway between their calibration points.

 Challenge and Extension, p. 211

A 106 Unit: Matter and Energy

Thermometers

Temperature is measured by using a device called a thermometer. A **thermometer** measures temperature through the regular variation of some physical property of the material inside the thermometer. A mercury or alcohol thermometer, for example, can measure temperature because the liquid inside the thermometer always expands or contracts by a certain amount in response to a change in temperature.

Liquid-filled thermometers measure how much the liquid expands in a narrow tube as the temperature increases. The distances along the tube are marked so that the temperature can be read. At one time, thermometers were filled with liquid mercury because it expands or contracts evenly at both high and low temperatures. This means that mercury expands or contracts by the same amount in response to a given change in temperature. However, mercury is dangerous to handle, so many thermometers today are filled with alcohol instead.

Some thermometers work in a different way—they use a material whose electrical properties change when the temperature changes. These thermometers can be read by computers. Some show the temperature on a display panel and are often used in cars and in homes.

CHECK YOUR READING How do liquid-filled thermometers work?

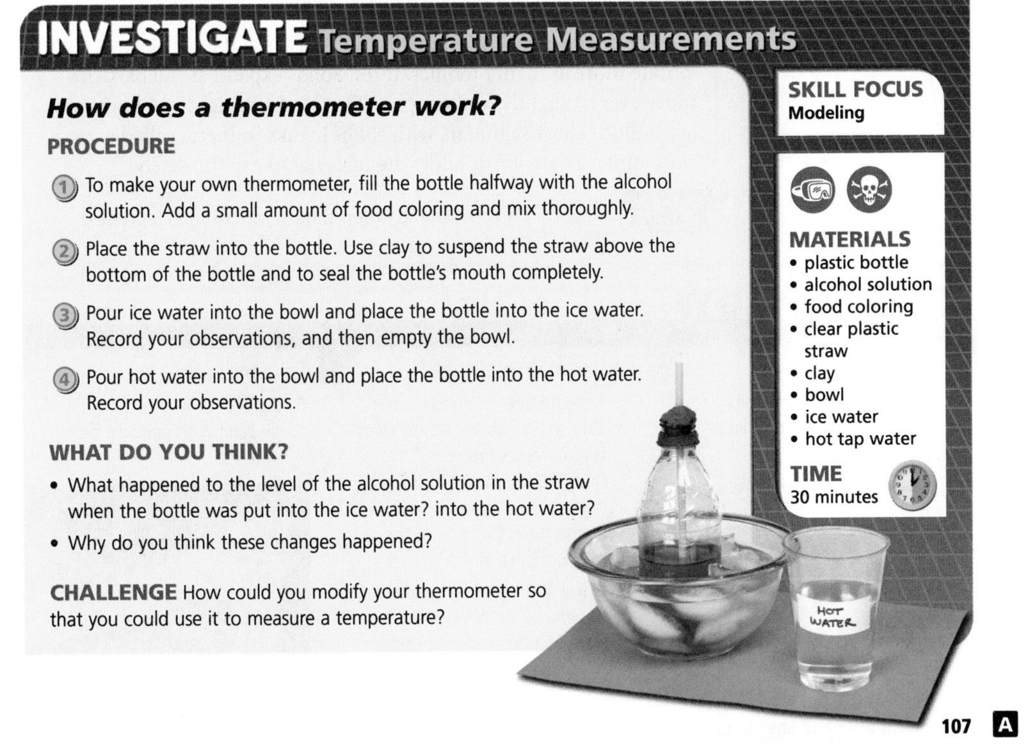

INVESTIGATE Temperature Measurements
How does a thermometer work?
PROCEDURE
1 To make your own thermometer, fill the bottle halfway with the alcohol solution. Add a small amount of food coloring and mix thoroughly.
2 Place the straw into the bottle. Use clay to suspend the straw above the bottom of the bottle and to seal the bottle's mouth completely.
3 Pour ice water into the bowl and place the bottle into the ice water. Record your observations, and then empty the bowl.
4 Pour hot water into the bowl and place the bottle into the hot water. Record your observations.
WHAT DO YOU THINK?
• What happened to the level of the alcohol solution in the straw when the bottle was put into the ice water? into the hot water?
• Why do you think these changes happened?
CHALLENGE How could you modify your thermometer so that you could use it to measure a temperature?
SKILL FOCUS
Modeling
MATERIALS
• plastic bottle
• alcohol solution
• food coloring
• clear plastic straw
• clay
• bowl
• ice water
• hot tap water
TIME
30 minutes
HOT WATER

INVESTIGATE Temperature Measurements

PURPOSE To observe how thermal expansion can be used to measure temperature

TIPS *30 min.*

- The mouth of the bottle must be completely sealed.
- You might want students to bring small plastic bottles from home.

WHAT DO YOU THINK? *The level dropped in ice water and rose in hot water. The liquid contracted and expanded in response to a change in temperature.*

CHALLENGE *Two known temperatures and levels on the straw are needed. The expansion and contraction would need to be observed several times.*

Datasheet, Temperature Measurements, p. 212

Technology Resources

Customize this student lab as needed or look for an alternative. Print rubrics to assess student lab reports.

Lab Generator CD-ROM

History of Science

Scientists developed tools to measure temperature changes in the 1500s and 1600s. These thermoscopes used air instead of liquid. At first they did not have temperature scales. Eventually, scientists developed many different scales and could not compare their measurements with those using other scales.

Ongoing Assessment

Describe how temperature is measured.

Ask: What are two things a thermometer needs to measure temperature? *a substance that varies regularly with changing temperatures and a scale with two known values.*

CHECK YOUR READING *Answer: The liquid expands or contracts by a consistent amount due to a change in temperature.*

DIFFERENTIATE INSTRUCTION

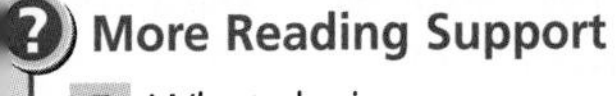

More Reading Support

F What device measures temperature? *thermometer*

Alternative Assessment Students can answer and extend the questions in "Investigate Temperature Measurements" by drawing and labeling their results and ideas.

Ongoing Assessment

Describe how changes in temperature can affect matter.

Ask: What happens to all gases, many liquids, and most solids when their temperature increases? *They expand.*

 *Answer: The particles move faster and move apart from each other very slightly.*

Real World Example

Thermal expansion can be used to solve problems. For example, a metal lid stuck on a jar will expand if held under hot tap water, and the jar will open more easily. Also, home thermostats contain a coil made of two metals with different rates of expansion. When the temperature drops, the coil bends one way, and when the temperature rises, it bends the other way.

Reinforce the BIG idea

Have students relate the section to the Big Idea.

 Reinforcing Key Concepts, p. 213

4.1 ASSESS & RETEACH

Assess

 Section 4.1 Quiz, p. 61

Reteach

Ask volunteers to explain each concept and give examples.

- how temperature depends on kinetic energy
- what is needed for a temperature scale
- how a thermometer works
- what thermal expansion is

Technology Resources

Have students visit **ClassZone.com** for reteaching of Key Concepts.

 CONTENT REVIEW

 CONTENT REVIEW CD-ROM

During construction of the Gateway Arch in St. Louis, engineers had to account for thermal expansion.

Thermal Expansion

The property that makes liquid-filled thermometers work is called thermal expansion. Thermal expansion affects many substances, not just alcohol and liquid mercury. All gases, many liquids, and most solids expand when their temperature increases.

Construction engineers often have to take thermal expansion into account because steel and concrete both expand with increasing temperature. An interesting example involves the construction of the Gateway Arch in St. Louis, which is built mostly of steel.

The final piece of the Arch to be put into place was the top segment joining the two legs. The Arch was scheduled to be completed in the middle of the day for its opening ceremony. However, engineers knew that the side of the Arch facing the Sun would get hot and expand due to thermal expansion.

This expansion would narrow the gap between the legs and prevent the last piece from fitting into place. In order to complete the Arch, workers sprayed water on the side facing the Sun. The water helped cool the Arch and decreased the amount of thermal expansion. Once the final segment was in place, engineers made the connection strong enough to withstand the force of the expanding material.

Thermal expansion occurs in solids because the particles of solids vibrate more at higher temperatures. Solids expand as the particles move ever so slightly farther apart. This is why bridges and highways are built in short segments with slight breaks in them, called expansion joints. These joints allow the material to expand safely.

 Why do objects expand when their temperatures increase?

4.1 Review

KEY CONCEPTS

1. Describe the relationship between temperature and kinetic energy.
2. Describe the way in which thermometers measure temperature.
3. How can you explain thermal expansion in terms of kinetic energy?

CRITICAL THINKING

4. **Synthesize** Suppose a mercury thermometer shows that the air temperature is 22°C (72°F). Do particles in the air have more average kinetic energy than particles in the mercury? Explain.
5. **Infer** If a puddle of water is frozen, do particles in the ice have kinetic energy? Explain.

CHALLENGE

6. **Apply** Why might a sidewalk be built with periodic breaks in it?

ANSWERS

1. Temperature is a measurement of average kinetic energy of particles in a substance. As the average kinetic energy increases, so does temperature.

2. through the regular variation of a physical property in response to a change in temperature

3. When the kinetic energy of particles in an object increases, they move faster and move farther apart from one another.

4. No; the temperatures are the same, so the particles of both have the same average kinetic energy.

5. Yes; particles in solids are always in motion.

6. As concrete expands and contracts with changes in temperature, it needs room to move.

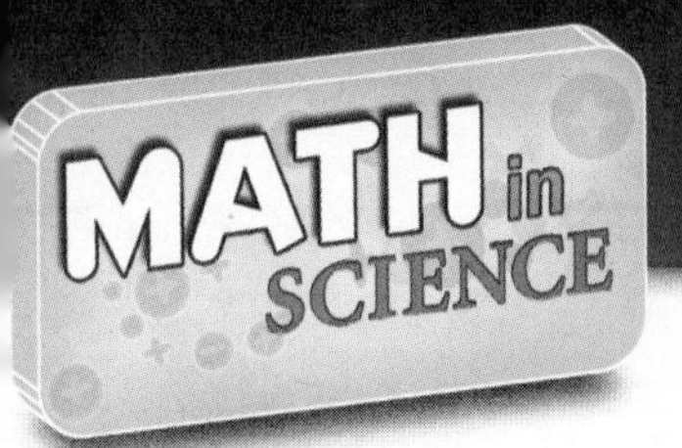

CLASSZONE.COM

Click on Math Tutorial for more help with temperature conversions.

SKILL: METRIC CONVERSIONS

How Hot Is Hot?

Temperatures on Earth can vary greatly, from extremely hot in some deserts to frigid in polar regions. The meaning of a temperature measurement depends on which temperature scale is being used. A very high temperature on the Fahrenheit scale is equal to a much lower temperature on the Celsius scale. The table shows the formulas used to convert temperatures between the two scales.

Conversion	Formula
Fahrenheit to Celsius	$°C = \frac{5}{9}(°F - 32)$
Celsius to Fahrenheit	$°F = \frac{9}{5}°C + 32$

Example

The boiling point of pure water is 212°F. Convert that temperature to a measurement on the Celsius scale.

(1) Use the correct conversion formula.

$$°C = \frac{5}{9}(°F - 32)$$

(2) Substitute the temperature given for the correct variable in the formula.

$$°C = \frac{5}{9}(212 - 32) = \frac{5}{9} \cdot 180 = 100$$

ANSWER °C = 100

Use the information in the table below to answer the questions that follow.

Highest and Lowest Temperatures Recorded on Earth			
Location	**Highest Temp. (°F)**	**Location**	**Lowest Temp. (°F)**
El Azizia, Libya	136	Vostok, Antarctica	–129
Death Valley, California	134	Oimekon, Russia	–90
Tirat Tsvi, Israel	129	Verkhoyansk, Russia	–90
Cloncurry, Australia	128	Northice, Greenland	–87
Seville, Spain	122	Snag, Yukon, Canada	–81

1. What is the highest temperature in °C?
2. What is the temperature difference in °C between the highest and second highest temperatures?
3. What is the difference between the highest and lowest temperatures in °F? in °C?

CHALLENGE **The surface of the Sun is approximately 5500°C. What is this temperature in °F?**

Temperatures on Earth, ranging from the extremes of frigid polar regions to the hottest deserts, can differ by more than 250°F.

ANSWERS

1. °C = 5/9 (136–32) = 5/9 (104) = 57.8

2. °C = 5/9 (134–32) = 5/9 (104) = 56.7; 57.8 – 56.7 = 1.1

3. 136 – (–129) = 265°F; 57.8 – (–89.4) = 147.2°C

CHALLENGE °F = 9/5 (5500) + 32 = 9900 + 32 = 9932

MATH IN SCIENCE
Math Skills Practice for Science

Set Learning Goal

To convert temperatures from Fahrenheit to Celsius and Celsius to Fahrenheit

Present the Science

Both common temperature scales have advantages. The Fahrenheit scale uses smaller degrees, so it is more accurate for weather reports. The Celsius scale is more closely tied to a physical constant (water's freezing point) and is more commonly used in science.

Develop Algebra Skills

- Remind students that the numerator of the fraction is above the line, and the denominator is below. To help them remember which part of the fraction to multiply, emphasize that they divide by the denominator.
- Point out that when converting from Fahrenheit to Celsius, the parentheses mean that the 32 is subtracted before multiplying by 5/9. When converting Celsius to Fahrenheit, the 32 is added after multiplying by 9/5.

DIFFERENTIATION TIP Use graph paper to model multiplying the fractions 1/5 and 1/9 by a number such as 45. Shade squares that show the fraction as a portion of a larger number.

Close

Ask: Why is knowing how to convert from one temperature scale to another useful? *Sample answer: Most people are more familiar with one scale than the other. Without conversion, temperatures given in an unfamiliar scale are meaningless.*

- Math Support, p. 240
- Math Practice, p. 241

Technology Resources

Students can visit **ClassZone.com** for practice with temperature conversions.

 **MATH TUTORIAL**

4.2 FOCUS

Set Learning Goals

Students will

- Compare heat and temperature.
- Describe how heat is measured.
- Understand why some substances change temperature more easily than others.
- Measure heat transfer in an experiment.

3-Minute Warm-Up

Display Transparency 28 or copy this exercise on the board:

Match the correct temperature scale to the descriptions.

Temperature Scale

1. Fahrenheit *b*
2. Celsius *a, c, d*

Description

a. freezing point of water is 0°
b. freezing point of water is 32°
c. used by scientists and most countries
d. 100 units between freezing and boiling points of water

3-Minute Warm-Up, p. T28

4.2 MOTIVATE

THINK ABOUT

PURPOSE To introduce that different substances warm up at different rates

DISCUSS Have students brainstorm answers to the question in the text. *More energy is needed to warm water.*

Ongoing Assessment

Explain how heat is different from temperature.

Ask: How does heat differ from temperature? *Heat is the flow of energy; temperature is a measure of it.*

Answer: the flow of energy from warmer to cooler objects

KEY CONCEPT

4.2 Energy flows from warmer to cooler objects.

BEFORE, you learned

- All matter is made of moving particles
- Temperature is the measurement of average kinetic energy of particles in an object
- Temperature can be measured

NOW, you will learn

- How heat is different from temperature
- How heat is measured
- Why some substances change temperature more easily than others

VOCABULARY

heat p. 110
thermal energy p. 111
calorie p. 112
joule p. 112
specific heat p. 113

THINK ABOUT

Why does water warm up so slowly?

If you have ever seen food being fried in oil or butter, you know that the metal frying pan heats up very quickly, as does the oil or butter used to coat the pan's surface. However, if you put the same amount of water as you put oil in the same pan, the water warms up more slowly. Why does water behave so differently from the metal, oil, or butter?

NOTE-TAKING STRATEGY
The mind map organizer would be a good choice for taking notes on heat.

Heat is different from temperature.

Heat and temperature are very closely related. As a result, people often confuse the concepts of heat and temperature. However, they are not the same. Temperature is a measurement of the average kinetic energy of particles in an object. **Heat** is a flow of energy from an object at a warmer temperature to an object at a cooler temperature.

If you add energy as heat to a pot of water, the water's temperature starts to increase. The added energy increases the average kinetic energy of the water molecules. Once the water starts to boil, however, adding energy no longer changes the temperature of the water. Instead, the heat goes into changing the physical state of the water from liquid to gas rather than increasing the kinetic energy of the water molecules. This fact is one demonstration that heat and temperature are not the same thing.

What is heat?

RESOURCES FOR DIFFERENTIATED INSTRUCTION

Below Level

UNIT RESOURCE BOOK

- Reading Study Guide A, pp. 216–217
- Decoding Support, p. 239

AUDIO CDS

Advanced

UNIT RESOURCE BOOK

Challenge and Extension, p. 222

English Learners

UNIT RESOURCE BOOK

Spanish Reading Study Guide, pp. 220–221

- Audio Readings in Spanish
- Audio Readings (English)

Heat and Thermal Energy

Learn more about thermal energy.

Suppose you place an ice cube in a bowl on a table. At first, the bowl and the ice cube have different temperatures. However, the ice cube melts, and the water that comes from the ice will eventually have the same temperature as the bowl. This temperature will be lower than the original temperature of the bowl but higher than the original temperature of the ice cube. The water and the bowl end up at the same temperature because the particles in the ice cube and the particles in the bowl continually bump into each other and energy is transferred from the bowl to the ice.

Heat is always the transfer of energy from an object at a higher temperature to an object at a lower temperature. So energy flows from the particles in the warmer bowl to the particles in the cold ice and, later, the cooler water. If energy flowed in the opposite direction—from cooler to warmer—the ice would get colder and the bowl would get hotter, and you know that never happens.

CHECK YOUR READING In which direction does heat always transfer energy?

When energy flows from a warmer object to a cooler object, the thermal energy of both of the objects changes. **Thermal energy** is the total random kinetic energy of particles in an object. Note that temperature and thermal energy are different from each other. Temperature is an average and thermal energy is a total. A glass of water can have the same temperature as Lake Superior, but the lake has far more thermal energy because the lake contains many more water molecules.

Another example of how energy is transferred through heat is shown on the right. Soon after you put ice cubes into a pitcher of lemonade, energy is transferred from the warmer lemonade to the colder ice. The lemonade's thermal energy decreases and the ice's thermal energy increases. Because the particles in the lemonade have transferred some of their energy to the particles in the ice, the average kinetic energy of the particles in the lemonade decreases. As a result, the temperature of the lemonade decreases.

lemonade
heat
ice

Energy is transferred from the warmer lemonade to the cold ice through heat until their temperatures are equal.

How are heat and thermal energy related to each other?

DIFFERENTIATE INSTRUCTION

A What does heat do? *transfers energy*

B What is the total random kinetic energy of all particles in an object? *thermal energy*

English Learners English learners rely on patterns and conventions in the English language, and may not recognize cause-and-effect relationships in sentences that do not follow *if/then* format. Point out the cause-and-effect relationship in the following sentence from this page: "Because the particles in the lemonade have transferred . . ., the average kinetic energy of the particles in the lemonade decreases." Encourage students to look for introductory clauses and phrases rather than a particular word, such as *if*.

4.2 INSTRUCT

Address Misconceptions

IDENTIFY Ask students how a flow of energy through heat differs from the flow of a fluid such as water. If students fail to recognize that heat is not a physical form of matter like water, they may hold the misconception that heat is a form of matter that literally flows from one object to another.

CORRECT Hold an ice cube in the palm of your hand. Ask students to describe the flow of heat. Point out that energy, not particles of matter, flows from your hand to the ice cube.

REASSESS Ask: What flows from one place to another when an object is heated? *energy*

Technology Resources

Visit **ClassZone.com** for background on common student misconceptions.

MISCONCEPTION DATABASE

Teach from Visuals

To help students interpret the diagram of heat transfer from lemonade to ice, ask:

- In what direction does heat transfer energy in the diagram? *from the warm lemonade to the cold ice*
- What happens to the ice's thermal energy? *It increases.*
- What happens to the lemonade's thermal energy? *It decreases.*

Teach Difficult Concepts

Heat is often thought of as both a form of energy and a way in which energy can be transferred. Point out that in this chapter, heat is not a form of energy but rather the flow of energy.

Ongoing Assessment

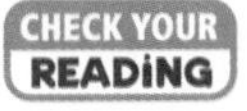

Answer: from warmer to cooler

Answer: As heat transfers energy, the thermal energy of the warmer object decreases, and the thermal energy of the cooler object increases.

INVESTIGATE Heat Transfer

PURPOSE To investigate the specific heat of different materials by measuring a change in temperature

TIPS *30 min.*

- You might want to have students bring pennies from home.
- The hot water should be at least 60°C for the best results.

WHAT DO YOU THINK? *The cup to which water is added will show the greatest change in temperature; the cup to which the pennies are added will show the smallest change in temperature. Different substances absorb different amounts of energy to show changes in temperature.*

CHALLENGE *Little energy is required to increase their temperatures.*

Datasheet, Heat Transfer, p. 223

Technology Resources

Customize this student lab as needed or look for an alternative. Print rubrics to assess student lab reports.

Lab Generator CD-ROM

Teaching with Technology

A probeware system with a temperature probe can be used to measure temperatures.

Metacognitive Strategy

Ask students what they could do to improve the accuracy of their results.

Ongoing Assessment

Describe how heat is measured.

Ask: What is a calorie? *energy to raise the temperature of 1 g of water 1°C*

CHECK YOUR READING *Answer: in calories or joules; the amount of energy transferred between substances*

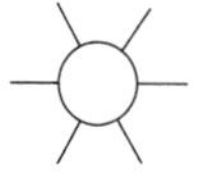

VOCABULARY
Remember to make description wheel diagrams for *calorie, joule,* and other vocabulary terms.

Measuring Heat

The most common units of heat measurement are the calorie and the joule (jool). One **calorie** is the amount of energy needed to raise the temperature of 1 gram of water by 1°C. The **joule** (J) is the standard scientific unit in which energy is measured. One calorie is equal to 4.18 joules.

D You probably think of calories in terms of food. However, in nutrition, one Calorie—written with a capital C—is actually one kilocalorie, or 1000 calories. This means that one Calorie in food contains enough energy to raise the temperature of 1 kilogram of water by 1°C. So, each Calorie in food contains 1000 calories of energy.

How do we know how many Calories are in a food, such as a piece of chocolate cake? The cake is burned inside an instrument called a calorimeter. The amount of energy released from the cake through heat is the number of Calories transferred from the cake to the calorimeter. The energy transferred to the calorimeter is equal to the amount of energy originally in the cake. A thermometer inside the calorimeter measures the increase in temperature from the burning cake, which is used to calculate how much energy is released.

CHECK YOUR READING How is heat measured?

INVESTIGATE Heat Transfer

Which substances change temperature faster?

SKILL FOCUS Measuring

MATERIALS
- graduated cylinder
- balance
- room-temperature water
- pennies
- aluminum foil
- hot tap water
- 100 mL beaker
- 3 plastic cups
- thermometer
- stopwatch

TIME 30 minutes

PROCEDURE

1. Using the graduated cylinder and the balance, separately measure 20 g of room-temperature water, 20 g of pennies, and 20 g of aluminum foil. Pour the water into a beaker until it is needed.
2. Using the graduated cylinder, pour 50 mL of hot water into each of the cups. Record the water temperature in each cup.
3. Pour the room-temperature water into one cup. Place the pennies in the second cup and the foil in the third. After 5 minutes, record the temperature of the water in each of the cups.

WHAT DO YOU THINK?
- How did the temperature changes in the three cups compare?
- What might account for the differences you observed?

CHALLENGE Why might items such as pots and pans be made of materials like copper, stainless steel, or iron?

DIFFERENTIATE INSTRUCTION

More Reading Support

C What are two common units for measuring heat? *calorie and joule*

D How many calories are in a Calorie, with a capital *C*? *1000*

Advanced Have students investigate and compare the number of Calories in fats, carbohydrates, and proteins; the number of Calories in foods they like; or the number of Calories that different people need. Have students make a table of their results to share with the class. What happens when too many Calories are consumed?

Challenge and Extension, p. 222

Some substances change temperature more easily than others.

Have you ever seen an apple pie taken right out of the oven? If you put a piece of pie on a plate to cool, you can touch the pie crust in a few minutes and it will feel only slightly warm. But if you try to take a bite, the hot pie filling will burn your mouth. The pie crust cools much more quickly than the filling, which is mostly water.

Specific Heat

The amount of energy required to raise the temperature of 1 gram of a substance by 1°C is the **specific heat** of that substance. Every substance has its own specific heat value. So, each substance absorbs a different amount of energy in order to show the same increase in temperature.

If you look back at the definition of a calorie, you will see that it is defined in terms of water—one calorie raises the temperature of 1 gram of water by 1°C. So, water has a specific heat of exactly 1.00 calorie per gram per °C. Because one calorie is equal to 4.18 J, it takes 4.18 J to raise the temperature of one gram of water by 1°C. In joules, water's specific heat is 4.18 J per gram per °C. If you look at the specific heat graph shown below, you will see that 4.18 is an unusually large value. For example, one gram of iron has to absorb only 0.45 joules for its temperature to increase by 1°C.

READING TIP Joules per gram per °C is shown as $\frac{J}{g°C}$.

A substance with a high specific heat value, like water, not only has to absorb a large quantity of energy for its temperature to increase, but it also must release a large quantity of energy for its temperature to decrease. This is why the apple pie filling can still be hot while the pie crust is cool. The liquid filling takes longer to cool. The high specific heat of water is also one reason it is used as a coolant in car radiators. The water can absorb a great deal of energy and protect the engine from getting too hot.

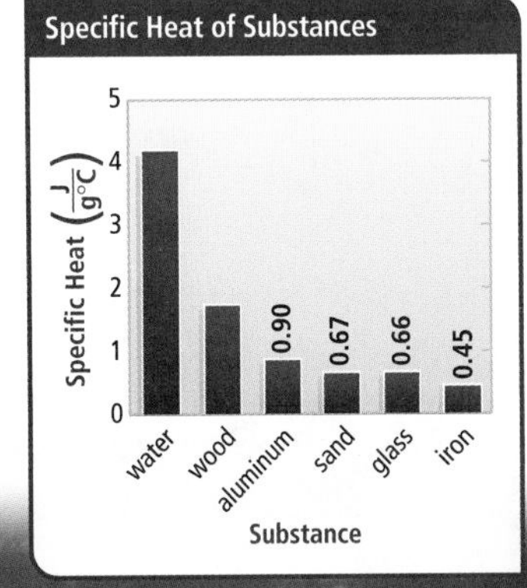

CHECK YOUR READING How is specific heat related to a change in temperature?

APPLY More energy is needed to warm water than many other substances. What materials in this photograph might be warmer than the water?

DIFFERENTIATE INSTRUCTION

More Reading Support

E What is the amount of energy needed to raise the temperature of 1 gram of a specific substance by 1°C? *the substance's specific heat*

Below Level If students have trouble understanding the concept of specific heat, ask them for examples of substances that get hot faster and stay hot longer than other substances. Possible examples include fruit or other fillings with a high water content inside muffins or pies. Explain that substances that stay hot longer have a higher specific heat value.

Teach from Visuals

To help students interpret the graph of specific heat, ask: What is the specific heat of wood? *1.76 J/g°C* What does this mean? *One gram of wood absorbs 1.76 J for its temperature to rise 1°C.*

Teach Difficult Concepts

As the atoms of elements become larger, less energy is required to produce an increase in temperature. Hydrogen has the greatest specific heat value (14.3 J/g°C) and is the smallest atom. The general relationship between atomic mass and specific heat was expressed by the Dulong-Petit law in 1819, about 50 years before Mendeleyev's modern periodic table.

Teacher Demo

To help students understand how substances absorb varying amounts of energy, do the following demonstration. Light a candle and ask students what will happen if you hold an index card over the flame. Hold the card over the flame; it will catch fire. (Keep a bucket of water nearby to put out the fire.) Gather some crayon shavings and place them on top of a second index card. Ask students what they think will happen this time. Hold the card and shavings over the flame. The shavings will melt, but the card will not burn. Remove the card after the wax is melted. Ask students why this second index card did not burn.

Ongoing Assessment

Understand why some substances change temperature more easily than others.

Ask: What determines the amount of energy a substance needs to absorb to increase its temperature? *specific heat*

PHOTO CAPTION Answer: metals, sand, wood, glass

CHECK YOUR READING *Answer: The greater a substance's specific heat, the more energy is required to raise that substance's temperature.*

Ongoing Assessment

CHECK YOUR READING *Answer: The more mass an object has, the more thermal energy it has.*

Teach Difficult Concepts

To help students understand that a substance with a low temperature can have more thermal energy than a substance with a higher temperature, place a gallon jug of hot water and a cup of hotter water on a table. Record the temperature of the water in both containers and have students carefully feel the outside of the containers. Wait 20 minutes and repeat the observations. The gallon of hot water should be warmer. Point out that the mass of a substance, in addition to its temperature, determines how much thermal energy it has.

Reinforce the BIG idea

Have students relate the section to the Big Idea.

 Reinforcing Key Concepts, p. 224

4.2 ASSESS & RETEACH

Assess

 Section 4.2 Quiz, p. 62

Reteach

Divide the class into groups. Make a three-column chart on the board to compare and contrast heat, thermal energy, and specific heat. Ask volunteers to fill in the chart with examples, characteristics, and misconceptions for each term.

Technology Resources

Have students visit **ClassZone.com** for reteaching of Key Concepts.

 CONTENT REVIEW

 CONTENT REVIEW CD-ROM

Specific Heat and Mass

Recall that thermal energy is the total kinetic energy of all particles in an object. So, thermal energy depends on the object's mass. Suppose you have a cup of water at a temperature of 90°C (194°F) and a bathtub full of water at a temperature of 40°C (104°F). Which mass of water has more thermal energy? There are many more water molecules in the bathtub, so the water in the tub has more thermal energy.

Specific Heat, Mass, and Weather

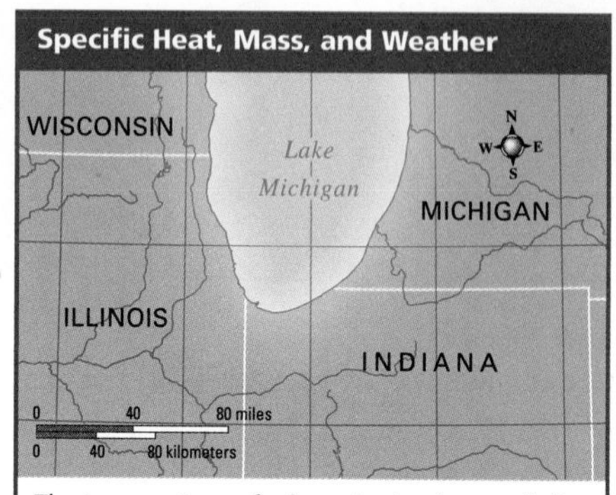

The temperature of a large body of water influences the temperature of nearby land. The green shading shows how far this effect extends.

The water in the cup has the same specific heat as the water in the tub. However, the cup of water will cool more quickly than the water in the bathtub. The tub of water has to release more thermal energy to its surroundings, through heat, to show a decrease in temperature because it has so much more mass.

This idea is particularly relevant to very large masses. For example, Lake Michigan holds 4.92 quadrillion liters (1.30 quadrillion gallons) of water. Because of the high specific heat of water and the mass of water in the lake, the temperature of Lake Michigan changes very slowly.

The temperature of the lake affects the temperatures on its shores. During spring and early summer, the lake warms slowly, which helps keep the nearby land cooler. During the winter, the lake cools slowly, which helps keep the nearby land warmer. Temperatures within about 15 miles of the lake can differ by as much as 10°F from areas farther away from the lake.

As you will read in the next section, the way in which a large body of water can influence temperatures on land depends on how energy is transferred through heat.

 How does an object's thermal energy depend on its mass?

4.2 Review

KEY CONCEPTS

1. How is temperature related to heat?
2. How do the units that are used to measure heat differ from the units that are used to measure temperature?
3. Describe specific heat in your own words.

CRITICAL THINKING

4. **Compare and Contrast** How are a calorie and a joule similar? How are they different?
5. **Synthesize** Describe the relationships among kinetic energy, temperature, heat, and thermal energy.

CHALLENGE

6. **Infer** Suppose you are spending a hot summer day by a pool. Why might the water in the pool cool the air near the pool?

ANSWERS

1. Heat transfers energy because a temperature difference exists.

2. Heat is measured in terms of temperature change.

3. the amount of energy that 1 gram of a substance needs to absorb to increase in temperature by 1°C

4. Both are used to measure heat. The joule is the standard unit for energy. One calorie equals 4.18 joules.

5. Temperature is the average kinetic energy of particles in an object; thermal energy is the total kinetic energy of particles in an object. Heat transfers thermal energy between objects due to temperature differences.

6. Water has a high specific heat, and heat transfers energy from the air to the cooler water.

CHEF

Cooking with Heat

A chef makes many decisions about cooking a meal based on heat and temperature. The appropriate temperature and cooking method must be used. A chef must calculate the cooking time of each part of the meal so that everything is finished at the same time. A chef also needs to understand how heat moves through food. For example, if an oven temperature is too hot, meat can be overcooked on the outside and undercooked on the inside.

Bread vs. Meat

Chefs have to understand how energy, as heat, is transferred to different foods. For example, the fluffy texture of bread comes from pockets of gas the separate its fibers. Because the gas is a poor conductor of energy, more energy, and a longer cooking time, is needed to cook bread than is needed to cook an equal amount of meat.

What Temperature?

Eggs cook very differently under different temperatures. For example, temperature is important when baking meringue, which is made of egg whites and sugar. A Key lime pie topped with meringue is baked at 400°F to make a meringue that is soft. However, meringue baked at 275°F makes light and crisp dessert shells.

Roasting and Heat

The shape of the food being roasted is just as important as what is being roasted. Heat moves more quickly through food with a thin shape than it will through food with a thicker shape.

EXPLORE

1. **COMPARE** Using a cookbook, find the oven temperatures for baking biscuits, potatoes, and beef. Could you successfully cook a roast and biscuits in the oven at the same time?
2. **CHALLENGE** Crack open three eggs. Lightly beat one egg in each of three separate bowls. Follow the steps below.
   1. Heat about two cups of water to 75°C in a small pan.
   2. Pour one of the eggs into the water in the pan.
   3. Observe the egg and record your observations.
   4. Repeat steps 1–3 twice, once with boiling water and then with room-temperature water.

   Describe the differences that you observed between the three eggs. What may account for these differences?

SCIENCE ON THE JOB

Relevance of Science to Non-science Jobs

Set Learning Goal

To understand why chefs need knowledge of heat and temperature

Present the Science

Many foods, such as beef, chicken, pork, and eggs, can contain microorganisms such as bacteria that cause food poisoning. Cooking foods until they reach a certain temperature kills these bacteria. Meat thermometers and other thermometers help ensure that foods reach the correct temperature. In addition, refrigerating leftovers and keeping foods cold help slow the growth of bacteria. Heat and temperature are also important to a chef because part of the job involves the presentation and appeal of food. The structure of protein changes when exposed to high temperatures, so a chef has to know how the texture of foods containing a large amount of protein will change during and after cooking.

Discussion Questions

- Ask: What can cause meat to be overcooked on the outside and undercooked on the inside? *an oven temperature that is too hot*
- Ask: How does the shape of a roast affect how fast it cooks? *Heat moves more quickly through a thin roast than a thick roast.*

Close

Ask: How do chefs use their knowledge of heat and temperature? *to choose the right temperature and cooking method; to finish cooking every part of the meal at the right time; to understand how heat moves through food*

EXPLORE

1. ***COMPARE*** *By changing the length of time they are in the oven, or by putting them in the oven at different times.*
2. ***CHALLENGE*** *The egg in room-temperature water doesn't visibly change. In boiling water it stays in one piece and turns white right away. In 75°C water it spreads out and slowly turns slightly white. The water temperature determines what happens to the eggs.*

4.3 FOCUS

Set Learning Goals

Students will

- Explain how energy is transferred through heat.
- Describe how materials are used to control the transfer of energy through heat.

3-Minute Warm-Up

Display Transparency 29 or copy this exercise on the board:

Match each definition with the correct term.

Definitions

1. the flow of energy from warmer objects to cooler objects *c*
2. the average kinetic energy of particles in an object *a*
3. the total kinetic energy of particles in an object *d*

Terms

a. temperature
b. specific heat
c. heat
d. thermal energy
e. kinetic energy

 3-Minute Warm-Up, p. T29

4.3 MOTIVATE

EXPLORE Conduction

PURPOSE To observe the transfer of energy between objects in direct contact

TIP *10 min.* Make sure the hot water in the large beaker does not overflow into the smaller beaker.

WHAT DO YOU THINK? *The temperature of the cold water increased, and the temperature of the hot water decreased. Energy flowed from warm to cooler; the changes in temperature are indicated by the two thermometers.*

KEY CONCEPT

4.3 The transfer of energy as heat can be controlled.

BEFORE, you learned

- Temperature is the average amount of kinetic energy of particles in an object
- Heat is the flow of energy from warmer objects to cooler objects

NOW, you will learn

- How energy is transferred through heat
- How materials are used to control the transfer of energy through heat

VOCABULARY

conduction p. 117
conductor p. 117
insulator p. 117
convection p. 118
radiation p. 119

EXPLORE Conduction

How can you observe a flow of energy?

PROCEDURE

1. Fill the large beaker halfway with hot tap water. Fill the small beaker halfway with cold water. Place a thermometer in each beaker. Record the temperature of the water in each beaker.
2. Without removing the water in either beaker, place the small beaker inside the large beaker. Record the temperature in each beaker every 30 seconds for 2 minutes.

MATERIALS

- 500 mL beaker
- hot tap water
- 200 mL beaker
- cold water
- 2 thermometers
- stopwatch

WHAT DO YOU THINK?

- How did the water temperature in each beaker change?
- In which direction did energy flow? How do you know?

NOTE-TAKING STRATEGY
Main idea and detail notes would be a useful strategy for taking notes on how heat transfers energy.

Energy moves as heat in three ways.

Think about what you do to keep warm on a cold day. You may wear several layers of clothing, sit next to a heater, or avoid drafty windows. On a hot day, you may wear light clothing and sit in the shade of a tree. In all of these situations, you are trying to control the transfer of energy between yourself and your surroundings.

Recall that heat is always a transfer of energy from objects at a higher temperature to objects at a lower temperature. How does energy get transferred from a warmer object to a cooler one? There are three different ways in which this transfer of energy can occur—by conduction, convection, and radiation. So, in trying to control heat, it is necessary to control conduction, convection, and radiation.

A 116 Unit: Matter and Energy

RESOURCES FOR DIFFERENTIATED INSTRUCTION

Below Level

UNIT RESOURCE BOOK

- Reading Study Guide A, pp. 227–228
- Decoding Support, p. 239

AUDIO CDS

 **Additional INVESTIGATION,** Observing Convection, A, B, & C, pp. 251–259; Teacher Instructions, pp. 262–263

Advanced

UNIT RESOURCE BOOK

- Challenge and Extension, p. 233
- Challenge Reading, pp. 235–236

English Learners

UNIT RESOURCE BOOK

Spanish Reading Study Guide, pp. 231–232

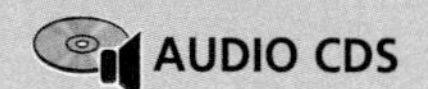

- Audio Readings in Spanish
- Audio Readings (English)

Conduction

One way in which energy is transferred as heat is through direct contact between objects. **Conduction** is the process that moves energy from one object to another when they are touching physically. If you have ever picked up a bowl of hot soup, you have experienced conduction.

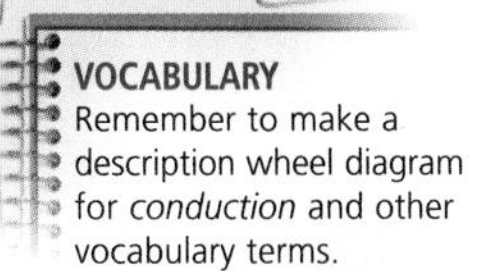

VOCABULARY
Remember to make a description wheel diagram for *conduction* and other vocabulary terms.

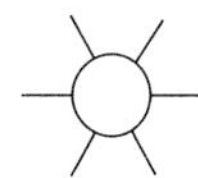

Conduction occurs any time that objects at different temperatures come into contact with each other. The average kinetic energy of particles in the warmer object is greater than that of the particles in the cooler object. When particles of the objects collide, some of the kinetic energy of the particles in the warmer object is transferred to the cooler object. As long as the objects are in contact, conduction continues until the temperatures of the objects are equal.

Conduction can also occur within a single object. In this case, energy is transferred from the warmer part of the object to the cooler part of the object by heat. Suppose you put a metal spoon into a cup of hot cocoa. Energy will be conducted from the warm end of the spoon to the cool end until the temperature of the entire spoon is the same.

Conduction transfers energy from the cocoa to the mug to the person's hands.

Some materials transfer the kinetic energy of particles better than others. **Conductors** are materials that transfer energy easily. Often, conductors also have a low specific heat. For example, metals are typically good conductors. You know that when one end of a metal object gets hot, the other end quickly becomes hot as well. Consider pots or pans that have metal handles. A metal handle becomes too hot to touch soon after the pan is placed on a stove that has been turned on.

Other materials, called **insulators,** are poor conductors. Insulators often have high specific heats. Some examples of insulators are wood, paper, and plastic foam. In fact, plastic foam is a good insulator because it contains many small spaces that are filled with air. A plastic foam cup will not easily transfer energy by conduction. As a result, plastic foam is often used to keep cold drinks cold or hot drinks hot. Think about the pan handle mentioned above. Often, the handle is made of a material that is an insulator, such as wood or plastic. Although a wood or plastic handle will get hot when the pan is on a stove, it takes a much longer time for wood or plastic to get hot as compared to a metal handle.

CHECK YOUR READING How are conductors and insulators different?

DIFFERENTIATE INSTRUCTION

More Reading Support

A What do two objects need for conduction to occur? *touch and a temperature difference*

B What are poor conductors of energy? *insulators*

English Learners Point out different word forms to English learners as they appear in the text.

conductor	insulator	radiate
conduction	insulation	radiation

Have students look up one of the pairs of words and compare their definitions. Ask students what they might conclude about the word ending *-tion.*

4.3 INSTRUCT

Teacher Demo

To help students understand heat conduction, place a small amount of wax at various spots on a copper rod and ask students to predict what will happen to the wax if the rod is heated on one end. Have students share their predictions about what will happen and their reasons for those predictions. Then use a candle to heat one end of the copper rod so students can test their predictions. The wax closest to the heat source will melt first, and the wax farthest away will melt last. Based on their observations, have them discuss the behavior of heat when a copper rod is heated.

Teach from Visuals

To help students interpret the photograph of conduction from the mug of cocoa, ask: What is happening in the photograph? *The hot cocoa transfers energy through the wall of the mug to the hand holding the mug and to the air.*

Develop Critical Thinking

PREDICT Have students predict what will happen in terms of particle movement when energy is conducted from the hot cocoa through the mug to a person's hands. *Particles that are moving faster will bump into slower particles and transfer energy.*

EXPLORE the BIG idea

Revisit "Does It Chill?" on p. 101. Have students explain their results.

Ongoing Assessment

CHECK YOUR READING *Answer: Conductors easily transfer energy, but insulators do not. Conductors often have low specific heats, and insulators often have high specific heats.*

Address Misconceptions

IDENTIFY Ask: In what direction, such as upward, downward, or sideways, is energy transferred through heat? If students say only upward, they may hold the misconception that heat only rises.

CORRECT Point out that the common phrase "Heat rises" is misleading. Have students look at the diagram again. Point out that warm air is pushed up by cooler, denser air. To show that heat can travel in any direction, use a radiant heat source. Have students put their hand close to the heat source, but not above it, to observe that heat travels in all directions. The demonstration on p. 117 also shows how heat travels in different directions.

REASSESS Ask students to describe what causes warm air to rise. *It is pushed up by cooler, denser air.*

Technology Resources

Visit **ClassZone.com** for background on common student misconceptions.

MISCONCEPTION DATABASE

Teach from Visuals

To help students interpret the diagram of convection in air, ask:

- What happens to air as it cools? *It becomes more dense and sinks.*
- What happens to cooler, denser air when it moves under warm air? *It is warmed by conduction from the ground.*

Integrate the Sciences

Convection is the process thought to be responsible for the movement of Earth's internal energy. As hot mantle near Earth's core rises toward the crust, it cools and is pushed aside by hot mantle rising beneath it. It continues to cool, becomes more dense, and sinks back toward Earth's core, creating convection currents.

Ongoing Assessment

more dense at 2 where air is cool; less dense at 1 where air is warmer

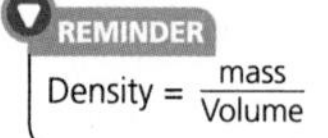

REMINDER

$\text{Density} = \frac{\text{mass}}{\text{Volume}}$

As you read about the cycle that occurs during convection, follow the steps in the illustration below.

Convection

Energy can also be transferred through the movement of gases or liquids. **Convection** is the process that transfers energy by the movement of large numbers of particles in the same direction within a liquid or gas. In most substances, as the kinetic energy of particles increases, the particles spread out over a larger area. An increased distance between particles causes a decrease in the density of the substance. Convection occurs when a cooler, denser mass of the gas or liquid replaces a warmer, less dense mass of the gas or liquid by pushing it upward.

Convection is a cycle in nature responsible for most winds and ocean currents. When the temperature of a region of air increases, the particles in the air spread out and the air becomes less dense.

1. Cooler, denser air flows in underneath the warmer, less dense air, and pushes the warmer air upward.
2. When this air cools, it becomes more dense than the warmer air beneath it.
3. The cooled air sinks and moves under the warmer air.

Convection in liquids is similar. Warm water is less dense than cold water, so the warm water is pushed upward as cooler, denser water moves underneath. When the warm water that has been pushed up cools, its density increases. The cycle continues when this more dense water sinks, pushing warmer water up again.

Recall that a large body of water, such as Lake Michigan, influences the temperature of the land nearby. This effect is due to convection. During the spring and early summer, the lake is cool and warms more slowly than the land. The air above the land gets warmer than the air over the water. The warmer air above the land is less dense than the cooler air above the water. The cooler, denser air moves onshore and pushes the warmer air up. The result is a cooling breeze from the lake.

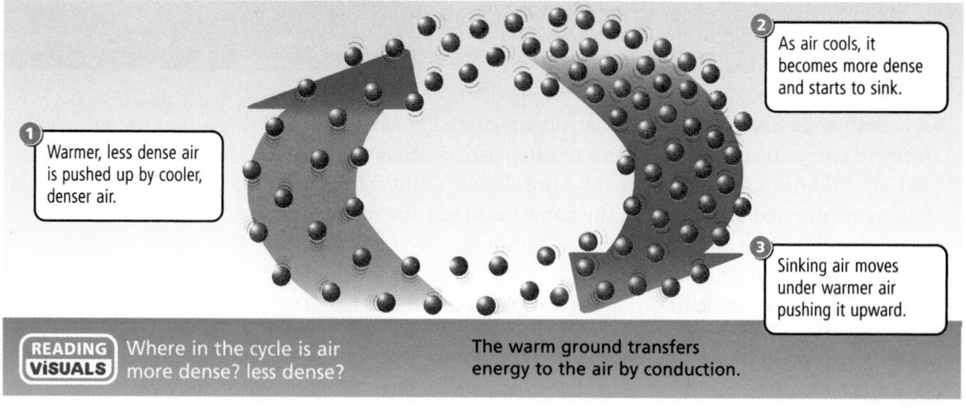

1 Warmer, less dense air is pushed up by cooler, denser air.
2 As air cools, it becomes more dense and starts to sink.
3 Sinking air moves under warmer air pushing it upward.
The warm ground transfers energy to the air by conduction.

READING VISUALS Where in the cycle is air more dense? less dense?

DIFFERENTIATE INSTRUCTION

More Reading Support

C What is the process that transfers energy by the motion of many particles in liquids or gases? *convection*

Additional Investigation To reinforce Section 4.3 learning goals, use the following full-period investigation:

Additional INVESTIGATION, Observing Convection, A, B, & C, pp. 251–259, 262–263

Below Level Remind students that density is the amount of mass of a substance in a certain volume. Because the particles in warm air are farther apart than the particles in cold air, a certain volume of warm air has fewer particles than the same volume of cold air. Warm air therefore is less dense than cold air.

Radiation

Radiation is another way in which energy can be transferred from one place to another. **Radiation** is energy that travels as electromagnetic waves, which include visible light, microwaves, and infrared (IHN-fruh-REHD) light. The Sun is the most significant source of radiation that you experience on a daily basis. However, all objects—even you—emit radiation and release energy to their surroundings.

Consider radiation from the Sun. You can feel radiation as heat when radiation from the Sun warms your skin. The radiation emitted from the Sun strikes the particles in your body and transfers energy. This transfer of energy increases the movement of particles in your skin, which you detect as an increase in temperature. Of course, you are not the only object on Earth that absorbs the Sun's radiation. Everything—from air to concrete sidewalks—absorbs radiation that increases particle motion and produces an increase in temperature.

When radiation from the Sun is absorbed, energy is transferred through heat.

When radiation is emitted from one object and then is absorbed by another, the result is often a transfer of energy through heat. Like both conduction and convection, radiation can transfer energy from warmer to cooler objects. However, radiation differs from conduction and convection in a very significant way. Radiation can travel through empty space, as it does when it moves from the Sun to Earth. If this were not the case, radiation from the Sun would have no effect on Earth.

SIMULATION CLASSZONE.COM

Identify examples of conduction, convection, or radiation.

CHECK YOUR READING How does radiation transfer energy?

Different materials are used to control the transfer of energy.

Energy is always being transferred between objects at different temperatures. It is often important to slow this movement of energy. For example, if energy were always transferred quickly and efficiently through heat, it would not be possible to keep a building warm during a cold day or to keep cocoa hot in a thermos.

DIFFERENTIATE INSTRUCTION

More Reading Support

D What is both a form of energy and a way that heat transfers energy? *radiation*

Advanced Have students research solar air heaters and try to construct one in a window. Solar air heaters typically have the following characteristics: clear covers, dark interior surfaces, and insulating materials.

R Challenge and Extension, p. 233

Have students who are interested in how heat is transferred on the sun read the following article:

R Challenge Reading, pp. 235–236

Teach from Visuals

To help students interpret the diagram and photograph of the person being warmed by radiation from the sun, ask:

- What is striking particles of this person's skin and transferring energy? *radiation*
- What happens to the particles of skin? *Their movement increases, and the person detects an increase in temperature.*

Real World Example

Radiation from the Sun is used in buildings designed for passive solar heating. In the Northern Hemisphere, these buildings have many windows facing south, so that more sunlight will shine into the building. The walls and floors are designed to absorb radiation, and their temperature increases when sunlight hits them. Conduction transfers the energy from the walls and floor to the air in the building.

Integrate the Sciences

Cold-blooded animals cannot maintain their own internal body temperature, so it remains about the same as the temperature of their environment. Some animals, such as turtles, frogs, lizards, and butterflies, lie in sunlight to absorb radiation from the Sun; others use muscle movement to increase their body temperature.

Ongoing Assessment

Explain how energy is transferred through heat.

Ask: What are three ways that energy can be transferred? *conduction, convection, radiation*

CHECK YOUR READING *Answer: When electromagnetic waves strike an object, they transfer energy to the object.*

Teach from Visuals

To help students interpret the diagrams and photographs of the polar bear and the vacuum flask, ask:

- Why is the polar bear's hollow hair an effective insulator? *Air inside the hair does not easily conduct energy from the warm bear to the cold air.*
- What is the insulator in the vacuum flask? *empty space*

 This visual is also available as T30 in the Unit Transparency Book.

Integrate the Sciences

In some homes, much of the energy used for heating is wasted because energy flows through leaks around windows and doors and in floors, walls, and ceilings. Using more fuel and electricity adds to the pollution of the environment and to the depletion of nonrenewable resources. Properly insulating a home makes it more energy efficient and saves money.

Develop Critical Thinking

EVALUATE In addition to hollow guard hairs and a thick layer of fat, polar bears have other characteristics that help keep them warm in their environment. Their fur is oily, and their skin is black. How might these characteristics help? *Radiation that penetrates a polar bear's fur is absorbed more readily by the black pigment of the skin; the oily fur repels water so a polar bear can easily shake off water before the water freezes into ice.*

Ongoing Assessment

READING VISUALS *Sample answer: It provides insulation and slows the transfer of energy through conduction. In the hair, air is the insulator; in the vacuum flask, empty space is the insulator.*

Insulation
Insulators used by people are similar to insulators in nature. Polar bears are so well insulated that they tend to overheat.
Vacuum Flask
The polar bear's hollow guard hair is an effective insulator because air inside the hair does not easily conduct energy.
hollow hair
hot liquid (inside flask)
air (outside flask)
inner reflective layer
empty space
outer case
The empty space between layers in a vacuum flask prevents the conduction of energy through heat.
Polar bears have several layers of insulation. They have a layer of fat up to 11 cm thick, a 2.5–5.0 cm thick layer of fur, and an outer layer of hollow guard hairs.
READING VISUALS How is the polar bear's hollow hair similar to the empty space in a vacuum flask? How is it different?
A 120 Unit: Matter and Energy

DIFFERENTIATE INSTRUCTION

Alternative Assessment Have students write a paragraph describing the insulators in the photograph and how they work.

Insulators are used to control and slow the transfer of energy from warmer objects to cooler objects because they are poor conductors of energy. You can think of an insulator as a material that keeps cold things cold or hot things hot.

Sometimes people say that insulation "keeps out the cold." An insulator actually works by trapping energy. During the winter, you use insulators such as wool to slow the loss of your body heat to cold air. The wool traps air against your body, and because both air and wool are poor conductors, you lose body heat at a slower rate. Fiberglass insulation in the outer walls of a building works in the same way. The fiberglass slows the movement of energy from a building to the outside during cold weather, and it slows the movement of energy into the building during hot weather.

A vacuum flask, or thermos, works in a slightly different way to keep liquids either hot or cold. Between two layers of the flask is an empty space. This space prevents conduction between the inside and outside walls of the flask. Also, the inside of the flask is covered with a shiny material that reflects much of the radiation that strikes it. This prevents radiation from either entering or leaving the flask.

Insulators that people use are often very similar to insulators in nature. Look at the photograph of the polar bear on page 120. Because of the arctic environment in which the polar bear lives, it needs several different types of insulation. The polar bear's fur helps to trap a layer of air against its body to keep warmth inside. Polar bears also have guard hairs that extend beyond the fur. These guard hairs are hollow and contain air. Because air is a poor conductor, the bear's body heat is not easily released into the air.

How does insulation keep a building warm?

KEY CONCEPTS

1. What are three ways in which energy can be transferred through heat? Provide an example of each.
2. Explain how convection is a cycle in nature.
3. Describe how an insulator can slow a transfer of energy.

CRITICAL THINKING

4. **Compare and Contrast** Describe the similarities and differences among conduction, convection, and radiation.
5. **Synthesize** Do you think solids can undergo convection? Why or why not? Explain.

CHALLENGE

6. **Infer** During the day, wind often blows from a body of water to the land. What do you think would happen at night? Explain.

ANSWERS

1. Sample answer: conduction, metal spoon in hot cocoa; convection, air moving in a cycle; radiation, sunlight transferring energy to skin

2. Warm air is pushed up by cooler, denser air. As air cools, it becomes more dense, sinks, and moves under warmer air.

3. It is a poor conductor, so energy is only slowly transferred.

4. All are ways that heat transfers energy. Conduction occurs only with direct contact. Convection occurs in a cycle in gases or liquids. Radiation travels as EM waves through empty space.

5. No; particles in a solid cannot move freely.

6. The land cools more quickly than the water, and wind blows from the land to the water.

Ongoing Assessment

Describe how different materials are used to control the transfer of energy through heat.

Ask: What property do materials that are used to control the transfer of energy as heat have in common? *They trap energy.*

CHECK YOUR READING *Answer: by slowing the transfer of energy from the building to the cooler outside air*

Reinforce 

Have students relate the section to the Big Idea.

 Reinforcing Key Concepts, p. 234

4.3 ASSESS & RETEACH

Assess

 Section 4.3 Quiz, p. 63

Reteach

Ask students to help you make a Venn diagram on the board that compares conduction, convection, and radiation. Ask volunteers to come forward and contribute examples of each.

Technology Resources

Have students visit **ClassZone.com** for reteaching of Key Concepts.

 **CONTENT REVIEW**

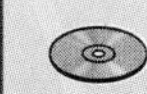 **CONTENT REVIEW CD-ROM**

CHAPTER INVESTIGATION

Focus

PURPOSE To design and test an insulated bottle that slows a change in the temperature of water as compared to a noninsulated control bottle

OVERVIEW Students will design and build an insulated bottle. They will measure the temperature change of water in the insulated bottle and a control bottle. Students will find the following:

- The temperature of the water in the control bottle should change more than that of the water in the insulated bottle.
- Insulation slows the transfer of energy through heat.

Lab Preparation

- Have students bring in small plastic bottles that hold over 200 mL. Each group should use two identical bottles.
- Prior to the investigation, have students read through the investigation and prepare their data tables. Or you may wish to copy and distribute datasheets and rubrics.

UNIT RESOURCE BOOK, pp. 242–250

SCIENCE TOOLKIT, F13

Lab Management

- Review with students how to put the thermometers in the bottles and hold them in place with clay so that they do not touch the bottom or sides of the bottle.
- Students should be ready to insert the thermometers and clay before getting the hot or cold water.

Teaching with Technology

A probeware system with a temperature probe can be used to measure and record temperatures. A CBL system can graph the data.

CHAPTER INVESTIGATION

Insulators

DESIGN — YOUR OWN —

OVERVIEW AND PURPOSE

To keep warm in cold weather, a person needs insulation. A down-filled coat, such as the one worn by the girl in the photograph, is a very effective insulator because it contains a great deal of air. Energy is transferred rapidly through some substances and quite slowly through others. In this investigation, you will

- design and build an insulator for a bottle to maintain the temperature of the water inside
- test an unchanged bottle and your experimental bottle to see which maintains the water's temperature more effectively

Problem

How can a bottle be insulated most effectively?

Procedure

1. Create a data table similar to the one shown on the sample notebook page to record your measurements.
2. Set aside plastic bottles, thermometers, modeling clay, and a graduated cylinder. Decide whether you will test hot or cold water in your bottles.
3. From the other materials available to you, design a way to modify one of the bottles so that it will keep the temperature of the water constant for a longer period of time than the control bottle.
4. Build your modified bottle by using one or more of the insulating materials available.

MATERIALS
- 2 small plastic bottles
- 2 thermometers
- modeling clay
- graduated cylinder
- tap water (hot or cold)
- foam packing peanuts
- plastic wrap
- aluminum foil
- soil
- sand
- rubber bands
- coffee can
- beaker
- stopwatch

INVESTIGATION RESOURCES

CHAPTER INVESTIGATION, Insulators
- Level A, pp. 242–245
- Level B, pp. 246–249
- Level C, p. 250

Advanced students should complete Levels B & C.

Writing a Lab Report, D12–13

Technology Resources

Customize this student lab as needed or look for an alternative. Print rubrics to assess student lab reports.

Lab Generator CD-ROM

5 Fill each bottle with 200 mL of hot or cold water. Make sure that the water in each bottle is the same temperature.

6 Place a thermometer into each bottle. The thermometers should touch only the water, not the bottom or sides of the bottles. Use modeling clay to hold the thermometers in place in the bottles.

7 Record the starting temperature of the water in both bottles. Continue to observe and record the temperature of the water in both bottles every 2 minutes for 30 minutes. Record these temperatures in your data table.

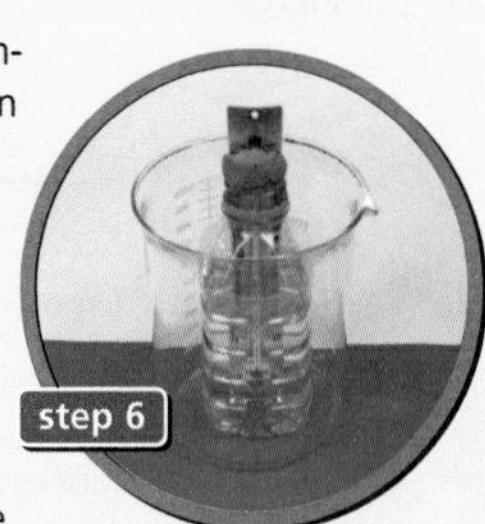
step 6

Observe and Analyze

Write It Up

1. **COMMUNICATE** Draw the setup of your experimental bottle in your notebook. Be sure to label the materials that you used to insulate your experimental bottle.
2. **RECORD OBSERVATIONS** Make sure you record all of your measurements and observations in the data table.
3. **GRAPH** Make a double line graph of the temperature data. Graph temperature versus time. Plot the temperature on the vertical axis, or *y*-axis, and the time on the horizontal axis, or *x*-axis. Use different colors to show the data from the different bottles.
4. **IDENTIFY VARIABLES, CONTROLS, AND CONSTANTS** Which bottle was the control? What was the variable? What were the constants in both setups?
5. **ANALYZE** Obtain the experimental results from two other groups that used a different insulator. Compare your results with the results from the other groups. Which bottle changed temperature most quickly?

Conclude

1. **EVALUATE** Explain why the materials used by different groups might have been more or less effective as insulators. How might you change your design to improve its insulating properties?
2. **IDENTIFY LIMITS** Describe possible sources of error in the procedure or any points at which errors might have occurred. Why is it important to use the same amount of water in both bottles?
3. **APPLY** Energy can be transferred as heat sby radiation, conduction, and convection. Which of these processes might be slowed by the insulation around your bottle? Explain.

INVESTIGATE Further

CHALLENGE We depend on our clothing to keep us from losing body heat when we go outside in cold weather. How might you determine the type of clothing that would provide the best insulation? Design an experiment that would test your hypothesis.

Insulators

Problem How can a bottle be insulated most effectively?

Observe and Analyze

Table 1. Water Temperature Measurements

Time (min)	Control Bottle Temperature (°C)	Experimental Bottle Temperature (°C)
0		
2		
4		
6		
8		
10		

Conclude

Observe and Analyze

1. *See students' drawings.*
2. *See students' tables.*
3. *The graph should indicate that the water in the insulated bottle has a more stable temperature than the water in the control bottle.*
4. *The control was the unmodified bottle. The variable was the insulating material used. The amount and the starting temperature of the water were the same for both setups.*
5. *Answers will vary depending on the different insulators being compared.*

Conclude

1. *Sample answer: The most effective designs used materials that have the highest specific heats and contain a large amount of air. Materials that are tightly packed are less effective. A design may be improved by using a material with more effective insulating properties.*
2. *Possible sources of error include different starting temperatures, inaccurate volume measurements, thermometers touching the bottom or sides of the bottles, inaccurate temperature measurements, different-sized bottles, and one bottle being disturbed more than the other. The same amount of water is important because both bottles need to start with the same thermal energy.*
3. *Conduction is the form of energy transfer that will be most affected by the materials available, although transfer by radiation will also be affected.*

INVESTIGATE Further

CHALLENGE Answer: Students' experiments should have a control and test different types of clothing materials.

Post-Lab Discussion

Ask: Why didn't it matter if you used hot or cold water? *Heat is the transfer of energy from a warmer substance to a colder substance. The insulating material slows down the transfer of energy either from the hot water to the cooler bottle and air or from the warm air and bottle to the colder water.*

BACK TO

the BIG idea

Have students look at the photograph on pp. 100–101. Ask them to summarize how heat affects temperatures in the photograph. *Radiation from the Sun is transferring energy to particles in the ground and the giraffe's skin. This increases the kinetic energy of the particles in the ground and skin, which increases their temperatures.*

KEY CONCEPTS SUMMARY

SECTION 4.1

Ask: Which liquid has particles with a higher average kinetic energy, and how do you know? *The particles in the hot liquid have a higher average kinetic energy and are moving faster, as shown by the motion lines.*

SECTION 4.2

Ask: What is happening to energy in the lemonade and ice? *Energy is transferred through heat from the lemonade to the cooler ice.*

Ask: When will the transfer of energy stop? *when there is no temperature difference*

SECTION 4.3

Ask: How does energy transferred differently by conduction, convection, and radiation? *Conduction requires direct contact, convection requires movement of gases or liquids, and radiation travels via electromagnetic waves.*

Review Concepts

- Big Idea Flow Chart, p. T25
- Chapter Outline, pp. T31–T32

4 Chapter Review

the BIG idea

Heat is a flow of energy due to temperature differences.

KEY CONCEPTS SUMMARY

4.1 Temperature depends on particle movement.

- All particles in matter have kinetic energy.
- Temperature is the measurement of the average kinetic energy of particles in an object.
- Temperature is commonly measured on the Fahrenheit or Celsius scales.

Particles in a warmer substance have a greater average kinetic energy than particles in a cooler substance.

VOCABULARY
kinetic theory of matter p. 104
temperature p. 105
degree p. 106
thermometer p. 107

4.2 Energy flows from warmer to cooler objects.

- Heat is a transfer of energy from an object at a higher temperature to an object at a lower temperature.
- Different materials require different amounts of energy to change temperature.

Energy is transferred from the warmer lemonade to the cold ice through heat.

VOCABULARY
heat p. 110
thermal energy p. 111
calorie p. 112
joule p. 112
specific heat p. 113

4.3 The transfer of energy as heat can be controlled.

- Energy can be transferred by conduction, convection, and radiation.
- Different materials are used to control the transfer of energy.

Types of Energy Transfer

Conduction	Convection	Radiation
• Energy transferred by direct contact • Energy flows directly from warmer object to cooler object • Can occur within one object • Continues until object temperatures are equal	• Occurs in gases and liquids • Movement of large number of particles in same direction • Occurs due to difference in density • Cycle occurs while temperature differences exist	• Energy transferred by electromagnetic waves such as light, microwaves, and infrared radiation • All objects radiate energy • Can transfer energy through empty space

VOCABULARY
conduction p. 117
conductor p. 117
insulator p. 117
convection p. 118
radiation p. 119

A 124 Unit: Matter and Energy

Technology Resources

Have students visit **ClassZone.com** or use the CD-ROM for a cumulative review of concepts.

 **CONTENT REVIEW**

 **CONTENT REVIEW CD-ROM**

Engage students in a whole-class interactive review of Key Concepts. Edit content as you wish.

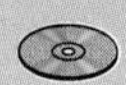 **POWER PRESENTATIONS**

Reviewing Vocabulary

Make a frame for each of the vocabulary terms listed below. Write the term in the center. Decide what information to frame it with. Use definitions, examples, descriptions, parts, or pictures.

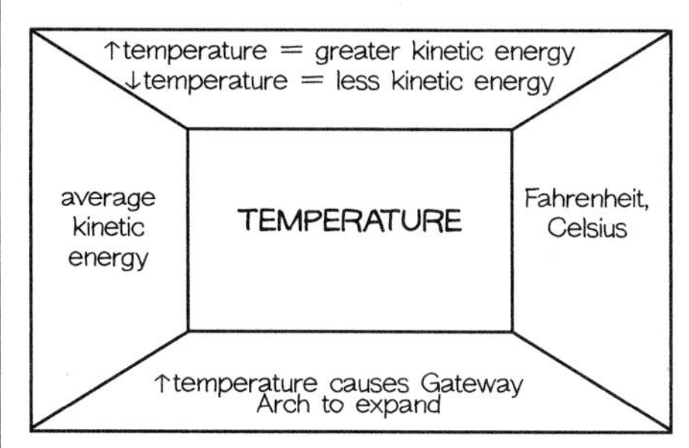

1. kinetic theory of matter
2. heat
3. thermal energy
4. conduction
5. convection
6. radiation

In two or three sentences, describe how the terms in the following pairs are related to each other. Underline each term in your answers.

7. calorie, joule
8. conductor, insulator

Reviewing Key Concepts

Multiple Choice *Choose the letter of the best answer.*

9. What is the zero point in the Celsius scale?
   a. the freezing point of pure water
   b. the boiling point of pure water
   c. the freezing point of mercury
   d. the boiling point of alcohol

10. Energy is always transferred through heat from?
   a. an object with a lower specific heat to one with a higher specific heat
   b. a cooler object to a warmer object
   c. an object with a higher specific heat to one with a lower specific heat
   d. a warmer object to a cooler object

11. The average kinetic energy of particles in an object can be measured by its
   a. heat
   b. thermal energy
   c. calories
   d. temperature

12. How is energy transferred by convection?
   a. by direct contact between objects
   b. by electromagnetic waves
   c. by movement of groups of particles in gases or liquids
   d. by movement of groups of particles in solid objects

13. The total kinetic energy of particles in an object is
   a. heat
   b. thermal energy
   c. calories
   d. temperature

14. Water requires more energy than an equal mass of iron for its temperature to increase because water has a greater
   a. thermal energy
   b. specific heat
   c. temperature
   d. kinetic energy

15. Energy from the Sun travels to Earth through which process?
   a. temperature
   b. conduction
   c. radiation
   d. convection

16. An insulator keeps a home warm by
   a. slowing the transfer of cold particles from outside to inside
   b. increasing the specific heat of the air inside
   c. slowing the transfer of energy from inside to outside
   d. increasing the thermal energy of the walls

17. Conduction is the transfer of energy from a warmer object to a cooler object through
   a. a vacuum
   b. a gas
   c. direct contact
   d. empty space

Short Answer *Write a short answer to each question.*

18. How are kinetic energy and temperature related to each other?

19. What is the difference between heat and temperature?

ASSESSMENT RESOURCES

UNIT ASSESSMENT BOOK
- Chapter Test A, pp. 64–67
- Chapter Test B, pp. 68–71
- Chapter Test C, pp. 72–75
- Alternative Assessment, pp. 76–77
- Unit Test, A, B, & C, pp. 78–89

SPANISH ASSESSMENT BOOK
- Spanish Chapter Test, pp. 225–228
- Spanish Unit Test, pp. 229–232

Technology Resources

Edit test items and answer choices.

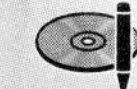 **Test Generator CD-ROM**

Visit **ClassZone.com** to extend test practice.

 **Test Practice**

Reviewing Vocabulary

1. *Frames should include that the kinetic theory of matter states that all particles in matter are in constant motion.*
2. *Frames should include that heat is the transfer of energy from an object at a higher temperature to one at a lower temperature.*
3. *Frames should include that thermal energy is the total amount of kinetic energy of the particles in an object.*
4. *Frames should include that conduction transfers energy through direct contact between objects.*
5. *Frames should include that convection transfers energy through the movement of many particles of a gas or liquid.*
6. *Frames should include that radiation transfers energy through electromagnetic waves.*
7. *Sample answer: The calorie and the joule are measures of heat. One calorie is the amount of energy needed to raise the temperature of 1 g of water by 1°C. One calorie is equal to 4.18 joules.*
8. *Sample answer: Conductors easily transfer energy, and insulators slow the transfer of energy. Conductors typically have a low specific heat, and insulators typically have a high specific heat.*

Reviewing Key Concepts

9. a
10. d
11. d
12. c
13. b
14. b
15. c
16. c
17. c

18. *Temperature measures the average kinetic energy of particles in a substance.*
19. *Heat is the transfer of energy between objects that differ in temperature. Temperature is a measurement of the average kinetic energy of particles in an object.*

Thinking Critically

20. *B has the higher temperature because its particles are moving faster.*
21. *If A were chilled, its particles would slow down and become more tightly packed. If B were warmed, its particles would speed up and have more space between them.*
22. *Energy would flow from B to A because B is warmer and heat always transfers energy from warm substances to cooler substances.*
23. *The illustrations would be identical because heat will transfer energy until the substances are the same temperature.*
24. *Both processes transfer energy. Convection occurs in large regions of gases and liquids, but not in solids. Conduction occurs by direct contact between substances.*
25. *Radiation directly from the Sun is being avoided. Conduction to any part of the body in contact with the ground is still felt. Convection might be felt if air is moving.*
26. *Check students' diagrams; answers should be similar in concept to the convection diagram on p. 118*

Using Math Skills in Science

27. *about 20°F*
28. *100 calories*
29. *418 joules*
30. *45 joules*

the BIG idea

31. *Answers will vary.*
32. *Answers should indicate that the kinetic theory of matter states that all particles of matter are in constant motion.*

UNIT PROJECTS

Have students present their projects. Use the appropriate rubrics from the URB to evaluate their work.

Unit Projects, pp. 5–10

Thinking Critically

The illustrations below show particle movement in a substance at two different temperatures. Use the illustrations to answer the next four questions.

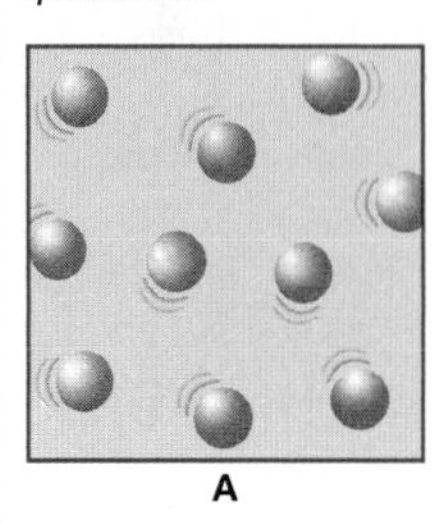
A

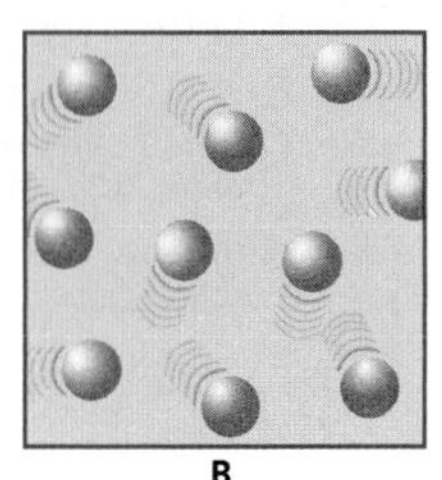
B

20. **OBSERVE** Which illustration represents the substance when it is at a higher temperature? Explain.
21. **PREDICT** What would happen to the particles in illustration A if the substance were chilled? What would happen if the particles in illustration B were warmed?
22. **PREDICT** If energy is transferred from one of the substances to the other through heat, in which direction would the energy flow (from A to B, or from B to A)? Why?
23. **COMMUNICATE** Suppose energy is transferred from one of the substances to the other through heat. Draw a sketch that shows what the particles of both substances would look like when the transfer of energy is complete. Explain.
24. **COMPARE AND CONTRAST** How are conduction and convection similar? How are they different?
25. **DRAW CONCLUSIONS** Suppose you are outdoors on a hot day and you move into the shade of a tree. Which form of energy transfer are you avoiding? Which type of energy transfer are you still feeling? Explain.
26. **COMMUNICATE** Draw a sketch that shows how convection occurs in a liquid. Label the sketch to indicate how the process occurs in a cycle.

Using Math Skills in Science

Use the illustrations of the two thermometers below to answer the next four questions.

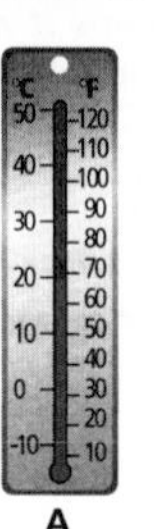

A

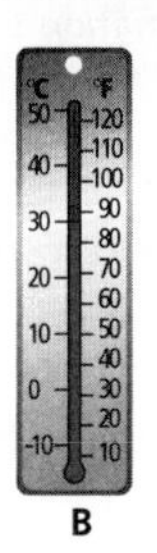

B

27. How much of a change in temperature occurred between A and B in the Fahrenheit scale?
28. Suppose the temperatures were measured in 10 g of water. How much energy, in calories, would have been added to cause that increase in temperature? (**Hint:** 1 calorie raises the temperature of 1 g of water by 1°C.)
29. Again, suppose the temperatures shown above were measured in 10 g of water. How much energy, in joules, would have been added? (**Hint:** 1 calorie = 4.18 joules.)
30. Suppose that the temperatures were measured for 10 g of iron. How much energy, in joules, would have been added to cause the increase in temperature? (**Hint:** see graph on p. 113.)

the BIG idea

31. **ANALYZE** Look back at the photograph and the question on pages 100 and 101. How has your understanding of temperature and heat changed after reading the chapter?
32. **COMMUNICATE** Explain the kinetic theory of matter in your own words. What, if anything, about the kinetic theory of matter surprised you?

UNIT PROJECTS

Evaluate all the data, results, and information from your project folder. Prepare to present your project.

MONITOR AND RETEACH

If students have trouble applying the concepts of heat transfer in items 24–26, suggest that they review the diagrams on pp. 117–119. Have them draw and label a diagram that includes all three processes (conduction, convection, and radiation) in a natural setting.

Students may benefit from summarizing one or more sections of the chapter.

Summarizing the Chapter, pp. 260–261

A 126 Unit: Matter and Energy

Standardized Test Practice

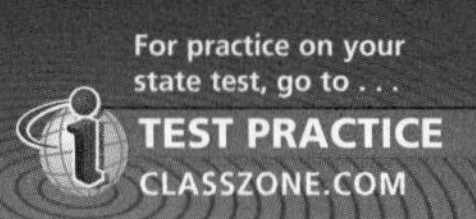

Interpreting Diagrams

The diagrams below illustrate the process that occurs in sea and land breezes.

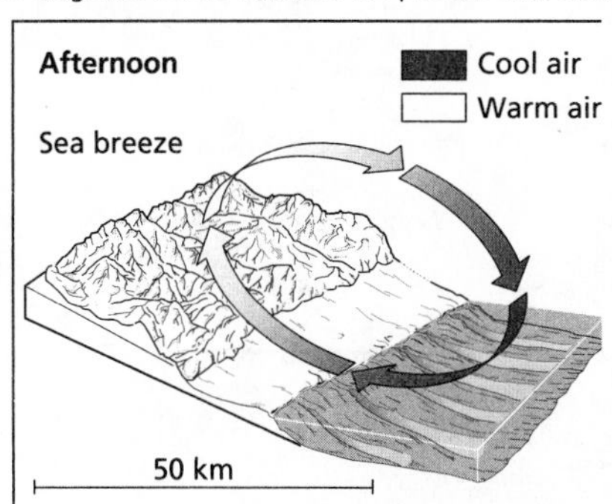

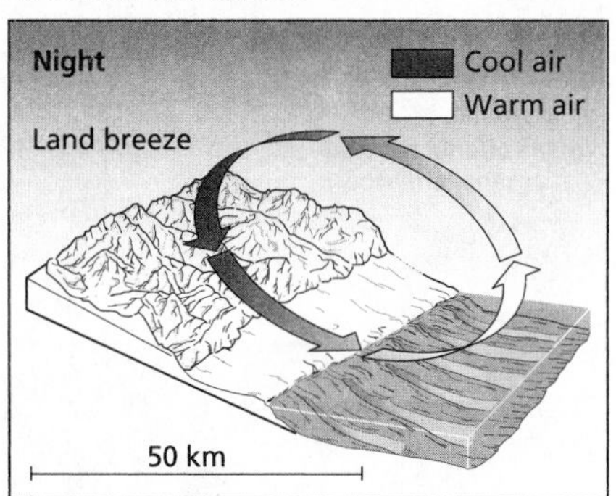

Use the diagrams above to answer the next five questions.

1. What happens during the day?
- **a.** Cool air from the land flows out to sea.
- **b.** Warm air from the land flows out to sea close to sea level.
- **c.** Cool air from the sea flows to the land.
- **d.** Warm air from the sea flows to the land.

2. What characteristic of large bodies of water explains why the seawater is cooler than the land in the hot afternoon sun?
- **a.** Water is liquid while the land is solid.
- **b.** Water has a higher specific heat than land.
- **c.** Land is a better insulator than water.
- **d.** Land has a higher specific heat than water.

3. What process causes the warm air to move upward over the land during the day?
- **a.** convection
- **b.** condensation
- **c.** evaporation
- **d.** radiation

4. Warm air is pushed upwards by cooler air during convection because the warm air
- **a.** is more dense
- **b.** has more mass
- **c.** is less dense
- **d.** has less mass

5. About how far over water does this land breeze extend?
- **a.** 1 kilometer
- **b.** 10 kilometers
- **c.** 25 kilometers
- **d.** 50 kilometers

Extended Response

Answer the two questions below in detail. Include some of the terms from the word box on the right. Underline each term that you use in your answer.

boiling point	convection	radiation
conduction	freezing point	zero point

6. What are the differences between the Fahrenheit and Celsius temperature scales? Which one is used in science? Why might this be the case?

7. Suppose you place three spoons—one metal, one plastic, and one wood—into a cup filled with hot water. The bowl end of the spoon is inside the cup and the handle is sticking up into the air. On each handle, you place a bead, held to the spoon by a dab of margarine. From which spoon will the bead fall first, and why?

METACOGNITIVE ACTIVITY

Have students answer the following questions in their **Science Notebook:**

1. What did you find the most challenging to understand about temperature and heat?
2. Which topics in this chapter would you like to learn more about?
3. What are the strongest pieces right now in your Unit Project?

Interpreting Diagrams

1. c *2. b* *3. a* *4. c* *5. c*

Extended Response

6. RUBRIC

4 points for a response that correctly answers the three questions and uses the following terms accurately:

- freezing point
- boiling point
- zero point

Sample: The Fahrenheit and Celsius scales have a different number of degrees between the freezing point and boiling point of water. On the Fahrenheit scale, the freezing point of water is 32° and the boiling point is 212°. On the Celsius scale, the freezing point of water is 0° and the boiling point is 100°. Scientists use the Celsius scale because it has a well-defined zero point.

3 points correctly answers two questions and uses two terms correctly
2 points correctly answers two questions and uses one term correctly
1 point correctly answers one of the questions

7. RUBRIC

4 points for a response that answers the question correctly and uses the following terms accurately:

- conduction
- specific heat
- heat

Sample: The bead will fall first from the metal spoon because metal has a higher conductivity and a lower specific heat than wood and plastic, so it warms up faster. Heat transfers energy from the hot water to the metal spoon by conduction. Heat transfers energy through the spoon and melts the margarine.

3 points correctly answers the question and uses one term accurately
2 points correctly answers the question
1 point does not correctly answer the question but uses the terms accurately

Student Resource Handbooks

Scientific Thinking Handbook

Making Observations

An **observation** is an act of noting and recording an event, characteristic, behavior, or anything else detected with an instrument or with the senses.

Observations allow you to make informed hypotheses and to gather data for experiments. Careful observations often lead to ideas for new experiments. There are two categories of observations:

- **Quantitative observations** can be expressed in numbers and include records of time, temperature, mass, distance, and volume.
- **Qualitative observations** include descriptions of sights, sounds, smells, and textures.

EXAMPLE

A student dissolved 30 grams of Epsom salts in water, poured the solution into a dish, and let the dish sit out uncovered overnight. The next day, she made the following observations of the Epsom salt crystals that grew in the dish.

To determine the mass, the student found the mass of the dish before and after growing the crystals and then used subtraction to find the difference.

The student measured several crystals and calculated the mean length. (To learn how to calculate the mean of a data set, see page R36.)

Table 1. Observations of Epsom Salt Crystals

Quantitative Observations	Qualitative Observations
• mass = 30 g • mean crystal length = 0.5 cm • longest crystal length = 2 cm	• Crystals are clear. • Crystals are long, thin, and rectangular. • White crust has formed around edge of dish.

Photographs or sketches are useful for recording qualitative observations.

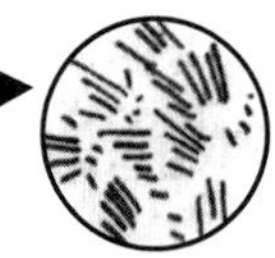

Epsom salt crystals

MORE ABOUT OBSERVING

- Make quantitative observations whenever possible. That way, others will know exactly what you observed and be able to compare their results with yours.
- It is always a good idea to make qualitative observations too. You never know when you might observe something unexpected.

Predicting and Hypothesizing

A **prediction** is an expectation of what will be observed or what will happen. A **hypothesis** is a tentative explanation for an observation or scientific problem that can be tested by further investigation.

SCIENTIFIC THINKING HANDBOOK

EXAMPLE

Suppose you have made two paper airplanes and you wonder why one of them tends to glide farther than the other one.

1. Start by asking a question.
2. Make an educated guess. After examination, you notice that the wings of the airplane that flies farther are slightly larger than the wings of the other airplane.
3. Write a prediction based upon your educated guess, in the form of an "If . . . , then . . ." statement. Write the independent variable after the word *if,* and the dependent variable after the word *then.*
4. To make a hypothesis, explain why you think what you predicted will occur. Write the explanation after the word *because*.

1. Why does one of the paper airplanes glide farther than the other?
2. The size of an airplane's wings may affect how far the airplane will glide.
3. Prediction: If I make a paper airplane with larger wings, then the airplane will glide farther.

To read about independent and dependent variables, see page R30.

4. Hypothesis: If I make a paper airplane with larger wings, then the airplane will glide farther, because the additional surface area of the wing will produce more lift.

Notice that the part of the hypothesis after *because* adds an explanation of why the airplane will glide farther.

MORE ABOUT HYPOTHESES

- The results of an experiment cannot prove that a hypothesis is correct. Rather, the results either support or do not support the hypothesis.
- Valuable information is gained even when your hypothesis is not supported by your results. For example, it would be an important discovery to find that wing size is not related to how far an airplane glides.
- In science, a hypothesis is supported only after many scientists have conducted many experiments and produced consistent results.

Inferring

An **inference** is a logical conclusion drawn from the available evidence and prior knowledge. Inferences are often made from observations.

EXAMPLE

A student observing a set of acorns noticed something unexpected about one of them. He noticed a white, soft-bodied insect eating its way out of the acorn.

The student recorded these observations.

Observations

- There is a hole in the acorn, about 0.5 cm in diameter, where the insect crawled out.
- There is a second hole, which is about the size of a pinhole, on the other side of the acorn.
- The inside of the acorn is hollow.

Here are some inferences that can be made on the basis of the observations.

Inferences

- The insect formed from the material inside the acorn, grew to its present size, and ate its way out of the acorn.
- The insect crawled through the smaller hole, ate the inside of the acorn, grew to its present size, and ate its way out of the acorn.
- An egg was laid in the acorn through the smaller hole. The egg hatched into a larva that ate the inside of the acorn, grew to its present size, and ate its way out of the acorn.

When you make inferences, be sure to look at all of the evidence available and combine it with what you already know.

MORE ABOUT INFERENCES

Inferences depend both on observations and on the knowledge of the people making the inferences. Ancient people who did not know that organisms are produced only by similar organisms might have made an inference like the first one. A student today might look at the same observations and make the second inference. A third student might have knowledge about this particular insect and know that it is never small enough to fit through the smaller hole, leading her to the third inference.

Identifying Cause and Effect

In a **cause-and-effect relationship,** one event or characteristic is the result of another. Usually an effect follows its cause in time.

There are many examples of cause-and-effect relationships in everyday life.

Cause	Effect
Turn off a light.	Room gets dark.
Drop a glass.	Glass breaks.
Blow a whistle.	Sound is heard.

Scientists must be careful not to infer a cause-and-effect relationship just because one event happens after another event. When one event occurs after another, you cannot infer a cause-and-effect relationship on the basis of that information alone. You also cannot conclude that one event caused another if there are alternative ways to explain the second event. A scientist must demonstrate through experimentation or continued observation that an event was truly caused by another event.

EXAMPLE

Make an Observation

Suppose you have a few plants growing outside. When the weather starts getting colder, you bring one of the plants indoors. You notice that the plant you brought indoors is growing faster than the others are growing. You cannot conclude from your observation that the change in temperature was the cause of the increased plant growth, because there are alternative explanations for the observation. Some possible explanations are given below.

- The humidity indoors caused the plant to grow faster.
- The level of sunlight indoors caused the plant to grow faster.
- The indoor plant's being noticed more often and watered more often than the outdoor plants caused it to grow faster.
- The plant that was brought indoors was healthier than the other plants to begin with.

To determine which of these factors, if any, caused the indoor plant to grow faster than the outdoor plants, you would need to design and conduct an experiment.

See pages R28–R35 for information about designing experiments.

Recognizing Bias

Television, newspapers, and the Internet are full of experts claiming to have scientific evidence to back up their claims. How do you know whether the claims are really backed up by good science?

SCIENTIFIC THINKING HANDBOOK

Bias is a slanted point of view, or personal prejudice. The goal of scientists is to be as objective as possible and to base their findings on facts instead of opinions. However, bias often affects the conclusions of researchers, and it is important to learn to recognize bias.

When scientific results are reported, you should consider the source of the information as well as the information itself. It is important to critically analyze the information that you see and read.

SOURCES OF BIAS

There are several ways in which a report of scientific information may be biased. Here are some questions that you can ask yourself:

1. **Who is sponsoring the research?**

   Sometimes, the results of an investigation are biased because an organization paying for the research is looking for a specific answer. This type of bias can affect how data are gathered and interpreted.

2. **Is the research sample large enough?**

   Sometimes research does not include enough data. The larger the sample size, the more likely that the results are accurate, assuming a truly random sample.

3. **In a survey, who is answering the questions?**

   The results of a survey or poll can be biased. The people taking part in the survey may have been specifically chosen because of how they would answer. They may have the same ideas or lifestyles. A survey or poll should make use of a random sample of people.

4. **Are the people who take part in a survey biased?**

   People who take part in surveys sometimes try to answer the questions the way they think the researcher wants them to answer. Also, in surveys or polls that ask for personal information, people may be unwilling to answer questions truthfully.

SCIENTIFIC BIAS

It is also important to realize that scientists have their own biases because of the types of research they do and because of their scientific viewpoints. Two scientists may look at the same set of data and come to completely different conclusions because of these biases. However, such disagreements are not necessarily bad. In fact, a critical analysis of disagreements is often responsible for moving science forward.

Identifying Faulty Reasoning

Faulty reasoning is wrong or incorrect thinking. It leads to mistakes and to wrong conclusions. Scientists are careful not to draw unreasonable conclusions from experimental data. Without such caution, the results of scientific investigations may be misleading.

EXAMPLE

Scientists try to make generalizations based on their data to explain as much about nature as possible. If only a small sample of data is looked at, however, a conclusion may be faulty. Suppose a scientist has studied the effects of the El Niño and La Niña weather patterns on flood damage in California from 1989 to 1995. The scientist organized the data in the bar graph below.

The scientist drew the following conclusions:

1. The La Niña weather pattern has no effect on flooding in California.
2. When neither weather pattern occurs, there is almost no flood damage.
3. A weak or moderate El Niño produces a small or moderate amount of flooding.
4. A strong El Niño produces a lot of flooding.

Flood and Storm Damage in California
Estimated damage (millions of dollars)
2000
1500
1000
500
0
1989 1992 1995
Starting year of season (July 1–June 30)
Weak–moderate El Niño
Strong El Niño
SOURCE: *Governor's Office of Emergency Services, California*

For the six-year period of the scientist's investigation, these conclusions may seem to be reasonable. However, a six-year study of weather patterns may be too small of a sample for the conclusions to be supported. Consider the following graph, which shows information that was gathered from 1949 to 1997.

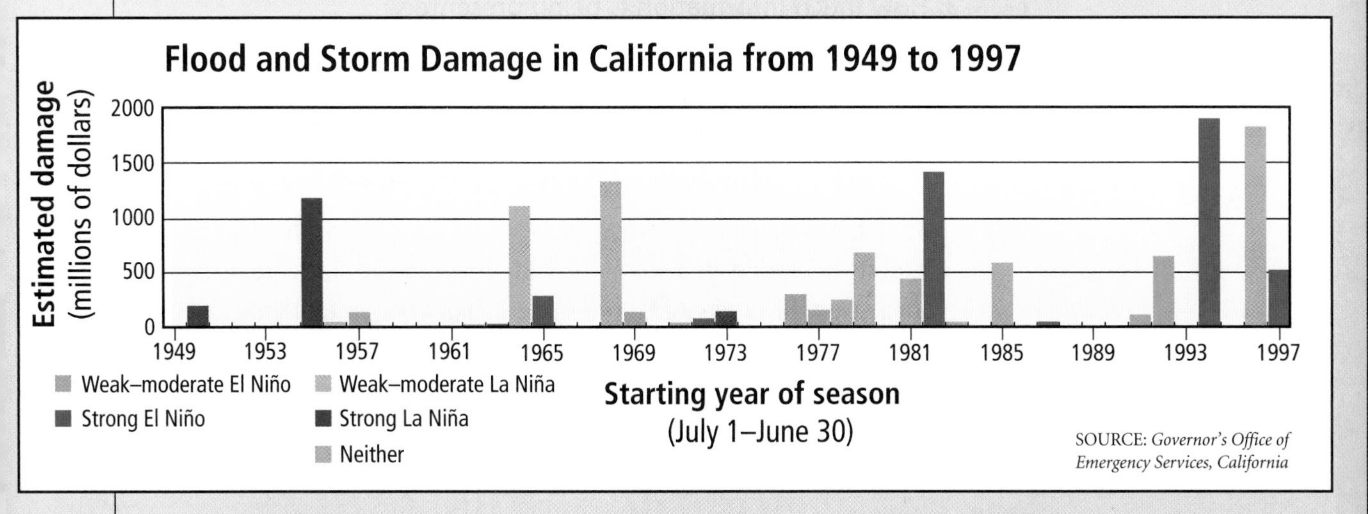

The only one of the conclusions that all of this information supports is number 3: a weak or moderate El Niño produces a small or moderate amount of flooding. By collecting more data, scientists can be more certain of their conclusions and can avoid faulty reasoning.

Analyzing Statements

To **analyze** a statement is to examine its parts carefully. Scientific findings are often reported through media such as television or the Internet. A report that is made public often focuses on only a small part of research. As a result, it is important to question the sources of information.

Evaluate Media Claims

To **evaluate** a statement is to judge it on the basis of criteria you've established. Sometimes evaluating means deciding whether a statement is true.

Reports of scientific research and findings in the media may be misleading or incomplete. When you are exposed to this information, you should ask yourself some questions so that you can make informed judgments about the information.

1. **Does the information come from a credible source?**

   Suppose you learn about a new product and it is stated that scientific evidence proves that the product works. A report from a respected news source may be more believable than an advertisement paid for by the product's manufacturer.

2. **How much evidence supports the claim?**

   Often, it may seem that there is new evidence every day of something in the world that either causes or cures an illness. However, information that is the result of several years of work by several different scientists is more credible than an advertisement that does not even cite the subjects of the experiment.

3. **How much information is being presented?**

   Science cannot solve all questions, and scientific experiments often have flaws. A report that discusses problems in a scientific study may be more believable than a report that addresses only positive experimental findings.

4. **Is scientific evidence being presented by a specific source?**

   Sometimes scientific findings are reported by people who are called experts or leaders in a scientific field. But if their names are not given or their scientific credentials are not reported, their statements may be less credible than those of recognized experts.

Differentiate Between Fact and Opinion

Sometimes information is presented as a fact when it may be an opinion. When scientific conclusions are reported, it is important to recognize whether they are based on solid evidence. Again, you may find it helpful to ask yourself some questions.

1. **What is the difference between a fact and an opinion?**
   A **fact** is a piece of information that can be strictly defined and proved true. An **opinion** is a statement that expresses a belief, value, or feeling. An opinion cannot be proved true or false. For example, a person's age is a fact, but if someone is asked how old they feel, it is impossible to prove the person's answer to be true or false.
2. **Can opinions be measured?**
   Yes, opinions can be measured. In fact, surveys often ask for people's opinions on a topic. But there is no way to know whether or not an opinion is the truth.

HOW TO DIFFERENTIATE FACT FROM OPINION

Opinions
Notice words or phrases that express beliefs or feelings. The words *unfortunately* and *careless* show that opinions are being expressed.

Opinion
Look for statements that speculate about events. These statements are opinions, because they cannot be proved.

Human Activities and the Environment

Unfortunately, human use of fossil fuels is one of the most significant developments of the past few centuries. Humans rely on fossil fuels, a non-renewable energy resource, for more than 90 percent of their energy needs.

This careless misuse of our planet's resources has resulted in pollution, global warming, and the destruction of fragile ecosystems. For example, oil pipelines carry more than one million barrels of oil each day across tundra regions. Transporting oil across such areas can only result in oil spills that poison the land for decades.

Facts
Statements that contain statistics tend to be facts. Writers often use facts to support their opinions.

Safety Rules

Before you work in the laboratory, read these safety rules twice. Ask your teacher to explain any rules that you do not completely understand. Refer to these rules later on if you have questions about safety in the science classroom.

Directions

- Read all directions and make sure that you understand them before starting an investigation or lab activity. If you do not understand how to do a procedure or how to use a piece of equipment, ask your teacher.
- Do not begin any investigation or touch any equipment until your teacher has told you to start.
- Never experiment on your own. If you want to try a procedure that the directions do not call for, ask your teacher for permission first.
- If you are hurt or injured in any way, tell your teacher immediately.

Dress Code

goggles

apron

gloves

- Wear goggles when
  — using glassware, sharp objects, or chemicals
  — heating an object
  — working with anything that can easily fly up into the air and hurt someone's eye
- Tie back long hair or hair that hangs in front of your eyes.
- Remove any article of clothing—such as a loose sweater or a scarf—that hangs down and may touch a flame, chemical, or piece of equipment.
- Observe all safety icons calling for the wearing of eye protection, gloves, and aprons.

Heating and Fire Safety

fire safety

heating safety

- Keep your work area neat, clean, and free of extra materials.
- Never reach over a flame or heat source.
- Point objects being heated away from you and others.
- Never heat a substance or an object in a closed container.
- Never touch an object that has been heated. If you are unsure whether something is hot, treat it as though it is. Use oven mitts, clamps, tongs, or a test-tube holder.
- Know where the fire extinguisher and fire blanket are kept in your classroom.
- Do not throw hot substances into the trash. Wait for them to cool or use the container your teacher puts out for disposal.

Electrical Safety

electrical safety

- Never use lamps or other electrical equipment with frayed cords.
- Make sure no cord is lying on the floor where someone can trip over it.
- Do not let a cord hang over the side of a counter or table so that the equipment can easily be pulled or knocked to the floor.
- Never let cords hang into sinks or other places where water can be found.
- Never try to fix electrical problems. Inform your teacher of any problems immediately.
- Unplug an electrical cord by pulling on the plug, not the cord.

Chemical Safety

chemical safety

poison

fumes

- If you spill a chemical or get one on your skin or in your eyes, tell your teacher right away.
- Never touch, taste, or sniff any chemicals in the lab. If you need to determine odor, waft. Wafting consists of holding the chemical in its container 15 centimeters (6 in.) away from your nose, and using your fingers to bring fumes from the container to your nose.
- Keep lids on all chemicals you are not using.
- Never put unused chemicals back into the original containers. Throw away extra chemicals where your teacher tells you to.
- Pour chemicals over a sink or your work area, not over the floor.
- If you get a chemical in your eye, use the eyewash right away.
- Always wash your hands after handling chemicals, plants, or soil.

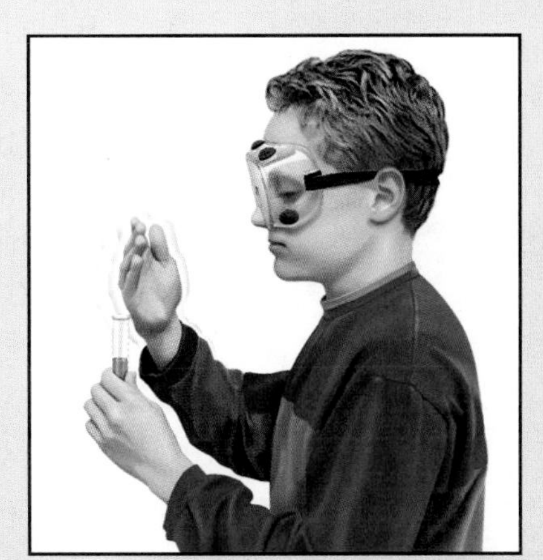

Wafting

Glassware and Sharp-Object Safety

sharp objects

- If you break glassware, tell your teacher right away.
- Do not use broken or chipped glassware. Give these to your teacher.
- Use knives and other cutting instruments carefully. Always wear eye protection and cut away from you.

Animal Safety

- Never hurt an animal.
- Touch animals only when necessary. Follow your teacher's instructions for handling animals.
- Always wash your hands after working with animals.

Cleanup

disposal

- Follow your teacher's instructions for throwing away or putting away supplies.
- Clean your work area and pick up anything that has dropped to the floor.
- Wash your hands.

R11

Using Lab Equipment

Different experiments require different types of equipment. But even though experiments differ, the ways in which the equipment is used are the same.

Beakers

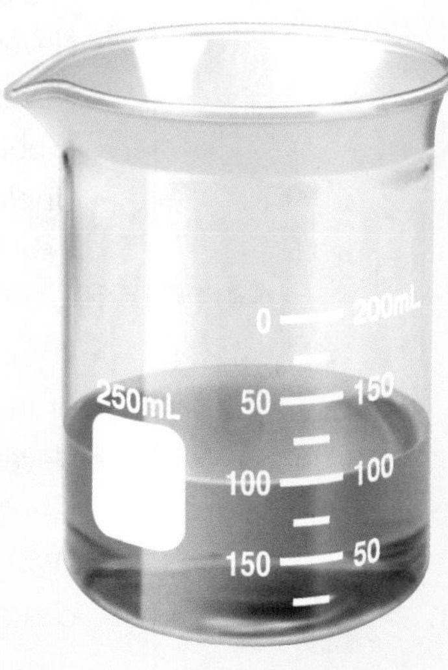

- Use beakers for holding and pouring liquids.
- Do not use a beaker to measure the volume of a liquid. Use a graduated cylinder instead. (See page R16.)
- Use a beaker that holds about twice as much liquid as you need. For example, if you need 100 milliliters of water, you should use a 200- or 250-milliliter beaker.

Test Tubes

- Use test tubes to hold small amounts of substances.
- Do not use a test tube to measure the volume of a liquid.
- Use a test tube when heating a substance over a flame. Aim the mouth of the tube away from yourself and other people.
- Liquids easily spill or splash from test tubes, so it is important to use only small amounts of liquids.

Test-Tube Holder

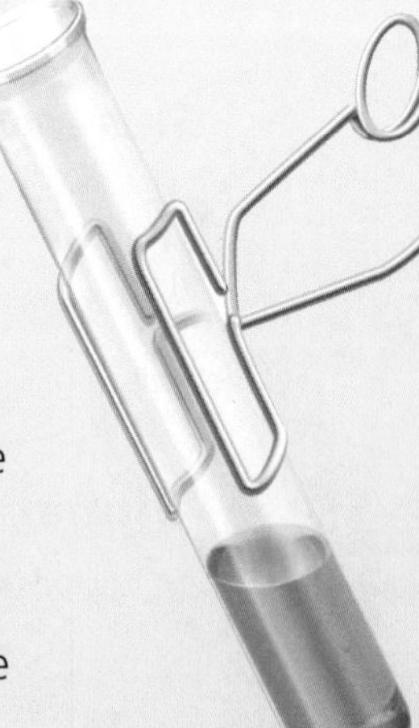

- Use a test-tube holder when heating a substance in a test tube.
- Use a test-tube holder if the substance in a test tube is dangerous to touch.
- Make sure the test-tube holder tightly grips the test tube so that the test tube will not slide out of the holder.
- Make sure that the test-tube holder is above the surface of the substance in the test tube so that you can observe the substance.

R12

Test-Tube Rack

- Use a test-tube rack to organize test tubes before, during, and after an experiment.
- Use a test-tube rack to keep test tubes upright so that they do not fall over and spill their contents.
- Use a test-tube rack that is the correct size for the test tubes that you are using. If the rack is too small, a test tube may become stuck. If the rack is too large, a test tube may lean over, and some of its contents may spill or splash.

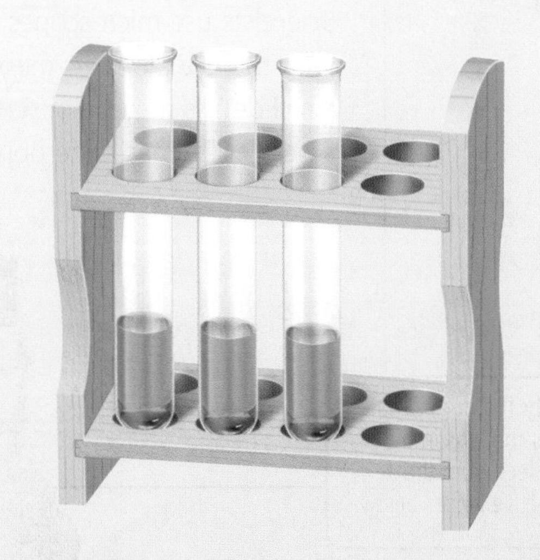

Forceps

- Use forceps when you need to pick up or hold a very small object that should not be touched with your hands.
- Do not use forceps to hold anything over a flame, because forceps are not long enough to keep your hand safely away from the flame. Plastic forceps will melt, and metal forceps will conduct heat and burn your hand.

Hot Plate

- Use a hot plate when a substance needs to be kept warmer than room temperature for a long period of time.
- Use a hot plate instead of a Bunsen burner or a candle when you need to carefully control temperature.
- Do not use a hot plate when a substance needs to be burned in an experiment.
- Always use "hot hands" safety mitts or oven mitts when handling anything that has been heated on a hot plate.

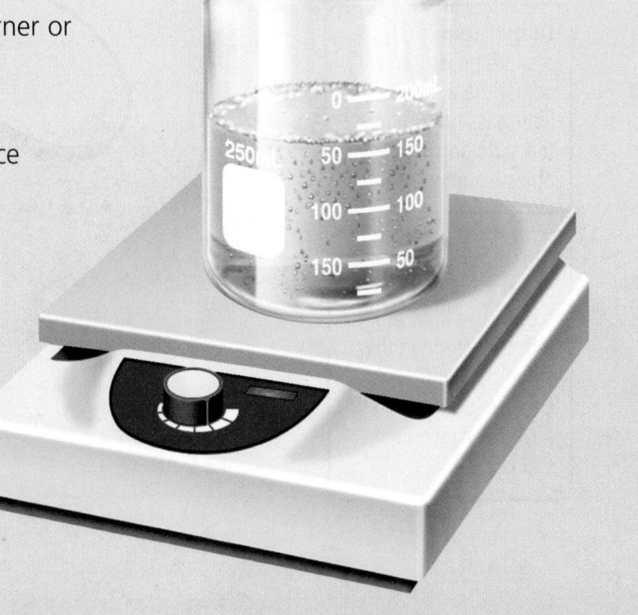

LAB HANDBOOK

Microscope

Scientists use microscopes to see very small objects that cannot easily be seen with the eye alone. A microscope magnifies the image of an object so that small details may be observed. A microscope that you may use can magnify an object 400 times—the object will appear 400 times larger than its actual size.

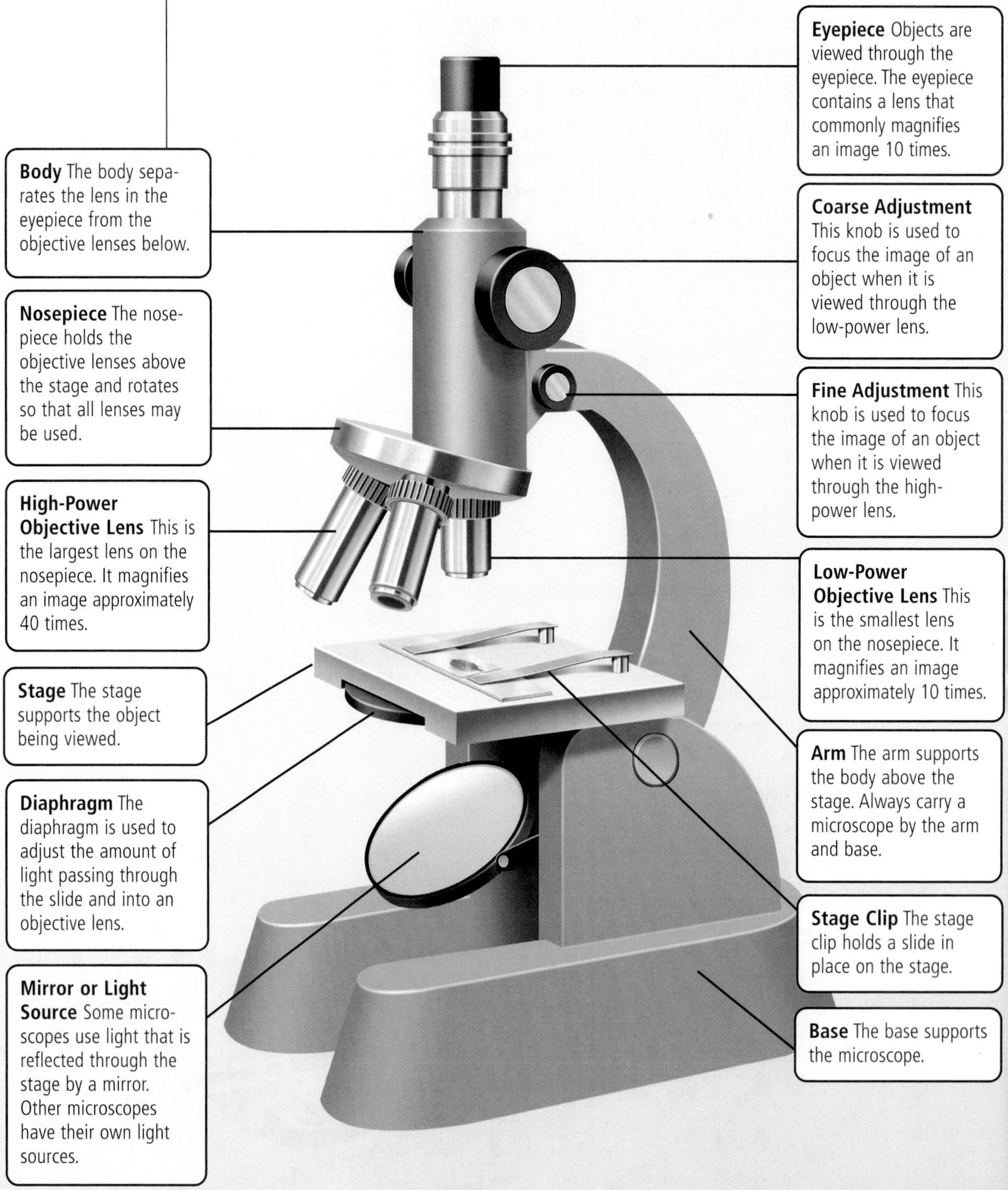

Eyepiece Objects are viewed through the eyepiece. The eyepiece contains a lens that commonly magnifies an image 10 times.

Body The body separates the lens in the eyepiece from the objective lenses below.

Coarse Adjustment This knob is used to focus the image of an object when it is viewed through the low-power lens.

Nosepiece The nosepiece holds the objective lenses above the stage and rotates so that all lenses may be used.

Fine Adjustment This knob is used to focus the image of an object when it is viewed through the high-power lens.

High-Power Objective Lens This is the largest lens on the nosepiece. It magnifies an image approximately 40 times.

Low-Power Objective Lens This is the smallest lens on the nosepiece. It magnifies an image approximately 10 times.

Stage The stage supports the object being viewed.

Arm The arm supports the body above the stage. Always carry a microscope by the arm and base.

Diaphragm The diaphragm is used to adjust the amount of light passing through the slide and into an objective lens.

Stage Clip The stage clip holds a slide in place on the stage.

Mirror or Light Source Some microscopes use light that is reflected through the stage by a mirror. Other microscopes have their own light sources.

Base The base supports the microscope.

VIEWING AN OBJECT

1. Use the coarse adjustment knob to raise the body tube.
2. Adjust the diaphragm so that you can see a bright circle of light through the eyepiece.
3. Place the object or slide on the stage. Be sure that it is centered over the hole in the stage.
4. Turn the nosepiece to click the low-power lens into place.
5. Using the coarse adjustment knob, slowly lower the lens and focus on the specimen being viewed. Be sure not to touch the slide or object with the lens.
6. When switching from the low-power lens to the high-power lens, first raise the body tube with the coarse adjustment knob so that the high-power lens will not hit the slide.
7. Turn the nosepiece to click the high-power lens into place.
8. Use the fine adjustment knob to focus on the specimen being viewed. Again, be sure not to touch the slide or object with the lens.

MAKING A SLIDE, OR WET MOUNT

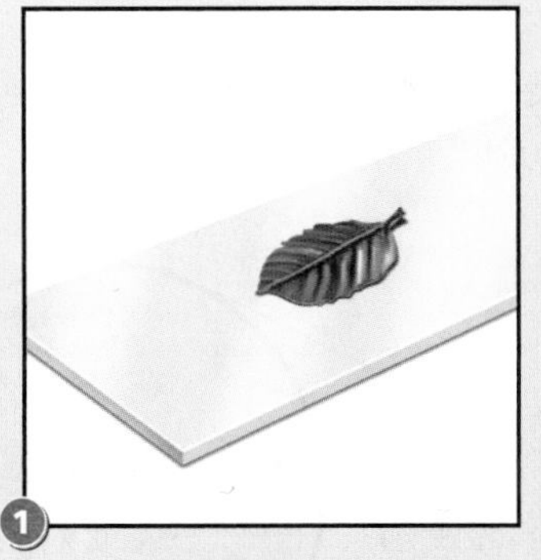

1. Place the specimen in the center of a clean slide.

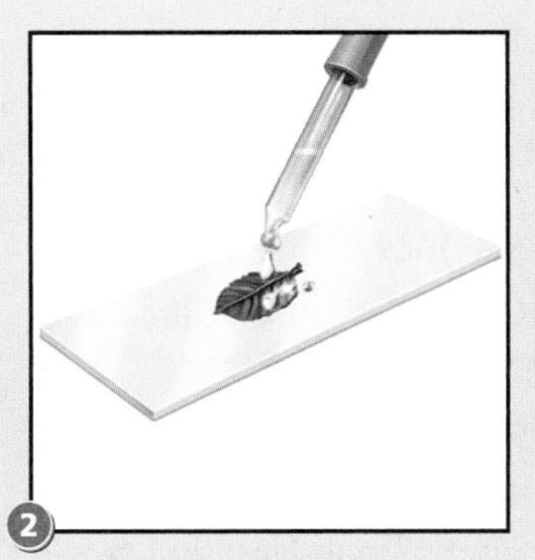

2. Place a drop of water on the specimen.

3. Place a cover slip on the slide. Put one edge of the cover slip into the drop of water and slowly lower it over the specimen.

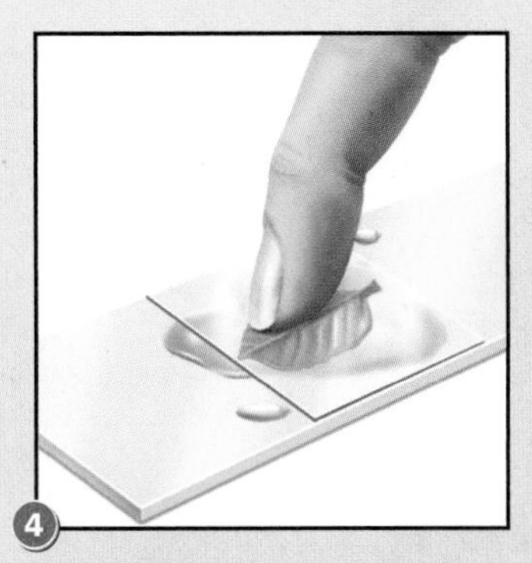

4. Remove any air bubbles from under the cover slip by gently tapping the cover slip.

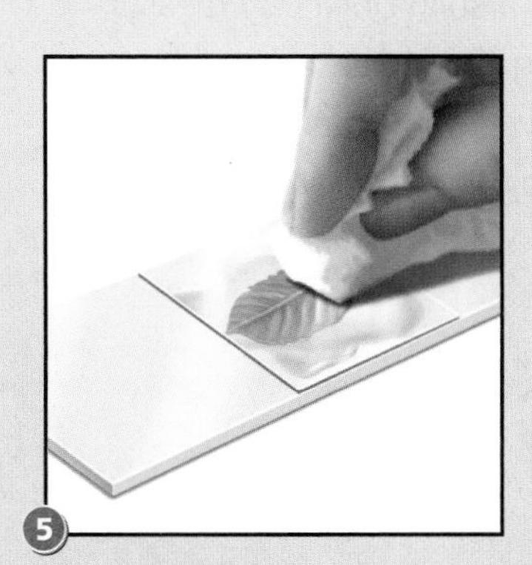

5. Dry any excess water before placing the slide on the microscope stage for viewing.

Spring Scale (Force Meter)

- Use a spring scale to measure a force pulling on the scale.
- Use a spring scale to measure the force of gravity exerted on an object by Earth.
- To measure a force accurately, a spring scale must be zeroed before it is used. The scale is zeroed when no weight is attached and the indicator is positioned at zero.
- Do not attach a weight that is either too heavy or too light to a spring scale. A weight that is too heavy could break the scale or exert too great a force for the scale to measure. A weight that is too light may not exert enough force to be measured accurately.

Graduated Cylinder

- Use a graduated cylinder to measure the volume of a liquid.
- Be sure that the graduated cylinder is on a flat surface so that your measurement will be accurate.
- When reading the scale on a graduated cylinder, be sure to have your eyes at the level of the surface of the liquid.
- The surface of the liquid will be curved in the graduated cylinder. Read the volume of the liquid at the bottom of the curve, or meniscus (muh-NIHS-kuhs).
- You can use a graduated cylinder to find the volume of a solid object by measuring the increase in a liquid's level after you add the object to the cylinder.

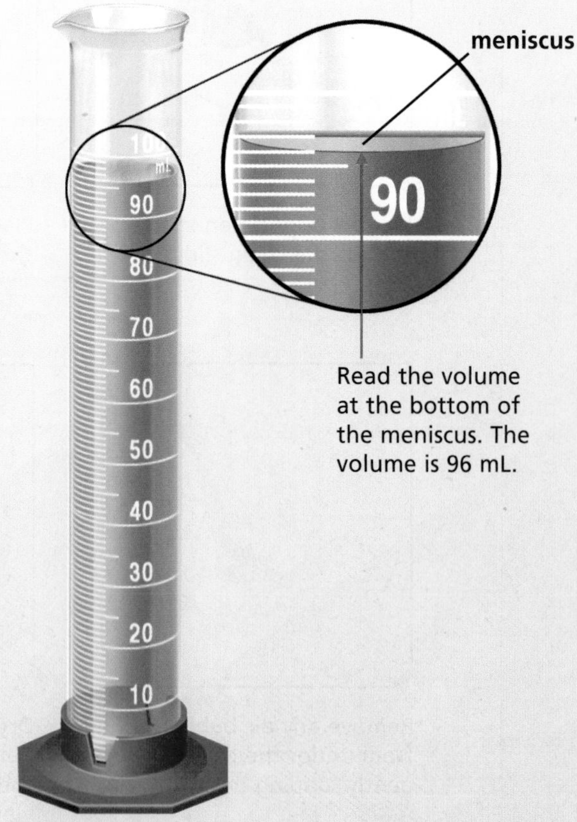

Read the volume at the bottom of the meniscus. The volume is 96 mL.

Metric Rulers

- Use metric rulers or meter sticks to measure objects' lengths.
- Do not measure an object from the end of a metric ruler or meter stick, because the end is often imperfect. Instead, measure from the 1-centimeter mark, but remember to subtract a centimeter from the apparent measurement.
- Estimate any lengths that extend between marked units. For example, if a meter stick shows centimeters but not millimeters, you can estimate the length that an object extends between centimeter marks to measure it to the nearest millimeter.
- **Controlling Variables** If you are taking repeated measurements, always measure from the same point each time. For example, if you're measuring how high two different balls bounce when dropped from the same height, measure both bounces at the same point on the balls—either the top or the bottom. Do not measure at the top of one ball and the bottom of the other.

EXAMPLE

How to Measure a Leaf

1. Lay a ruler flat on top of the leaf so that the 1-centimeter mark lines up with one end. Make sure the ruler and the leaf do not move between the time you line them up and the time you take the measurement.
2. Look straight down on the ruler so that you can see exactly how the marks line up with the other end of the leaf.
3. Estimate the length by which the leaf extends beyond a marking. For example, the leaf below extends about halfway between the 4.2-centimeter and 4.3-centimeter marks, so the apparent measurement is about 4.25 centimeters.
4. Remember to subtract 1 centimeter from your apparent measurement, since you started at the 1-centimeter mark on the ruler and not at the end. The leaf is about 3.25 centimeters long (4.25 cm – 1 cm = 3.25 cm).

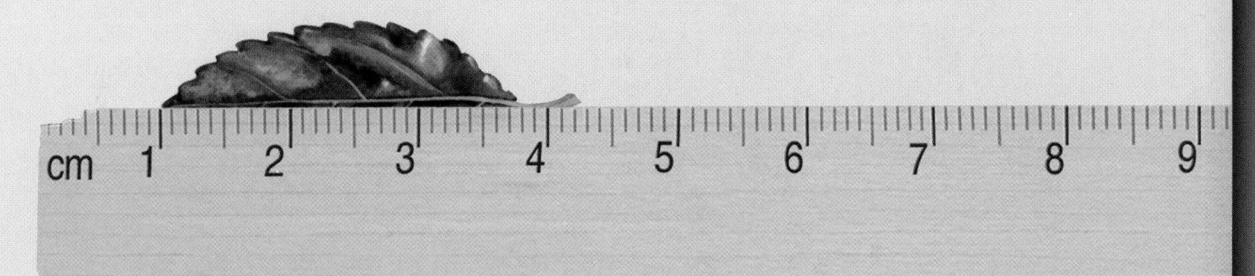

LAB HANDBOOK

Triple-Beam Balance

This balance has a pan and three beams with sliding masses, called riders. At one end of the beams is a pointer that indicates whether the mass on the pan is equal to the masses shown on the beams.

1. Make sure the balance is zeroed before measuring the mass of an object. The balance is zeroed if the pointer is at zero when nothing is on the pan and the riders are at their zero points. Use the adjustment knob at the base of the balance to zero it.
2. Place the object to be measured on the pan.
3. Move the riders one notch at a time away from the pan. Begin with the largest rider. If moving the largest rider one notch brings the pointer below zero, begin measuring the mass of the object with the next smaller rider.
4. Change the positions of the riders until they balance the mass on the pan and the pointer is at zero. Then add the readings from the three beams to determine the mass of the object.

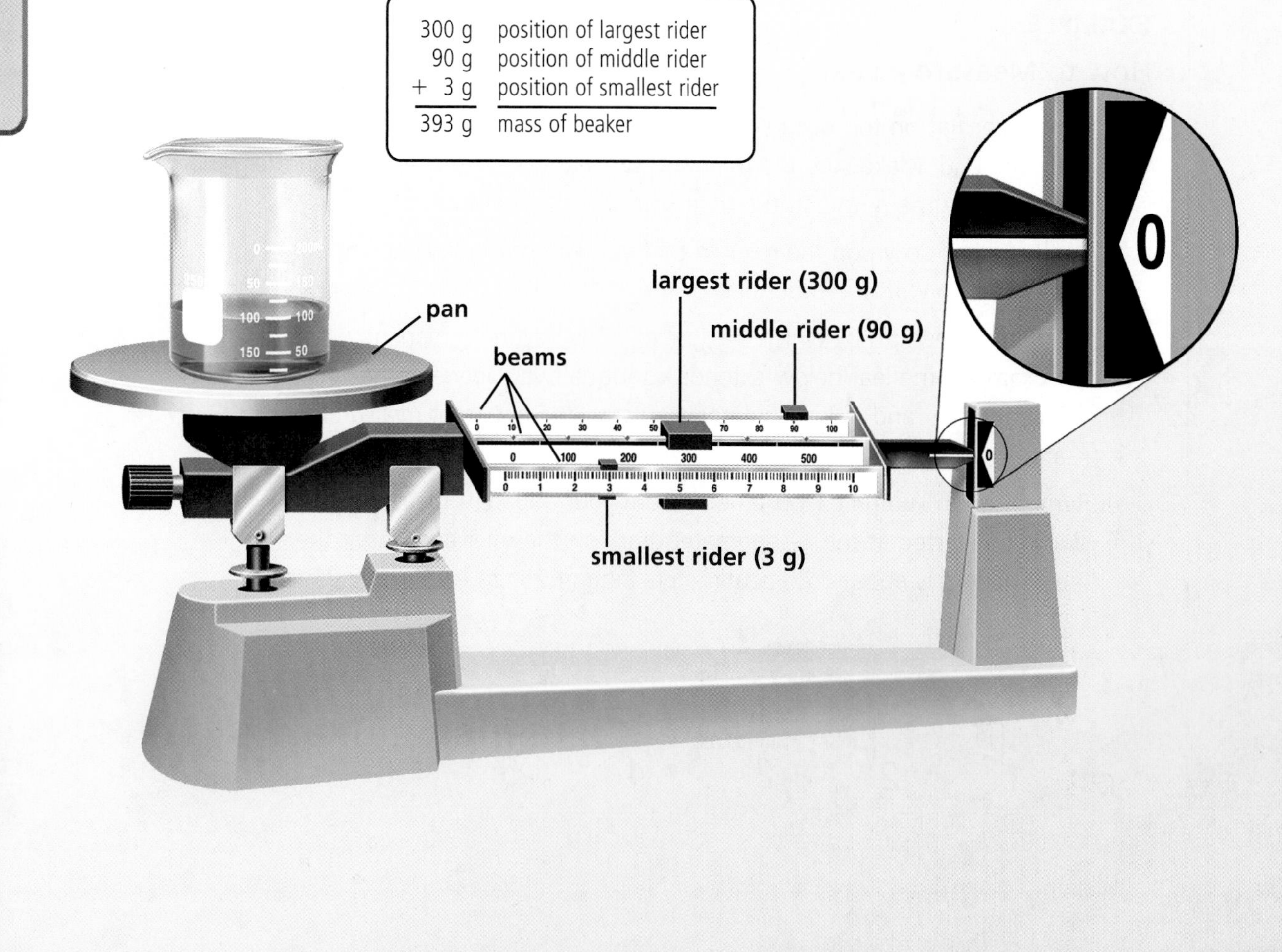

Double-Pan Balance

This type of balance has two pans. Between the pans is a pointer that indicates whether the masses on the pans are equal.

1. Make sure the balance is zeroed before measuring the mass of an object. The balance is zeroed if the pointer is at zero when there is nothing on either of the pans. Many double-pan balances have sliding knobs that can be used to zero them.
2. Place the object to be measured on one of the pans.
3. Begin adding standard masses to the other pan. Begin with the largest standard mass. If this adds too much mass to the balance, begin measuring the mass of the object with the next smaller standard mass.
4. Add standard masses until the masses on both pans are balanced and the pointer is at zero. Then add the standard masses together to determine the mass of the object being measured.

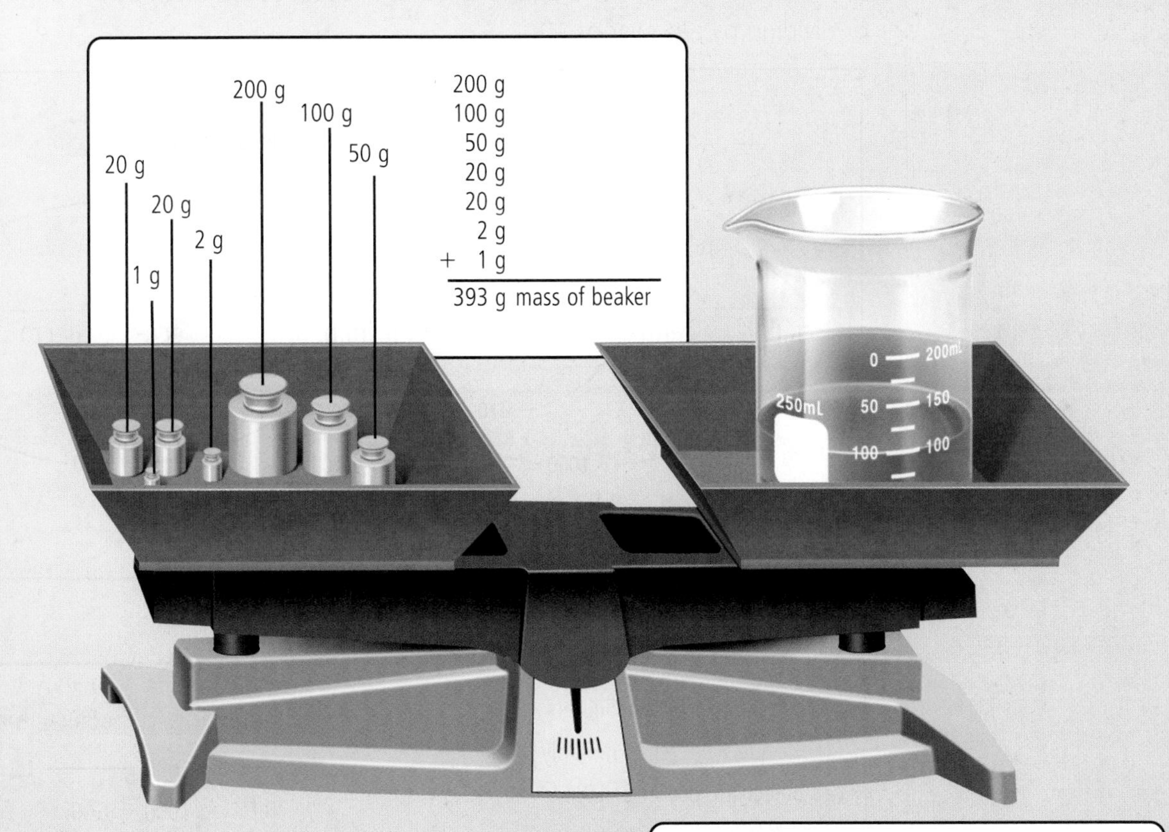

Never place chemicals or liquids directly on a pan. Instead, use the following procedure:

1. Determine the mass of an empty container, such as a beaker.
2. Pour the substance into the container, and measure the total mass of the substance and the container.
3. Subtract the mass of the empty container from the total mass to find the mass of the substance.

The Metric System and SI Units

Scientists use International System (SI) units for measurements of distance, volume, mass, and temperature. The International System is based on multiples of ten and the metric system of measurement.

Basic SI Units		
Property	**Name**	**Symbol**
length	meter	m
volume	liter	L
mass	kilogram	kg
temperature	kelvin	K

SI Prefixes		
Prefix	**Symbol**	**Multiple of 10**
kilo-	k	1000
hecto-	h	100
deca-	da	10
deci-	d	0.1 $(\frac{1}{10})$
centi-	c	0.01 $(\frac{1}{100})$
milli-	m	0.001 $(\frac{1}{1000})$

Changing Metric Units

You can change from one unit to another in the metric system by multiplying or dividing by a power of 10.

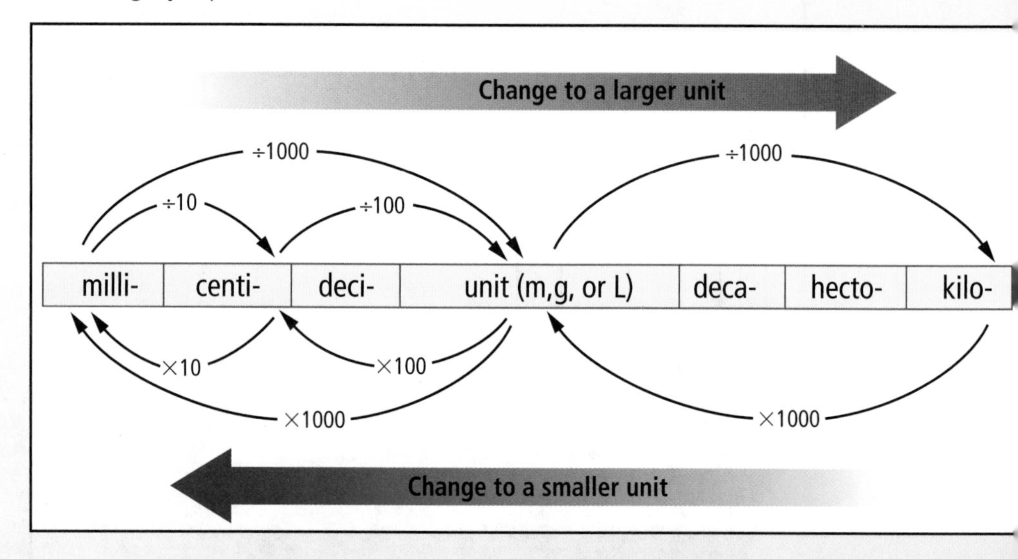

Example

Change 0.64 liters to milliliters.

(1) Decide whether to multiply or divide.

(2) Select the power of 10.

ANSWER 0.64 L = 640 mL

Change to a smaller unit by multiplying.

mL ← × 1000 — **L**

0.64 × 1000 = **640.**

Example

Change 23.6 grams to kilograms.

(1) Decide whether to multiply or divide.

(2) Select the power of 10.

ANSWER 23.6 g = 0.0236 kg

Change to a larger unit by dividing.

g — ÷ 1000 → **kg**

23.6 ÷ 1000 = **0.0236**

LAB HANDBOOK

Temperature Conversions

Even though the kelvin is the SI base unit of temperature, the degree Celsius will be the unit you use most often in your science studies. The formulas below show the relationships between temperatures in degrees Fahrenheit (°F), degrees Celsius (°C), and kelvins (K).

$$°C = \frac{5}{9}(°F - 32)$$

$$°F = \frac{9}{5}°C + 32$$

$$K = °C + 273$$

See page R42 for help with using formulas.

Examples of Temperature Conversions		
Condition	Degrees Celsius	Degrees Fahrenheit
Freezing point of water	0	32
Cool day	10	50
Mild day	20	68
Warm day	30	86
Normal body temperature	37	98.6
Very hot day	40	104
Boiling point of water	100	212

Converting Between SI and U.S. Customary Units

Use the chart below when you need to convert between SI units and U.S. customary units.

SI Unit	From SI to U.S. Customary			From U.S. Customary to SI		
Length	**When you know**	**multiply by**	**to find**	**When you know**	**multiply by**	**to find**
kilometer (km) = 1000 m	kilometers	0.62	miles	miles	1.61	kilometers
meter (m) = 100 cm	meters	3.28	feet	feet	0.3048	meters
centimeter (cm) = 10 mm	centimeters	0.39	inches	inches	2.54	centimeters
millimeter (mm) = 0.1 cm	millimeters	0.04	inches	inches	25.4	millimeters
Area	**When you know**	**multiply by**	**to find**	**When you know**	**multiply by**	**to find**
square kilometer (km^2)	square kilometers	0.39	square miles	square miles	2.59	square kilometers
square meter (m^2)	square meters	1.2	square yards	square yards	0.84	square meters
square centimeter (cm^2)	square centimeters	0.155	square inches	square inches	6.45	square centimeters
Volume	**When you know**	**multiply by**	**to find**	**When you know**	**multiply by**	**to find**
liter (L) = 1000 mL	liters	1.06	quarts	quarts	0.95	liters
	liters	0.26	gallons	gallons	3.79	liters
	liters	4.23	cups	cups	0.24	liters
	liters	2.12	pints	pints	0.47	liters
milliliter (mL) = 0.001 L	milliliters	0.20	teaspoons	teaspoons	4.93	milliliters
	milliliters	0.07	tablespoons	tablespoons	14.79	milliliters
	milliliters	0.03	fluid ounces	fluid ounces	29.57	milliliters
Mass	**When you know**	**multiply by**	**to find**	**When you know**	**multiply by**	**to find**
kilogram (kg) = 1000 g	kilograms	2.2	pounds	pounds	0.45	kilograms
gram (g) = 1000 mg	grams	0.035	ounces	ounces	28.35	grams

Precision and Accuracy

When you do an experiment, it is important that your methods, observations, and data be both precise and accurate.

low precision

precision, but not accuracy

precision and accuracy

Precision

In science, **precision** is the exactness and consistency of measurements. For example, measurements made with a ruler that has both centimeter and millimeter markings would be more precise than measurements made with a ruler that has only centimeter markings. Another indicator of precision is the care taken to make sure that methods and observations are as exact and consistent as possible. Every time a particular experiment is done, the same procedure should be used. Precision is necessary because experiments are repeated several times and if the procedure changes, the results will change.

EXAMPLE

Suppose you are measuring temperatures over a two-week period. Your precision will be greater if you measure each temperature at the same place, at the same time of day, and with the same thermometer than if you change any of these factors from one day to the next.

Accuracy

In science, it is possible to be precise but not accurate. **Accuracy** depends on the difference between a measurement and an actual value. The smaller the difference, the more accurate the measurement.

EXAMPLE

Suppose you look at a stream and estimate that it is about 1 meter wide at a particular place. You decide to check your estimate by measuring the stream with a meter stick, and you determine that the stream is 1.32 meters wide. However, because it is hard to measure the width of a stream with a meter stick, it turns out that you didn't do a very good job. The stream is actually 1.14 meters wide. Therefore, even though your estimate was less precise than your measurement, your estimate was actually more accurate.

Making Data Tables and Graphs

Data tables and graphs are useful tools for both recording and communicating scientific data.

Making Data Tables

You can use a **data table** to organize and record the measurements that you make. Some examples of information that might be recorded in data tables are frequencies, times, and amounts.

EXAMPLE

Suppose you are investigating photosynthesis in two elodea plants. One sits in direct sunlight, and the other sits in a dimly lit room. You measure the rate of photosynthesis by counting the number of bubbles in the jar every ten minutes.

1. Title and number your data table.
2. Decide how you will organize the table into columns and rows.
3. Any units, such as seconds or degrees, should be included in column headings, not in the individual cells.

Table 1. Number of Bubbles from Elodea

Time (min)	Sunlight	Dim Light
0	0	0
10	15	5
20	25	8
30	32	7
40	41	10
50	47	9
60	42	9

Always number and title data tables.

The data in the table above could also be organized in a different way.

Table 1. Number of Bubbles from Elodea

Light Condition	Time (min)						
	0	10	20	30	40	50	60
Sunlight	0	15	25	32	41	47	42
Dim light	0	5	8	7	10	9	9

Put units in column heading.

LAB HANDBOOK

Making Line Graphs

You can use a **line graph** to show a relationship between variables. Line graphs are particularly useful for showing changes in variables over time.

EXAMPLE

Suppose you are interested in graphing temperature data that you collected over the course of a day.

Table 1. Outside Temperature During the Day on March 7

	Time of Day						
	7:00 A.M.	9:00 A.M.	11:00 A.M.	1:00 P.M.	3:00 P.M.	5:00 P.M.	7:00 P.M.
Temp (°C)	8	9	11	14	12	10	6

1. Use the vertical axis of your line graph for the variable that you are measuring—temperature.
2. Choose scales for both the horizontal axis and the vertical axis of the graph. You should have two points more than you need on the vertical axis, and the horizontal axis should be long enough for all of the data points to fit.
3. Draw and label each axis.
4. Graph each value. First find the appropriate point on the scale of the horizontal axis. Imagine a line that rises vertically from that place on the scale. Then find the corresponding value on the vertical axis, and imagine a line that moves horizontally from that value. The point where these two imaginary lines intersect is where the value should be plotted.
5. Connect the points with straight lines.

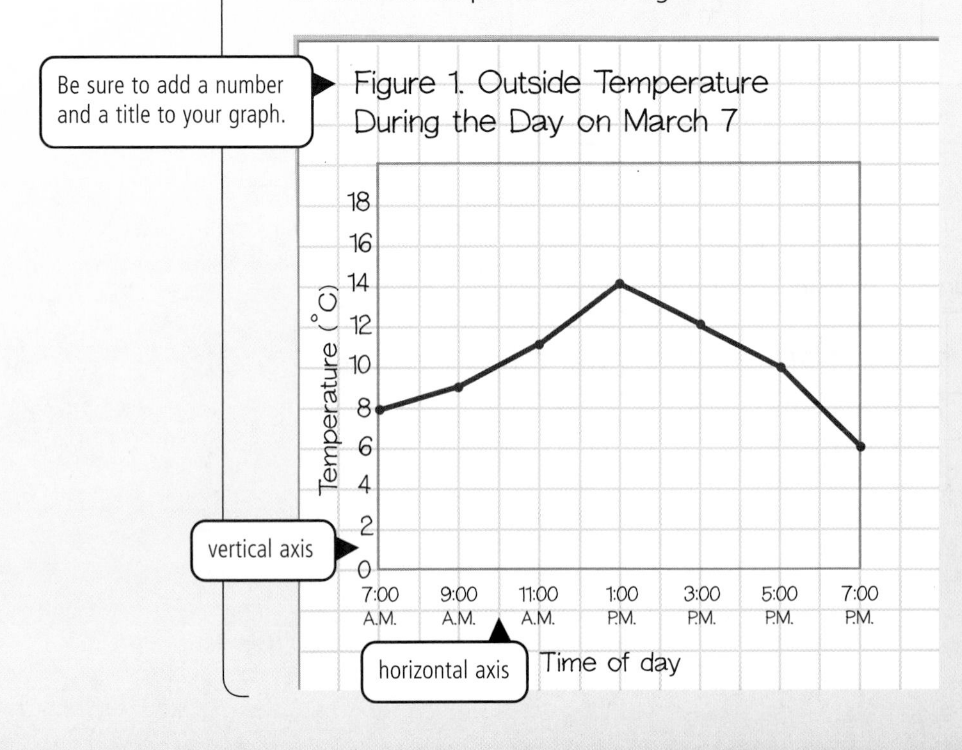

LAB HANDBOOK

Making Circle Graphs

You can use a **circle graph,** sometimes called a pie chart, to represent data as parts of a circle. Circle graphs are used only when the data can be expressed as percentages of a whole. The entire circle shown in a circle graph is equal to 100 percent of the data.

EXAMPLE

Suppose you identified the species of each mature tree growing in a small wooded area. You organized your data in a table, but you also want to show the data in a circle graph.

1. To begin, find the total number of mature trees.

   $56 + 34 + 22 + 10 + 28 = 150$

2. To find the degree measure for each sector of the circle, write a fraction comparing the number of each tree species with the total number of trees. Then multiply the fraction by 360°.

   Oak: $\frac{56}{150} \times 360° = 134.4°$

3. Draw a circle. Use a protractor to draw the angle for each sector of the graph.
4. Color and label each sector of the graph.
5. Give the graph a number and title.

Table 1. Tree Species in Wooded Area

Species	Number of Specimens
Oak	56
Maple	34
Birch	22
Willow	10
Pine	28

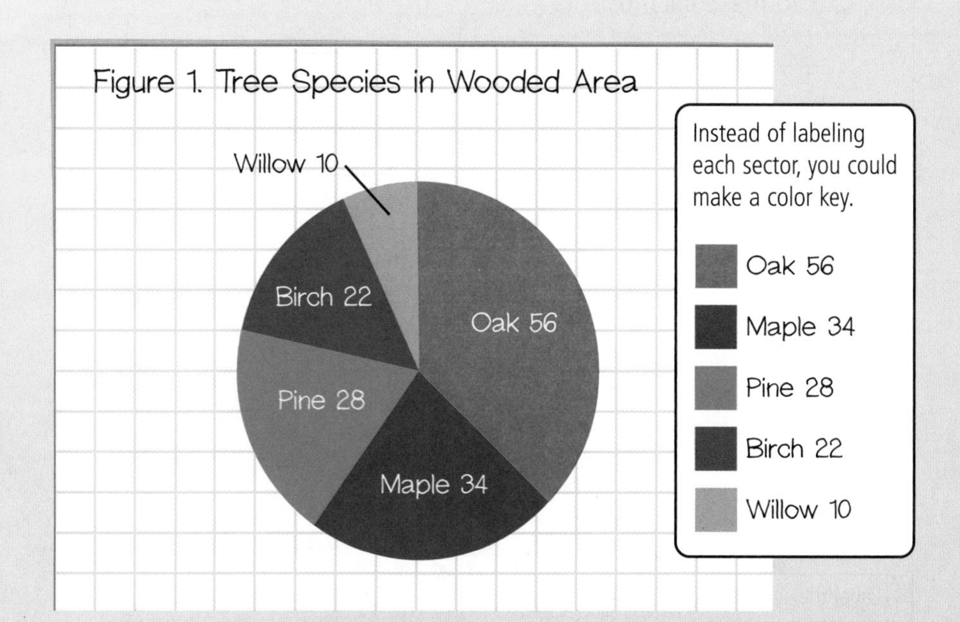

Figure 1. Tree Species in Wooded Area

Bar Graph

A **bar graph** is a type of graph in which the lengths of the bars are used to represent and compare data. A numerical scale is used to determine the lengths of the bars.

EXAMPLE

To determine the effect of water on seed sprouting, three cups were filled with sand, and ten seeds were planted in each. Different amounts of water were added to each cup over a three-day period.

Table 1. Effect of Water on Seed Sprouting

Daily Amount of Water (mL)	Number of Seeds That Sprouted After 3 Days in Sand
0	1
10	4
20	8

1. Choose a numerical scale. The greatest value is 8, so the end of the scale should have a value greater than 8, such as 10. Use equal increments along the scale, such as increments of 2.
2. Draw and label the axes. Mark intervals on the vertical axis according to the scale you chose.
3. Draw a bar for each data value. Use the scale to decide how long to make each bar.

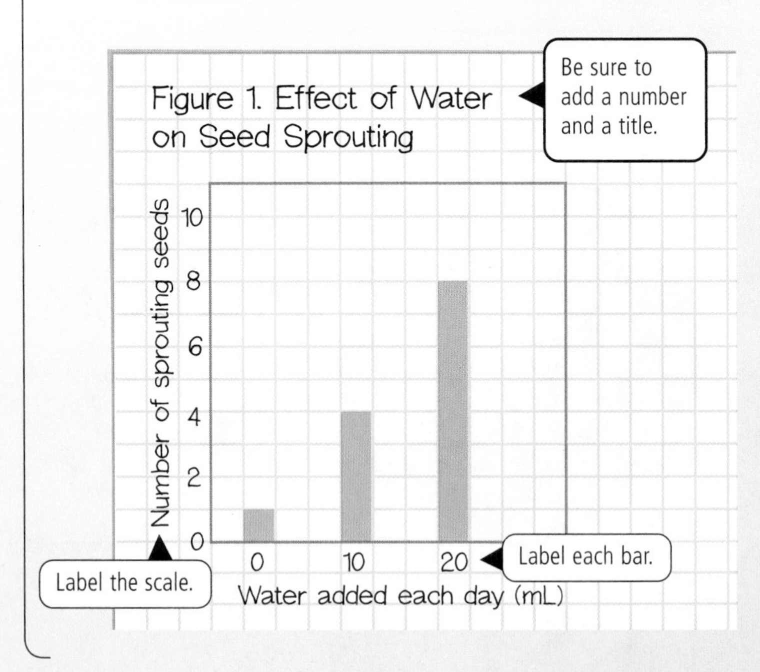

Double Bar Graph

A **double bar graph** is a bar graph that shows two sets of data. The two bars for each measurement are drawn next to each other.

EXAMPLE

The same seed-sprouting experiment was repeated with potting soil. The data for sand and potting soil can be plotted on one graph.

1. Draw one set of bars, using the data for sand, as shown below.
2. Draw bars for the potting-soil data next to the bars for the sand data. Shade them a different color. Add a key.

Table 2. Effect of Water and Soil on Seed Sprouting

Daily Amount of Water (mL)	Number of Seeds That Sprouted After 3 Days in Sand	Number of Seeds That Sprouted After 3 Days in Potting Soil
0	1	2
10	4	5
20	8	9

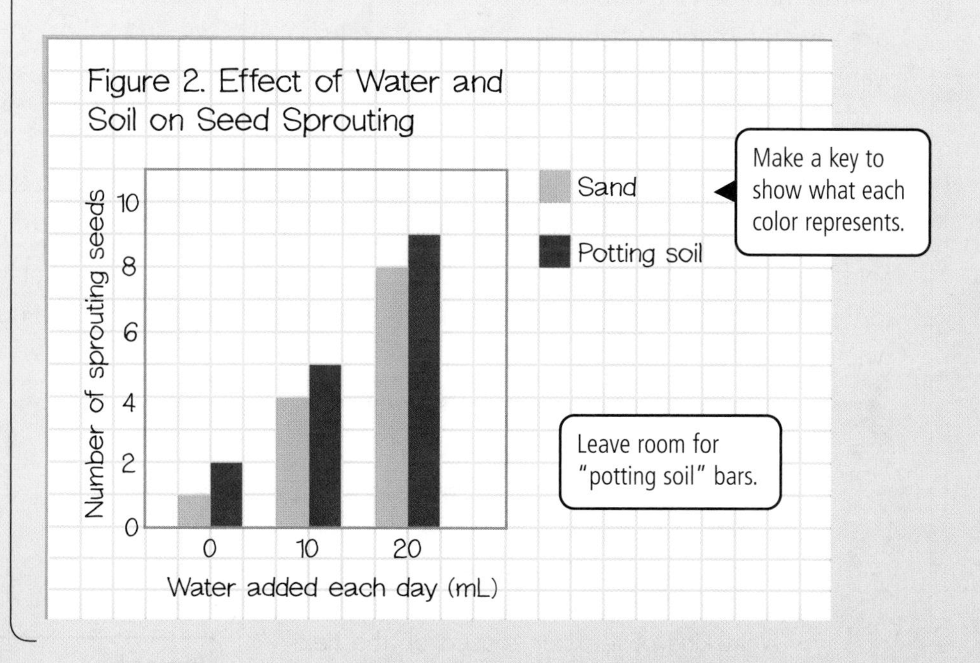

LAB HANDBOOK

Designing an Experiment

Use this section when designing or conducting an experiment.

Determining a Purpose

You can find a purpose for an experiment by doing research, by examining the results of a previous experiment, or by observing the world around you. An **experiment** is an organized procedure to study something under controlled conditions.

1. Write the purpose of your experiment as a question or problem that you want to investigate.
2. Write down research questions and begin searching for information that will help you design an experiment. Consult the library, the Internet, and other people as you conduct your research.

Don't forget to learn as much as possible about your topic before you begin.

EXAMPLE

Middle school students observed an odor near the lake by their school. They also noticed that the water on the side of the lake near the school was greener than the water on the other side of the lake. The students did some research to learn more about their observations. They discovered that the odor and green color in the lake came from algae. They also discovered that a new fertilizer was being used on a field nearby. The students inferred that the use of the fertilizer might be related to the presence of the algae and designed a controlled experiment to find out whether they were right.

Problem

How does fertilizer affect the presence of algae in a lake?

Research Questions

- Have other experiments been done on this problem? If so, what did those experiments show?
- What kind of fertilizer is used on the field? How much?
- How do algae grow?
- How do people measure algae?
- Can fertilizer and algae be used safely in a lab? How?

Research
As you research, you may find a topic that is more interesting to you than your original topic, or learn that a procedure you wanted to use is not practical or safe. It is OK to change your purpose as you research.

LAB HANDBOOK

Writing a Hypothesis

A **hypothesis** is a tentative explanation for an observation or scientific problem that can be tested by further investigation. You can write your hypothesis in the form of an "If . . . , then . . . , because . . ." statement.

Hypothesis

If the amount of fertilizer in lake water is increased, then the amount of algae will also increase, because fertilizers provide nutrients that algae need to grow.

Hypotheses
For help with hypotheses, refer to page R3.

Determining Materials

Make a list of all the materials you will need to do your experiment. Be specific, especially if someone else is helping you obtain the materials. Try to think of everything you will need.

Materials

- 1 large jar or container
- 4 identical smaller containers
- rubber gloves that also cover the arms
- sample of fertilizer-and-water solution
- eyedropper
- clear plastic wrap
- scissors
- masking tape
- marker
- ruler

Determining Variables and Constants

EXPERIMENTAL GROUP AND CONTROL GROUP

An experiment to determine how two factors are related always has two groups—a control group and an experimental group.

1. Design an experimental group. Include as many trials as possible in the experimental group in order to obtain reliable results.
2. Design a control group that is the same as the experimental group in every way possible, except for the factor you wish to test.

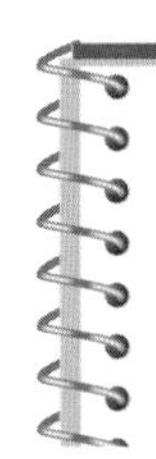

Experimental Group: two containers of lake water with one drop of fertilizer solution added to each

Control Group: two containers of lake water with no fertilizer solution added

Go back to your materials list and make sure you have enough items listed to cover both your experimental group and your control group.

VARIABLES AND CONSTANTS

Identify the variables and constants in your experiment. In a controlled experiment, a **variable** is any factor that can change. **Constants** are all of the factors that are the same in both the experimental group and the control group.

Hypothesis
If the amount of fertilizer in lake water is increased, then the amount of algae will also increase, because fertilizers provide nutrients that algae need to grow.

1. Read your hypothesis. The **independent variable** is the factor that you wish to test and that is manipulated or changed so that it can be tested. The independent variable is expressed in your hypothesis after the word *if*. Identify the independent variable in your laboratory report.
2. The **dependent variable** is the factor that you measure to gather results. It is expressed in your hypothesis after the word *then*. Identify the dependent variable in your laboratory report.

Table 1. Variables and Constants in Algae Experiment

Independent Variable	Dependent Variable	Constants
Amount of fertilizer in lake water	Amount of algae that grow	• Where the lake water is obtained • Type of container used • Light and temperature conditions where water will be stored

Set up your experiment so that you will test only one variable.

LAB HANDBOOK

MEASURING THE DEPENDENT VARIABLE

Before starting your experiment, you need to define how you will measure the dependent variable. An **operational definition** is a description of the one particular way in which you will measure the dependent variable.

Your operational definition is important for several reasons. First, in any experiment there are several ways in which a dependent variable can be measured. Second, the procedure of the experiment depends on how you decide to measure the dependent variable. Third, your operational definition makes it possible for other people to evaluate and build on your experiment.

EXAMPLE 1

An operational definition of a dependent variable can be qualitative. That is, your measurement of the dependent variable can simply be an observation of whether a change occurs as a result of a change in the independent variable. This type of operational definition can be thought of as a "yes or no" measurement.

Table 2. Qualitative Operational Definition of Algae Growth

Independent Variable	Dependent Variable	Operational Definition
Amount of fertilizer in lake water	Amount of algae that grow	Algae grow in lake water

A qualitative measurement of a dependent variable is often easy to make and record. However, this type of information does not provide a great deal of detail in your experimental results.

EXAMPLE 2

An operational definition of a dependent variable can be quantitative. That is, your measurement of the dependent variable can be a number that shows how much change occurs as a result of a change in the independent variable.

Table 3. Quantitative Operational Definition of Algae Growth

Independent Variable	Dependent Variable	Operational Definition
Amount of fertilizer in lake water	Amount of algae that grow	Diameter of largest algal growth (in mm)

A quantitative measurement of a dependent variable can be more difficult to make and analyze than a qualitative measurement. However, this type of data provides much more information about your experiment and is often more useful.

Writing a Procedure

Write each step of your procedure. Start each step with a verb, or action word, and keep the steps short. Your procedure should be clear enough for someone else to use as instructions for repeating your experiment.

If necessary, go back to your materials list and add any materials that you left out.

Procedure

1. Put on your gloves. Use the large container to obtain a sample of lake water.
2. Divide the sample of lake water equally among the four smaller containers.
3. Use the eyedropper to add one drop of fertilizer solution to two of the containers.
4. Use the masking tape and the marker to label the containers with your initials, the date, and the identifiers "Jar 1 with Fertilizer," "Jar 2 with Fertilizer," "Jar 1 without Fertilizer," and "Jar 2 without Fertilizer."
5. Cover the containers with clear plastic wrap. Use the scissors to punch ten holes in each of the covers.
6. Place all four containers on a window ledge. Make sure that they all receive the same amount of light.
7. Observe the containers every day for one week.
8. Use the ruler to measure the diameter of the largest clump of algae in each container, and record your measurements daily.

Controlling Variables
The same amount of fertilizer solution must be added to two of the four containers.

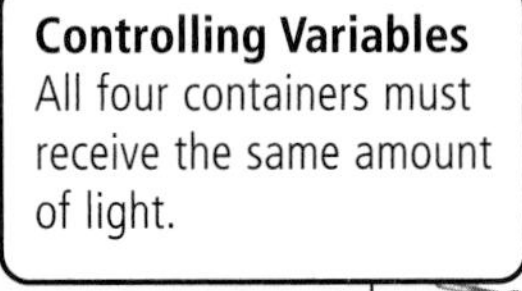

Controlling Variables
All four containers must receive the same amount of light.

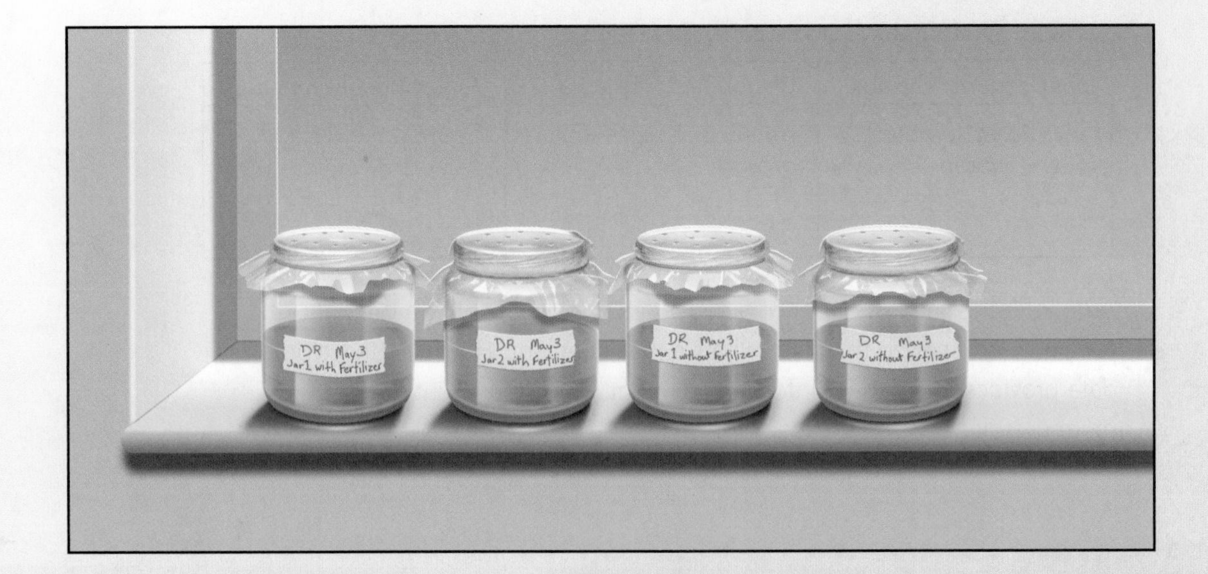

R32

Recording Observations

Once you have obtained all of your materials and your procedure has been approved, you can begin making experimental observations. Gather both quantitative and qualitative data. If something goes wrong during your procedure, make sure you record that too.

Observations
For help with making qualitative and quantitative observations, refer to page R2.

For more examples of data tables, see page R23.

Table 4. Fertilizer and Algae Growth

	Experimental Group		Control Group		
Date and Time	Jar 1 with Fertilizer (diameter of algae in mm)	Jar 2 with Fertilizer (diameter of algae in mm)	Jar 1 without Fertilizer (diameter of algae in mm)	Jar 2 without Fertilizer (diameter of algae in mm)	Observations
5/3 4:00 P.M.	0	0	0	0	condensation in all containers
5/4 4:00 P.M.	0	3	0	0	tiny green blobs in jar 2 with fertilizer
5/5 4:15 P.M.	4	5	0	3	green blobs in jars 1 and 2 with fertilizer and jar 2 without fertilizer
5/6 4:00 P.M.	5	6	0	4	water light green in jar 2 with fertilizer
5/7 4:00 P.M.	8	10	0	6	water light green in jars 1 and 2 with fertilizer and in jar 2 without fertilizer
5/8 3:30 P.M.	10	18	0	6	cover off jar 2 with fertilizer
5/9 3:30 P.M.	14	23	0	8	drew sketches of each container

Notice that on the sixth day, the observer found that the cover was off one of the containers. It is important to record observations of unintended factors because they might affect the results of the experiment.

Drawings of Samples Viewed Under Microscope on 5/9 at 100x

Use technology, such as a microscope, to help you make observations when possible.

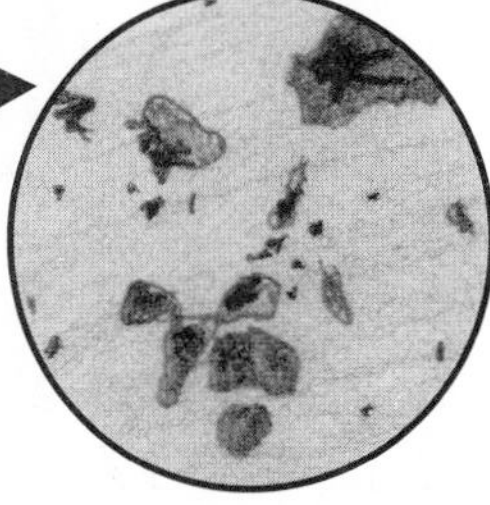
Jar 1 with Fertilizer

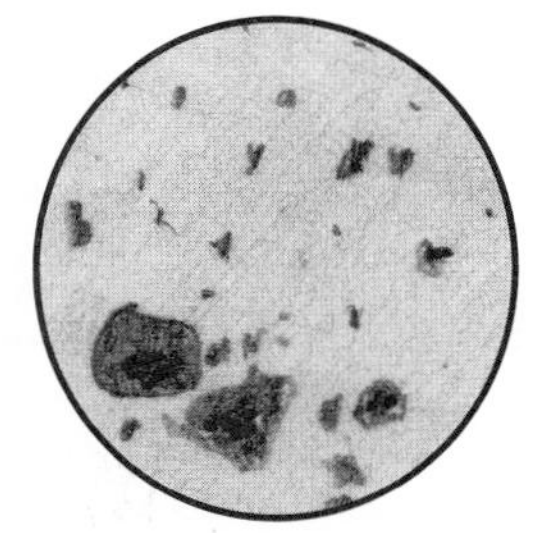
Jar 2 with Fertilizer

Jar 1 without Fertilizer

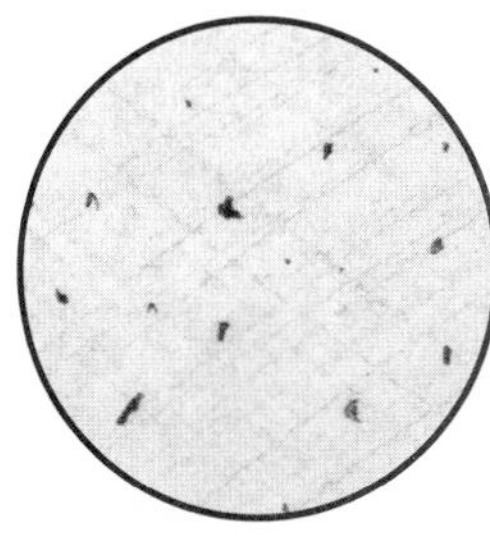
Jar 2 without Fertilizer

LAB HANDBOOK

Summarizing Results

To summarize your data, look at all of your observations together. Look for meaningful ways to present your observations. For example, you might average your data or make a graph to look for patterns. When possible, use spreadsheet software to help you analyze and present your data. The two graphs below show the same data.

EXAMPLE 1

Figure 1. Fertilizer and Algae Growth

Diameter of algae (mm)
25
20
15
10
5
0
5/3 5/4 5/5 5/6 5/7 5/8 5/9
Date
Jar 1 with fertilizer
Jar 1 without fertilizer
Jar 2 with fertilizer
Jar 2 without fertilizer

Always include a number and a title with a graph.

Line graphs are useful for showing changes over time. For help with line graphs, refer to page R24.

EXAMPLE 2

Bar graphs are useful for comparing different data sets. This bar graph has four bars for each day. Another way to present the data would be to calculate averages for the tests and the controls, and to show one test bar and one control bar for each day.

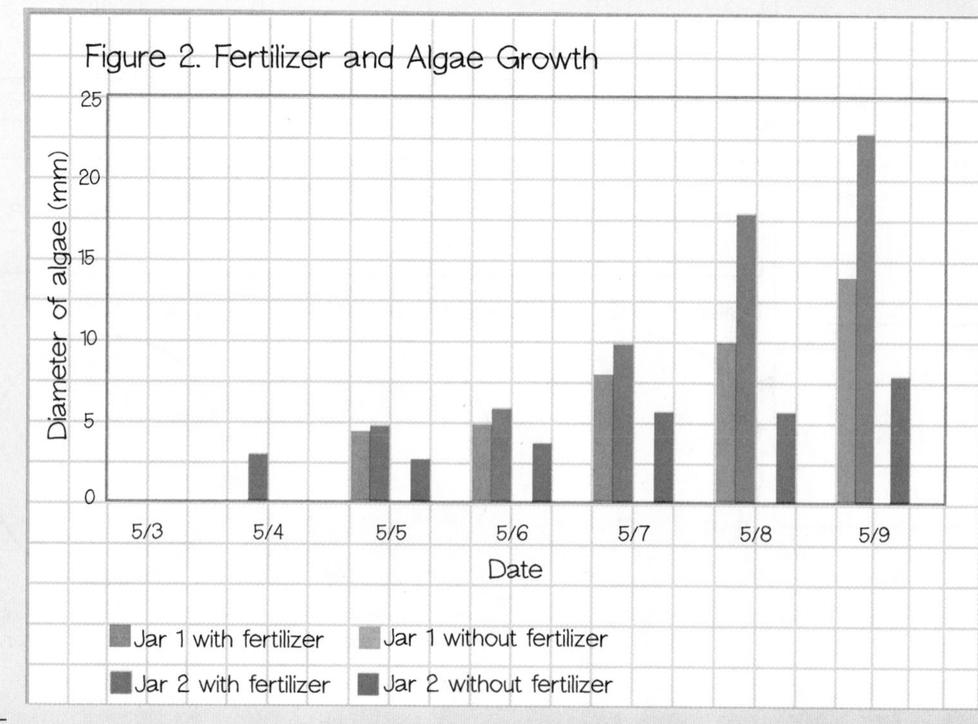

Drawing Conclusions

RESULTS AND INFERENCES

To draw conclusions from your experiment, first write your results. Then compare your results with your hypothesis. Do your results support your hypothesis? Be careful not to make inferences about factors that you did not test.

For help with making inferences, see page R4.

Results and Inferences

The results of my experiment show that more algae grew in lake water to which fertilizer had been added than in lake water to which no fertilizer had been added. My hypothesis was supported. I infer that it is possible that the growth of algae in the lake was caused by the fertilizer used on the field.

Notice that you cannot conclude from this experiment that the presence of algae in the lake was due only to the fertilizer.

QUESTIONS FOR FURTHER RESEARCH

Write a list of questions for further research and investigation. Your ideas may lead you to new experiments and discoveries.

Questions for Further Research

- What is the connection between the amount of fertilizer and algae growth?
- How do different brands of fertilizer affect algae growth?
- How would algae growth in the lake be affected if no fertilizer were used on the field?
- How do algae affect the lake and the other life in and around it?
- How does fertilizer affect the lake and the life in and around it?
- If fertilizer is getting into the lake, how is it getting there?

Describing a Set of Data

Means, medians, modes, and ranges are important math tools for describing data sets such as the following widths of fossilized clamshells.

13 mm   25 mm   14 mm   21 mm   16 mm   23 mm   14 mm

Mean

The **mean** of a data set is the sum of the values divided by the number of values.

Example

To find the mean of the clamshell data, add the values and then divide the sum by the number of values.

$$\frac{13\text{ mm} + 25\text{ mm} + 14\text{ mm} + 21\text{ mm} + 16\text{ mm} + 23\text{ mm} + 14\text{ mm}}{7} = \frac{126\text{ mm}}{7} = 18\text{ mm}$$

ANSWER The mean is 18 mm.

Median

The **median** of a data set is the middle value when the values are written in numerical order. If a data set has an even number of values, the median is the mean of the two middle values.

Example

To find the median of the clamshell data, arrange the values in order from least to greatest. The median is the middle value.

13 mm   14 mm   14 mm   16 mm   21 mm   23 mm   25 mm

ANSWER The median is 16 mm.

Mode

The **mode** of a data set is the value that occurs most often.

Example

To find the mode of the clamshell data, arrange the values in order from least to greatest and determine the value that occurs most often.

13 mm 14 mm 14 mm 16 mm 21 mm 23 mm 25 mm

ANSWER The mode is 14 mm.

A data set can have more than one mode or no mode. For example, the following data set has modes of 2 mm and 4 mm:

2 mm 2 mm 3 mm 4 mm 4 mm

The data set below has no mode, because no value occurs more often than any other.

2 mm 3 mm 4 mm 5 mm

Range

The **range** of a data set is the difference between the greatest value and the least value.

Example

To find the range of the clamshell data, arrange the values in order from least to greatest.

13 mm 14 mm 14 mm 16 mm 21 mm 23 mm 25 mm

Subtract the least value from the greatest value.

13 mm is the least value.
25 mm is the greatest value.

25 mm – 13 mm = 12 mm

ANSWER The range is 12 mm.

MATH HANDBOOK

Using Ratios, Rates, and Proportions

You can use ratios and rates to compare values in data sets. You can use proportions to find unknown values.

Ratios

A **ratio** uses division to compare two values. The ratio of a value a to a nonzero value b can be written as $\frac{a}{b}$.

Example

The height of one plant is 8 centimeters. The height of another plant is 6 centimeters. To find the ratio of the height of the first plant to the height of the second plant, write a fraction and simplify it.

$$\frac{8\text{ cm}}{6\text{ cm}} = \frac{4 \times \overset{1}{\cancel{2}}}{3 \times \underset{1}{\cancel{2}}} = \frac{4}{3}$$

ANSWER The ratio of the plant heights is $\frac{4}{3}$.

You can also write the ratio $\frac{a}{b}$ as "a to b" or as $a:b$. For example, you can write the ratio of the plant heights as "4 to 3" or as 4:3.

MATH HANDBOOK

Rates

A **rate** is a ratio of two values expressed in different units. A unit rate is a rate with a denominator of 1 unit.

Example

A plant grew 6 centimeters in 2 days. The plant's rate of growth was $\frac{6\text{ cm}}{2\text{ days}}$. To describe the plant's growth in centimeters per day, write a unit rate.

Divide numerator and denominator by 2: $\frac{6\text{ cm}}{2\text{ days}} = \frac{6\text{ cm} \div 2}{2\text{ days} \div 2}$

Simplify: $= \frac{3\text{ cm}}{1\text{ day}}$

You divide 2 days by 2 to get 1 day, so divide 6 cm by 2 also.

ANSWER The plant's rate of growth is 3 centimeters per day.

Proportions

A **proportion** is an equation stating that two ratios are equivalent. To solve for an unknown value in a proportion, you can use cross products.

Example

If a plant grew 6 centimeters in 2 days, how many centimeters would it grow in 3 days (if its rate of growth is constant)?

Write a proportion: $\frac{6 \text{ cm}}{2 \text{ days}} = \frac{x \text{ cm}}{3 \text{ days}}$

Set cross products: $6 \cdot 3 = 2x$

Multiply 6 and 3: $18 = 2x$

Divide each side by 2: $\frac{18}{2} = \frac{2x}{2}$

Simplify: $9 = x$

ANSWER The plant would grow 9 centimeters in 3 days.

Using Decimals, Fractions, and Percents

Decimals, fractions, and percentages are all ways of recording and representing data.

Decimals

A **decimal** is a number that is written in the base-ten place value system, in which a decimal point separates the ones and tenths digits. The values of each place is ten times that of the place to its right.

Example

A caterpillar traveled from point *A* to point *C* along the path shown.

A —— 36.9 cm —— B —— 52.4 cm —— C

ADDING DECIMALS To find the total distance traveled by the caterpillar, add the distance from *A* to *B* and the distance from *B* to *C*. Begin by lining up the decimal points. Then add the figures as you would whole numbers and bring down the decimal point.

$$\begin{array}{r} 36.9 \text{ cm} \\ +\ 52.4 \text{ cm} \\ \hline 89.3 \text{ cm} \end{array}$$

ANSWER The caterpillar traveled a total distance of 89.3 centimeters.

Example *continued*

SUBTRACTING DECIMALS To find how much farther the caterpillar traveled on the second leg of the journey, subtract the distance from *A* to *B* from the distance from *B* to *C*.

$$\begin{array}{r} 52.4 \text{ cm} \\ -\ 36.9 \text{ cm} \\ \hline 15.5 \text{ cm} \end{array}$$

ANSWER The caterpillar traveled 15.5 centimeters farther on the second leg of the journey.

Example

A caterpillar is traveling from point *D* to point *F* along the path shown. The caterpillar travels at a speed of 9.6 centimeters per minute.

D E 33.6 cm F

MULTIPLYING DECIMALS You can multiply decimals as you would whole numbers. The number of decimal places in the product is equal to the sum of the number of decimal places in the factors.

For instance, suppose it takes the caterpillar 1.5 minutes to go from *D* to *E*. To find the distance from *D* to *E*, multiply the caterpillar's speed by the time it took.

$$\begin{array}{r} 9.6 \\ \times\ 1.5 \\ \hline 480 \\ 96\ \ \\ \hline 14.40 \end{array} \qquad \begin{array}{l} 1 \text{ decimal place} \\ +\ 1 \text{ decimal place} \\ \\ \\ \\ 2 \text{ decimal places} \end{array}$$

Align as shown.

ANSWER The distance from *D* to *E* is 14.4 centimeters.

DIVIDING DECIMALS When you divide by a decimal, move the decimal points the same number of places in the divisor and the dividend to make the divisor a whole number.

For instance, to find the time it will take the caterpillar to travel from *E* to *F*, divide the distance from *E* to *F* by the caterpillar's speed.

$9.6\overline{)33.6}$

Move each decimal point one place to the right.

$$\begin{array}{r} 3.5 \\ 96\overline{)336.} \\ 288\ \ \\ \hline 480 \\ 480 \\ \hline 0 \end{array}$$

Line up decimal points.

ANSWER The caterpillar will travel from *E* to *F* in 3.5 minutes.

MATH HANDBOOK

Fractions

A **fraction** is a number in the form $\frac{a}{b}$, where b is not equal to 0. A fraction is in **simplest form** if its numerator and denominator have a greatest common factor (GCF) of 1. To simplify a fraction, divide its numerator and denominator by their GCF.

Example

A caterpillar is 40 millimeters long. The head of the caterpillar is 6 millimeters long. To compare the length of the caterpillar's head with the caterpillar's total length, you can write and simplify a fraction that expresses the ratio of the two lengths.

Write the ratio of the two lengths: $\frac{\text{Length of head}}{\text{Total length}} = \frac{6 \text{ mm}}{40 \text{ mm}}$

Write numerator and denominator as products of numbers and the GCF: $= \frac{3 \times 2}{20 \times 2}$

Divide numerator and denominator by the GCF: $= \frac{3 \times \cancel{2}^{1}}{20 \times \cancel{2}_{1}}$

Simplify: $= \frac{3}{20}$

ANSWER In simplest form, the ratio of the lengths is $\frac{3}{20}$.

Percents

A **percent** is a ratio that compares a number to 100. The word *percent* means "per hundred" or "out of 100." The symbol for *percent* is %.

For instance, suppose 43 out of 100 caterpillars are female. You can represent this ratio as a percent, a decimal, or a fraction.

Percent	Decimal	Fraction
43%	0.43	$\frac{43}{100}$

Example

In the preceding example, the ratio of the length of the caterpillar's head to the caterpillar's total length is $\frac{3}{20}$. To write this ratio as a percent, write an equivalent fraction that has a denominator of 100.

Multiply numerator and denominator by 5: $\frac{3}{20} = \frac{3 \times 5}{20 \times 5}$

$= \frac{15}{100}$

Write as a percent: $= 15\%$

ANSWER The caterpillar's head represents 15 percent of its total length.

MATH HANDBOOK

Using Formulas

A mathematical **formula** is a statement of a fact, rule, or principle. It is usually expressed as an equation.

The term *variable* is also used in science to refer to a factor that can change during an experiment.

In science, a formula often has a word form and a symbolic form. The formula below expresses Ohm's law.

Word Form	**Symbolic Form**
$\text{Current} = \frac{\text{voltage}}{\text{resistance}}$	$I = \frac{V}{R}$

In this formula, *I*, *V*, and *R* are variables. A mathematical **variable** is a symbol or letter that is used to represent one or more numbers.

Example

Suppose that you measure a voltage of 1.5 volts and a resistance of 15 ohms. You can use the formula for Ohm's law to find the current in amperes.

Write the formula for Ohm's law: $I = \frac{V}{R}$

Substitute 1.5 volts for V and 15 ohms for R: $I = \frac{1.5 \text{ volts}}{15 \text{ ohms}}$

Simplify: $I = 0.1 \text{ amp}$

ANSWER The current is 0.1 ampere.

If you know the values of all variables but one in a formula, you can solve for the value of the unknown variable. For instance, Ohm's law can be used to find a voltage if you know the current and the resistance.

Example

Suppose that you know that a current is 0.2 amperes and the resistance is 18 ohms. Use the formula for Ohm's law to find the voltage in volts.

Write the formula for Ohm's law: $I = \frac{V}{R}$

Substitute 0.2 amp for I and 18 ohms for R: $0.2 \text{ amp} = \frac{V}{18 \text{ ohms}}$

Multiply both sides by 18 ohms: $0.2 \text{ amp} \cdot 18 \text{ ohms} = V$

Simplify: $3.6 \text{ volts} = V$

ANSWER The voltage is 3.6 volts.

MATH HANDBOOK

Finding Areas

The area of a figure is the amount of surface the figure covers.

Area is measured in square units, such as square meters (m^2) or square centimeters (cm^2). Formulas for the areas of three common geometric figures are shown below.

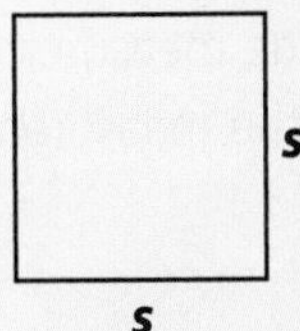

Area = (side length)^2
$A = s^2$

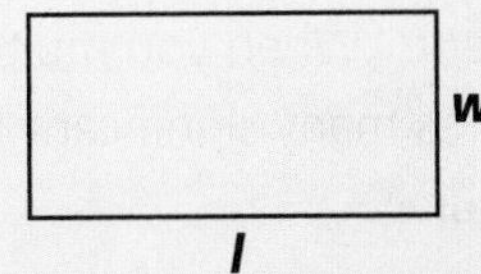

Area = length × width
$A = lw$

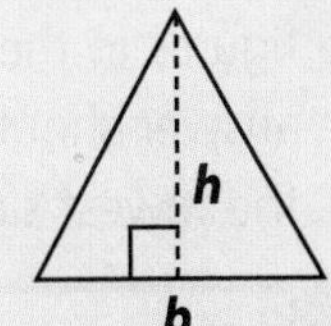

Area = $\frac{1}{2}$ × base × height
$A = \frac{1}{2}bh$

Example

Each face of a halite crystal is a square like the one shown. You can find the area of the square by using the steps below.

3 mm
3 mm

Write the formula for the area of a square: $A = s^2$

Substitute 3 mm for s: $= (3\text{ mm})^2$

Simplify: $= 9\text{ mm}^2$

ANSWER The area of the square is 9 square millimeters.

MATH HANDBOOK

Finding Volumes

The volume of a solid is the amount of space contained by the solid.

Volume is measured in cubic units, such as cubic meters (m^3) or cubic centimeters (cm^3). The volume of a rectangular prism is given by the formula shown below.

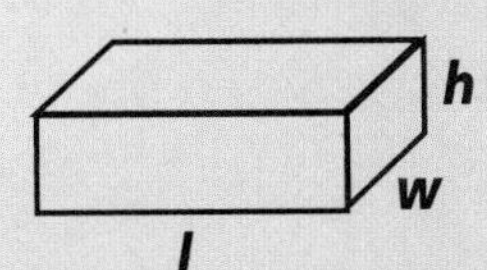

Volume = length × width × height
$V = lwh$

Example

A topaz crystal is a rectangular prism like the one shown. You can find the volume of the prism by using the steps below.

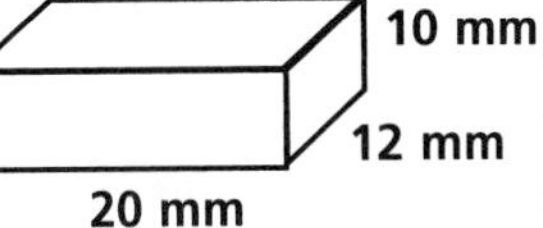

Write the formula for the volume of a rectangular prism: $V = lwh$

Substitute dimensions: $= 20\text{ mm} \times 12\text{ mm} \times 10\text{ mm}$

Simplify: $= 2400\text{ mm}^3$

ANSWER The volume of the rectangular prism is 2400 cubic millimeters.

Using Significant Figures

The **significant figures** in a decimal are the digits that are warranted by the accuracy of a measuring device.

When you perform a calculation with measurements, the number of significant figures to include in the result depends in part on the number of significant figures in the measurements. When you multiply or divide measurements, your answer should have only as many significant figures as the measurement with the fewest significant figures.

Example

Using a balance and a graduated cylinder filled with water, you determined that a marble has a mass of 8.0 grams and a volume of 3.5 cubic centimeters. To calculate the density of the marble, divide the mass by the volume.

Write the formula for density: $\text{Density} = \frac{\text{mass}}{\text{Volume}}$

Substitute measurements: $= \frac{8.0 \text{ g}}{3.5 \text{ cm}^3}$

Use a calculator to divide: $\approx 2.285714286 \text{ g/cm}^3$

ANSWER Because the mass and the volume have two significant figures each, give the density to two significant figures. The marble has a density of 2.3 grams per cubic centimeter.

MATH HANDBOOK

Using Scientific Notation

Scientific notation is a shorthand way to write very large or very small numbers. For example, 73,500,000,000,000,000,000,000 kg is the mass of the Moon. In scientific notation, it is 7.35×10^{22} kg.

Example

You can convert from standard form to scientific notation.

Standard Form	Scientific Notation
720,000	7.2×10^5
5 decimal places left	Exponent is 5.
0.000291	2.91×10^{-4}
4 decimal places right	Exponent is –4.

You can convert from scientific notation to standard form.

Scientific Notation	Standard Form
4.63×10^7	46,300,000
Exponent is 7.	7 decimal places right
1.08×10^{-6}	0.00000108
Exponent is –6.	6 decimal places left

Note-Taking Handbook

Note-Taking Strategies

Taking notes as you read helps you understand the information. The notes you take can also be used as a study guide for later review. This handbook presents several ways to organize your notes.

Content Frame

1. Make a chart in which each column represents a category.
2. Give each column a heading.
3. Write details under the headings.

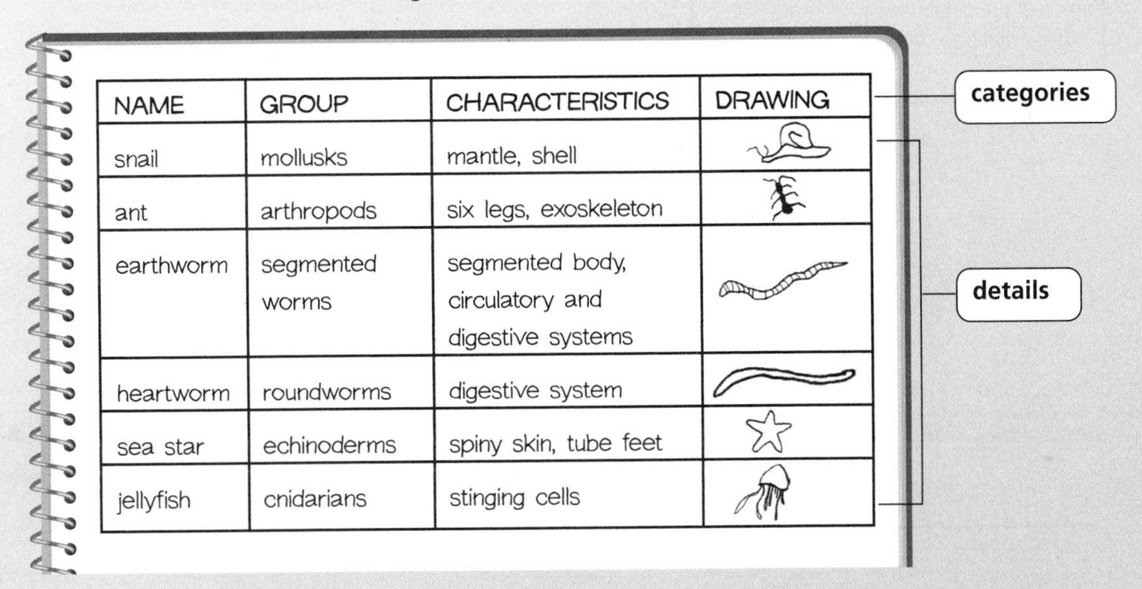

NAME	GROUP	CHARACTERISTICS	DRAWING
snail	mollusks	mantle, shell	
ant	arthropods	six legs, exoskeleton	
earthworm	segmented worms	segmented body, circulatory and digestive systems	
heartworm	roundworms	digestive system	
sea star	echinoderms	spiny skin, tube feet	
jellyfish	cnidarians	stinging cells	

Combination Notes

1. For each new idea or concept, write an informal outline of the information.
2. Make a sketch to illustrate the concept, and label it.

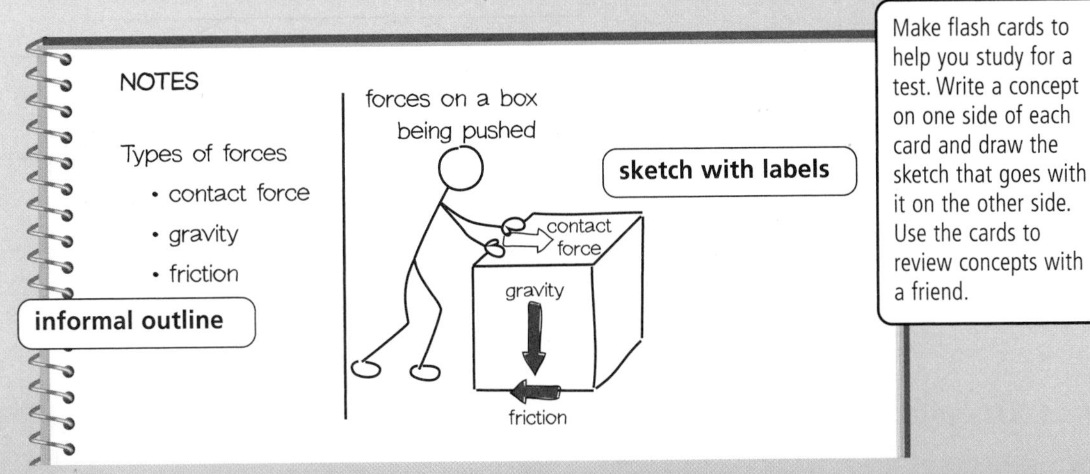

Make flash cards to help you study for a test. Write a concept on one side of each card and draw the sketch that goes with it on the other side. Use the cards to review concepts with a friend.

Main Idea and Detail Notes

1. In the left-hand column of a two-column chart, list main ideas. The blue headings express main ideas throughout this textbook.
2. In the right-hand column, write details that expand on each main idea.

You can shorten the headings in your chart. Be sure to use the most important words.

When studying for tests, cover up the detail notes column with a sheet of paper. Then use each main idea to form a question—such as "How does latitude affect climate?" Answer the question, and then uncover the detail notes column to check your answer.

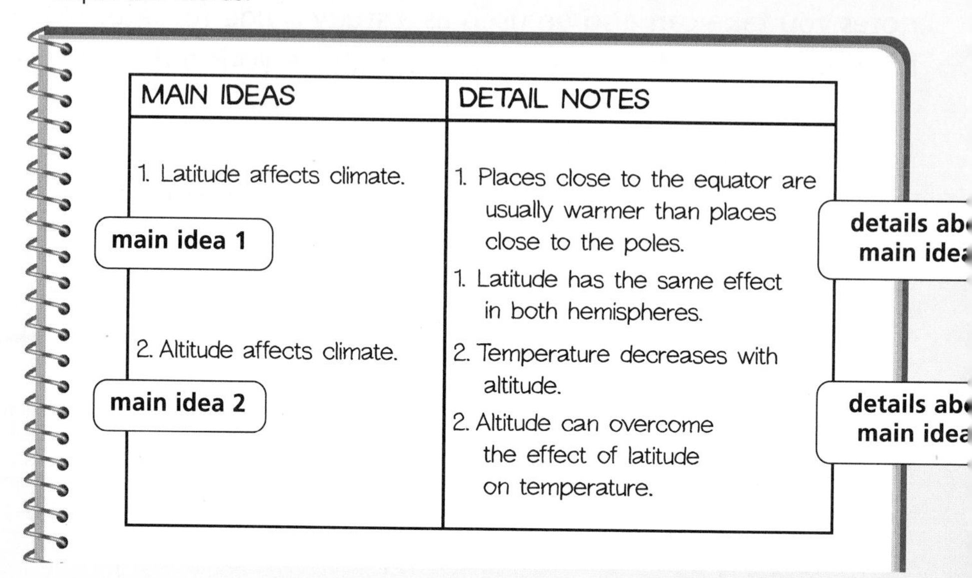

MAIN IDEAS	DETAIL NOTES
1. Latitude affects climate.	1. Places close to the equator are usually warmer than places close to the poles. 1. Latitude has the same effect in both hemispheres.
2. Altitude affects climate.	2. Temperature decreases with altitude. 2. Altitude can overcome the effect of latitude on temperature.

Main Idea Web

1. Write a main idea in a box.
2. Add boxes around it with related vocabulary terms and important details.

You can find definitions near highlighted terms.

definition of *work*
Work is the use of force to move an object.

formula
Work = force · distance

main idea
Force is necessary to do work.

definition of *joule*
The joule is the unit used to measure work.

important detail
Work depends on the size of a force.

Mind Map

1. Write a main idea in the center.
2. Add details that relate to one another and to the main idea.

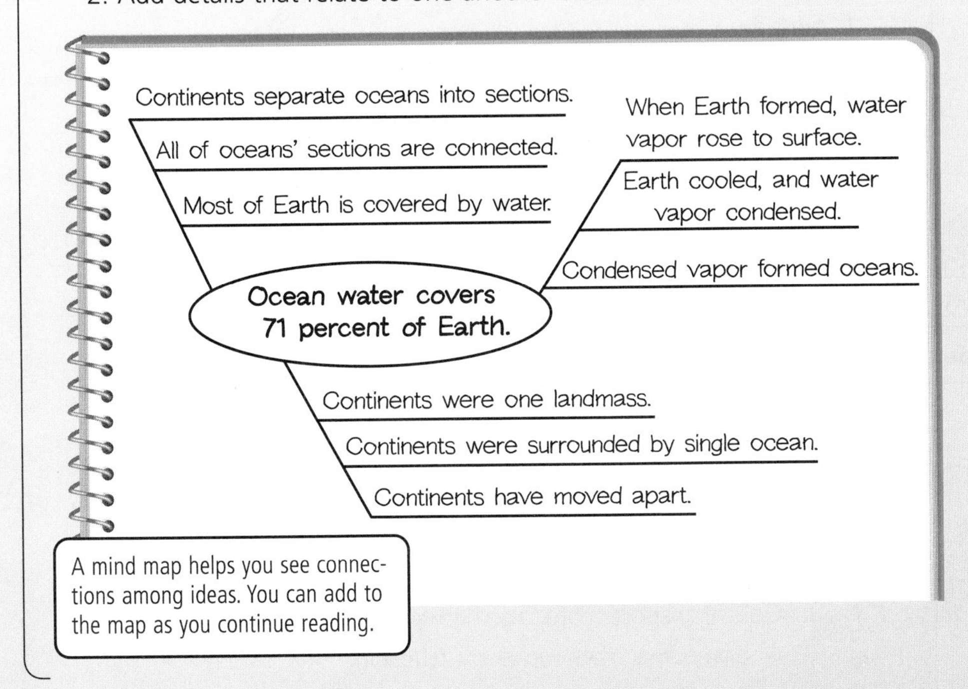

Continents separate oceans into sections.
All of oceans' sections are connected.
Most of Earth is covered by water.
When Earth formed, water vapor rose to surface.
Earth cooled, and water vapor condensed.
Condensed vapor formed oceans.
Ocean water covers 71 percent of Earth.
Continents were one landmass.
Continents were surrounded by single ocean.
Continents have moved apart.
A mind map helps you see connections among ideas. You can add to the map as you continue reading.

Supporting Main Ideas

1. Write a main idea in a box.
2. Add boxes underneath with information—such as reasons, explanations, and examples—that supports the main idea.

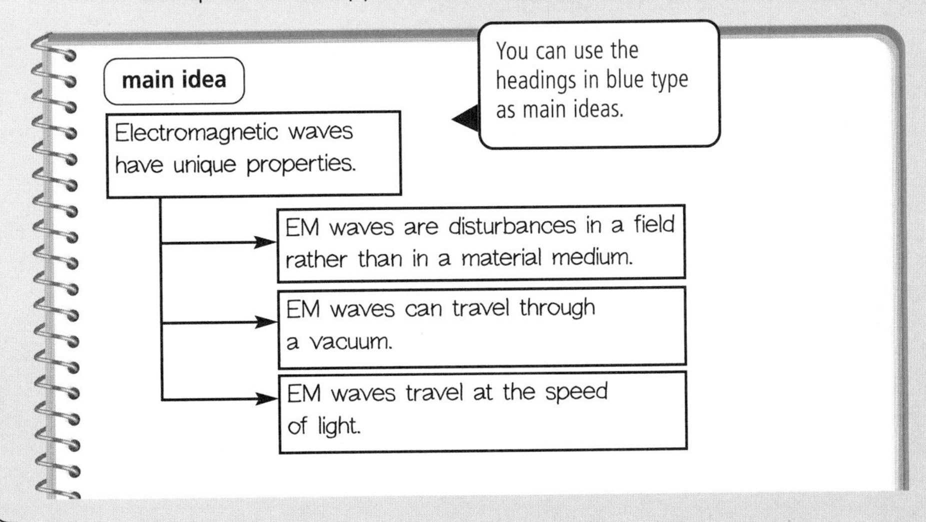

main idea
Electromagnetic waves have unique properties.
You can use the headings in blue type as main ideas.
EM waves are disturbances in a field rather than in a material medium.
EM waves can travel through a vacuum.
EM waves travel at the speed of light.

NOTE-TAKING HANDBOOK

Outline

1. Copy the chapter title and headings from the book in the form of an outline.
2. Add notes that summarize in your own words what you read.

Cell Processes

I. Cells capture and release energy. — **1st key idea**
  A. All cells need energy. — **1st subpoint of I**
  B. Some cells capture light energy. — **2nd subpoint of I**
    1. Process of photosynthesis — **1st detail about B**
    2. Chloroplasts (site of photosynthesis) — **2nd detail about B**
    3. Carbon dioxide and water as raw materials
    4. Glucose and oxygen as products
  C. All cells release energy.
    1. Process of cellular respiration
    2. Fermentation of sugar to carbon dioxide
    3. Bacteria that carry out fermentation

II. Cells transport materials through membranes.
  A. Some materials move by diffusion.
    1. Particle movement from higher to lower concentrations
    2. Movement of water through membrane (osmosis)
  B. Some transport requires energy.
    1. Active transport
    2. Examples of active transport

Correct Outline Form

Include a title.

Arrange key ideas, subpoints, and details as shown.

Indent the divisions of the outline as shown.

Use the same grammatical form for items of the same rank. For example, if A is a sentence, B must also be a sentence.

You must have at least two main ideas or subpoints. That is, every A must be followed by a B, and every 1 must be followed by a 2.

R48

Concept Map

1. Write an important concept in a large oval.
2. Add details related to the concept in smaller ovals.
3. Write linking words on arrows that connect the ovals.

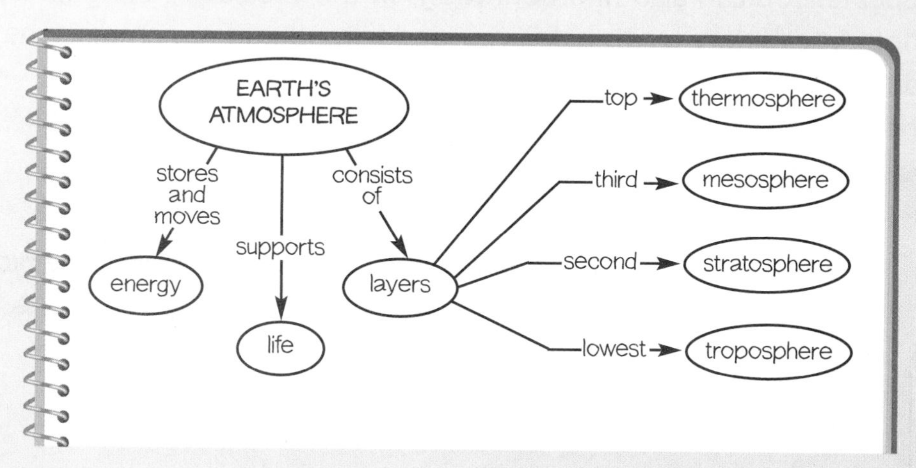

The main ideas or concepts can often be found in the blue headings. An example is "The atmosphere stores and moves energy." Use nouns from these concepts in the ovals, and use the verb or verbs on the lines.

Venn Diagram

1. Draw two overlapping circles, one for each item that you are comparing.
2. In the overlapping section, list the characteristics that are shared by both items.
3. In the outer sections, list the characteristics that are peculiar to each item.
4. Write a summary that describes the information in the Venn diagram.

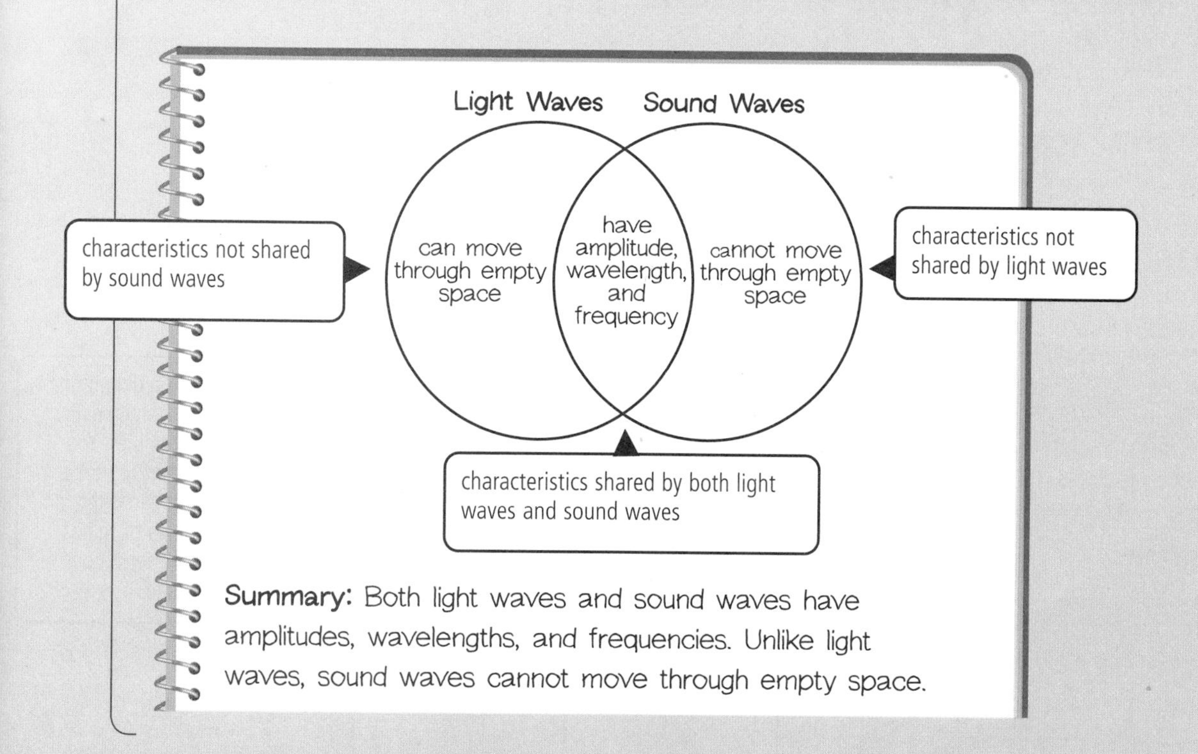

NOTE-TAKING HANDBOOK

Vocabulary Strategies

Important terms are highlighted in this book. A definition of each term can be found in the sentence or paragraph where the term appears. You can also find definitions in the Glossary. Taking notes about vocabulary terms helps you understand and remember what you read.

Description Wheel

1. Write a term inside a circle.
2. Write words that describe the term on "spokes" attached to the circle.

When studying for a test with a friend, read the phrases on the spokes one at a time until your friend identifies the correct term.

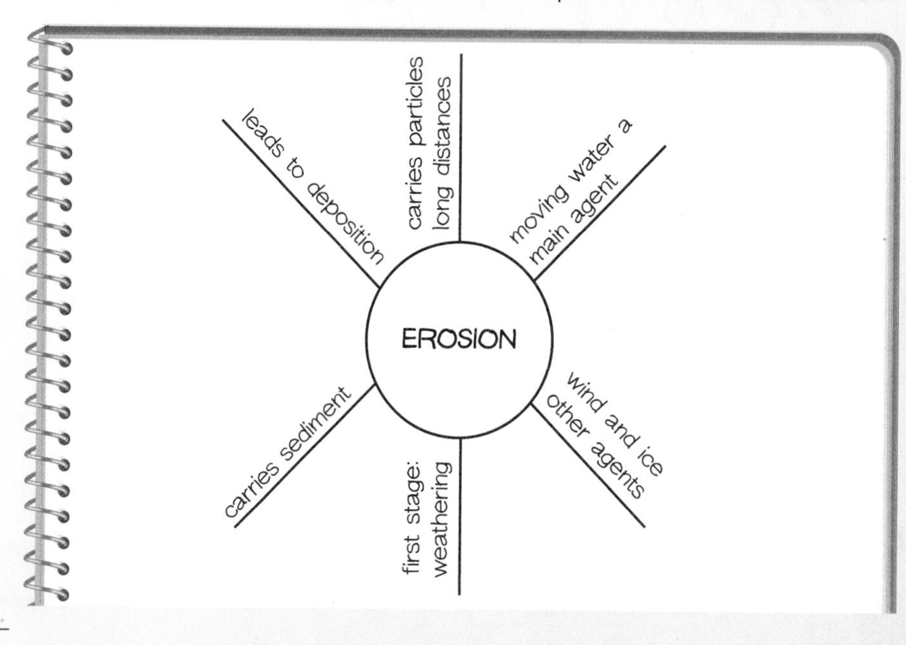

Four Square

1. Write a term in the center.
2. Write details in the four areas around the term.

ORGANISM

Definition	Characteristics
any living thing	needs food, water, air; needs energy; grows, develops, reproduces
Examples	**Nonexamples**
dogs, cats, birds, insects, flowers, trees	rocks, water, dirt

Include a definition, some characteristics, and examples. You may want to add a formula, a sketch, or examples of things that the term does *not* name.

NOTE-TAKING HANDBOOK

Frame Game

1. Write a term in the center.
2. Frame the term with details.

Include examples, descriptions, sketches, or sentences that use the term in context. Change the frame to fit each new term.

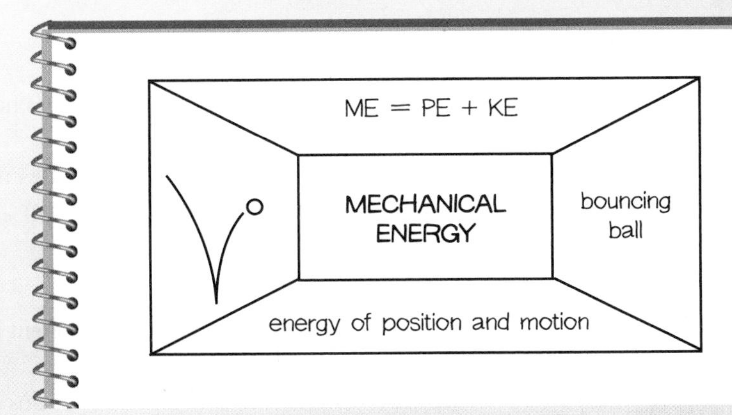

Magnet Word

1. Write a term on the magnet.
2. On the lines, add details related to the term.

You can also use phrases or sentences on the lines.

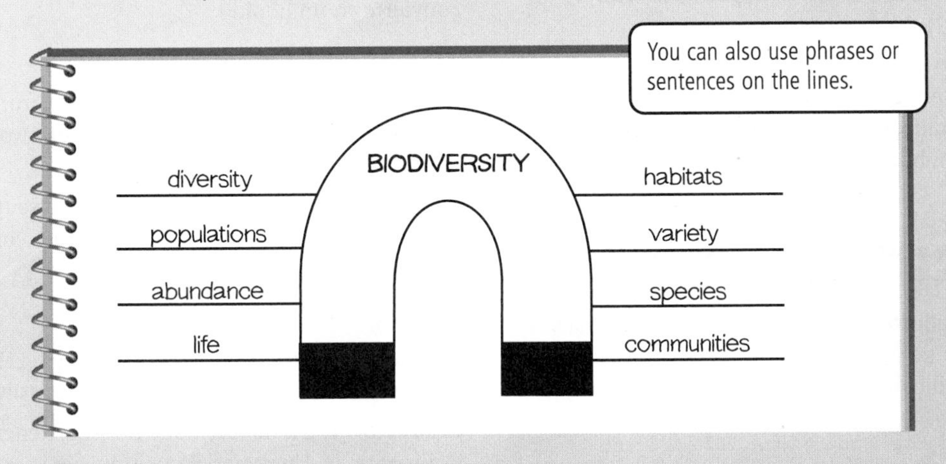

Word Triangle

1. Write a term and its definition in the bottom section.
2. In the middle section, write a sentence in which the term is used correctly.
3. In the top section, draw a small picture to illustrate the term.

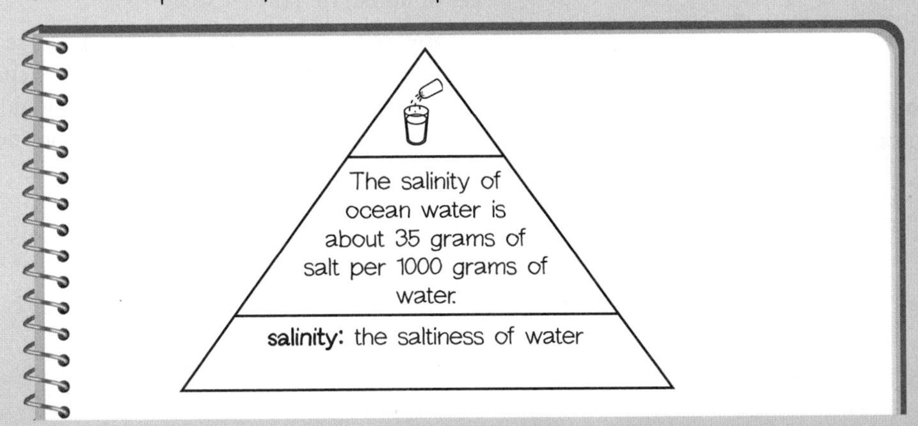

NOTE-TAKING HANDBOOK

Glossary

GLOSSARY

A

atom
The smallest particle of an element that has the chemical properties of that element. (p. 16)

átomo La partícula más pequeña de un elemento que tiene las propiedades químicas de ese elemento.

B

boiling
A process by which a substance changes from its liquid state to its gas state. The liquid is heated to a specific temperature at which bubbles of vapor form within the liquid. (p. 54)

ebullición Un proceso mediante el cual una sustancia cambia de su estado líquido a su estado gaseoso se calienta el líquido a una determinada temperatura a la cual se forman burbujas de vapor dentro del líquido.

boiling point
The temperature at which a substance changes from its liquid state to its gas state through boiling. (p. 54)

punto de ebullición La temperatura a la cual una sustancia cambia de su estado líquido a su estado gaseoso mediante ebullición.

C

calorie
The amount of energy needed to increase the temperature of one gram of water by one degree Celsius. (p. 112)

caloría La cantidad de energía que se necesita para aumentar la temperatura de un gramo de agua un grado centígrado.

chemical change
A change of one substance into another substance. (p. 46)

cambio químico La transformación de una sustancia a otra sustancia.

chemical property
A characteristic of a substance that describes how it can form a new substance. (p. 46)

propiedad química Una característica de una sustancia que describe como puede formar una nueva sustancia.

compound
A substance made up of two or more different types of atoms bonded together. (p. 23)

compuesto Una sustancia formada por dos o más diferentes tipos de átomos enlazados.

condensation
The process by which a gas becomes a liquid. (p. 55)

condensación El proceso mediante el cual un gas se convierte en un líquido.

conduction
The process by which energy is transferred from a warmer object to a cooler object by means of physical contact. (p. 117)

conducción El proceso mediante el cual se transfiere energía de un objeto más caliente a un objeto más frío por medio de contacto físico.

conductor
1. A material that transfers energy easily. (p. 117)
2. A material that transfers electric charge easily.

conductor 1. Un material que transfiere energía fácilmente. 2. Un material que transfiere cargas eléctricas fácilmente.

convection
A process by which energy is transferred in gases and liquids, occurring when a warmer, less dense area of gas or liquid is pushed up by a cooler, more dense area of the gas or liquid. (p. 118)

convección Un proceso mediante el cual se transfiere energía en los gases y los líquidos; ocurre cuando un área más fría y más densa del gas o del líquido empuja hacia arriba un área más caliente y menos densa de gas o de líquido.

cycle

n. A series of events or actions that repeat themselves regularly; a physical and/or chemical process in which one material continually changes locations and/or forms. Examples include the water cycle, the carbon cycle, and the rock cycle.

v. To move through a repeating series of events or actions.

ciclo *s.* Una serie de eventos o acciones que se repiten regularmente; un proceso físico y/o químico en el cual un material cambia continuamente de lugar y/o forma. Ejemplos: el ciclo del agua, el ciclo del carbono y el ciclo de las rocas.

D

data

Information gathered by observation or experimentation that can be used in calculating or reasoning. *Data* is a plural word; the singular is datum.

datos Información reunida mediante observación o experimentación y que se puede usar para calcular o para razonar.

degree

Evenly divided units of a temperature scale. (p. 106)

grado Unidades de una escala de temperatura distribuidas uniformemente.

density

A property of matter representing the mass per unit volume. (p. 43)

densidad Una propiedad de la materia que representa la masa por unidad de volumen.

E

element

A substance that cannot be broken down into a simpler substance by ordinary chemical changes. An element consists of atoms of only one type. (p. 22)

elemento Una sustancia que no puede descomponerse en otra sustancia más simple por medio de cambios químicos normales. Un elemento consta de átomos de un solo tipo.

energy

The ability to do work or to cause a change. For example, the energy of a moving bowling ball knocks over pins; energy from food allows animals to move and to grow; and energy from the Sun heats Earth's surface and atmosphere, which causes air to move. (p. 72)

energía La capacidad para trabajar o causar un cambio. Por ejemplo, la energía de una bola de boliche en movimiento tumba los pinos; la energía proveniente de su alimento permite a los animales moverse y crecer; la energía del Sol calienta la superficie y la atmósfera de la Tierra, lo que ocasiona que el aire se mueva.

energy efficiency

A measurement of usable energy after an energy conversion. (p. 83)

eficiencia energética Una medida de la energía utilizable después de una conversión energética.

evaporation

A process by which a substance changes from its liquid state to its gas state by random particle movement. Evaporation usually occurs at the surface of a liquid over a wide range of temperatures. (p. 53)

evaporación Un proceso mediante el cual una sustancia cambia de su estado líquido a su estado gaseoso por medio del movimiento aleatorio de las partículas. La evaporación normalmente ocurre en la superficie de un líquido en una amplia gama de temperaturas.

experiment

An organized procedure to study something under controlled conditions. (p. xxiv)

experimento Un procedimiento organizado para estudiar algo bajo condiciones controladas.

F

force

A push or a pull; something that changes the motion of an object. (p. xxi)

fuerza Un empuje o un jalón; algo que cambia el movimiento de un objeto.

freezing

The process by which a substance changes from its liquid state into its solid state. (p. 52)

congelación El proceso mediante el cual una sustancia cambia de su estado líquido a su estado sólido.

R53

freezing point
The temperature at which a substance changes from its liquid state to its solid state through freezing. (p. 52)

> **punto de congelación** La temperatura a la cual una sustancia cambia de su estado líquido a su estado sólido mediante congelación.

friction
A force that resists the motion between two surfaces in contact. (p. xxi)

> **fricción** Una fuerza que resiste el movimiento entre dos superficies en contacto.

gas
Matter with no definite volume and no definite shape. The molecules in a gas are very far apart, and the amount of space between them can change easily. (p. 28)

> **gas** Materia sin volumen definido ni forma definida. Las moléculas en un gas están muy separadas unas de otras, y la cantidad de espacio entre ellas puede cambiar fácilmente.

gravity
The force that objects exert on each other because of their mass. (p. xxi)

> **gravedad** La fuerza que los objetos ejercen entre sí debido a su masa.

heat
1. The flow of energy from an object at a higher temperature to an object at a lower temperature. (p. 110)
2. Energy that is transferred from a warmer object to a cooler object.

> **calor** 1. El flujo de energía de un objeto a mayor temperatura a un objeto a menor temperatura. 2. Energía que se transfiere de un objeto más caliente a un objeto más frío.

hypothesis
A tentative explanation for an observation or phenomenon. A hypothesis is used to make testable predictions. (p. xxiv)

> **hipótesis** Una explicación provisional de una observación o de un fenómeno. Una hipótesis se usa para hacer predicciones que se pueden probar.

insulator
1. A material that does not transfer energy easily. (p. 117)
2. A material that does not transfer electric charge easily.

> **aislante** 1. Un material que no transfiere energía fácilmente. 2. Un material que no transfiere cargas eléctricas fácilmente.

joule (jool) J
A unit used to measure energy and work. One calorie is equal to 4.18 joules of energy; one joule of work is done when a force of one newton moves an object one meter. (p. 112)

> **julio** Una unidad que se usa para medir la energía y el trabajo. Una caloría es igual a 4.18 julios de energía; se hace un joule de trabajo cuando una fuerza de un newton mueve un objeto un metro.

kinetic energy
The energy of motion. A moving object has the most kinetic energy at the point where it moves the fastest. (p. 74)

> **energía cinética** La energía del movimiento. Un objeto que se mueve tiene su mayor energía cinética en el punto en el cual se mueve con mayor rapidez.

kinetic theory of matter
A theory stating that all matter is made of particles in motion. (p. 104)

> **teoría cinética de la materia** Una teoría que establece que toda materia está compuesta de partículas en movimiento.

L

law
In science, a rule or principle describing a physical relationship that always works in the same way under the same conditions. The law of conservation of energy is an example.

ley En las ciencias, una regla o un principio que describe una relación física que siempre funciona de la misma manera bajo las mismas condiciones. La ley de la conservación de la energía es un ejemplo.

law of conservation of energy
A law stating that no matter how energy is transferred or transformed, it continues to exist in one form or another. (p. 82)

ley de la conservación de la energía Una ley que establece que no importa cómo se transfiere o transforma la energía, toda la energía sigue presente en alguna forma u otra.

liquid
Matter that has a definite volume but does not have a definite shape. The molecules in a liquid are close together but not bound to one another. (p. 28)

líquido Materia que tiene un volumen definido pero no tiene una forma definida. Las moléculas en un líquido están cerca unas de otras pero no están ligadas.

M, N, O

mass
A measure of how much matter an object is made of. (p. 10)

masa Una medida de la cantidad de materia de la que está compuesto un objeto.

matter
Anything that has mass and volume. Matter exists ordinarily as a solid, a liquid, or a gas. (p. 9)

materia Todo lo que tiene masa y volumen. Generalmente la materia existe como sólido, líquido o gas.

melting
The process by which a substance changes from its solid state to its liquid state. (p. 51)

fusión El proceso mediante el cual una sustancia cambia de su estado sólido a su estado líquido.

melting point
The temperature at which a substance changes from its solid state to its liquid state through melting. (p. 51)

punto de fusión La temperatura a la cual una sustancia cambia de su estado sólido a su estado líquido mediante fusión.

mixture
A combination of two or more substances that do not combine chemically but remain the same individual substances. Mixtures can be separated by physical means. (p. 23)

mezcla Una combinación de dos o más sustancias que no se combinan químicamente sino que permanecen como sustancias individuales. Las mezclas se pueden separar por medios físicos.

molecule
A group of atoms that are held together by covalent bonds so that they move as a single unit. (p. 18)

molécula Un grupo de átomos que están unidos mediante enlaces covalentes de tal manera que se mueven como una sola unidad.

P, Q

particle
A very small piece of matter, such as an atom, molecule, or ion.

partícula Una cantidad muy pequeña de materia, como un átomo, una molécula o un ión.

physical change
A change in a substance that does not change the substance into a different one. (p. 44)

cambio físico Un cambio en una sustancia que no transforma la sustancia a otra sustancia.

physical property
A characteristic of a substance that can be observed without changing the identity of the substance. (p. 41)

propiedad física Una característica de una sustancia que se puede observar sin cambiar la identidad de la sustancia.

potential energy
Stored energy; the energy an object has due to its position, molecular arrangement, or chemical composition. (p. 75)

energía potencial Energía almacenada; o la energía que tiene un objeto debido a su posición, arreglo molecular o composición química.

R

radiation
Energy that travels across distances in the form of electromagnetic waves. (p. 119)

radiación Energía que viaja a través de la distancia en forma de ondas electromagnéticas.

S

solar cell
A type of technology in which light-sensitive materials convert sunlight into electrical energy. (p. 88)

celda solar Un tipo de tecnología en el cual materiales sensibles a la luz convierten luz solar a energía eléctrica.

solid
Matter that has a definite shape and a definite volume. The molecules in a solid are in fixed positions and are close together. (p. 28)

sólido La materia que tiene una forma definida y un volumen definido. Las moléculas en un sólido están en posiciones fijas y cercanas unas a otras.

specific heat
The amount of energy required to raise the temperature of one gram of a substance by one degree Celsius. (p. 113)

calor específico La cantidad de energía que se necesita para aumentar la temperatura de un gramo de una sustancia un grado centígrado.

states of matter
The different forms in which matter can exist. Three familiar states are solid, liquid, and gas. (p. 27)

estados de la materia Las diferentes formas en las cuales puede existir la materia. Los tres estados conocidos son sólido, líquido y gas.

sublimation
The process by which a substance changes directly from its solid state to its gas state without becoming a liquid first. (p. 53)

sublimación El proceso mediante el cual una sustancia cambia directamente de su estado sólido a su estado gaseoso sin convertirse primero en líquido.

substance
Matter of a particular type. Elements, compounds, and mixtures are all substances.

sustancia La materia de cierto tipo. Los elementos, los compuestos y las mezclas son sustancias.

system
A group of objects or phenomena that interact. A system can be as simple as a rope, a pulley, and a mass. It also can be as complex as the interaction of energy and matter in the four spheres of the Earth system.

sistema Un grupo de objetos o fenómenos que interactúan. Un sistema puede ser algo tan sencillo como una cuerda, una polea y una masa. También puede ser algo tan complejo como la interacción de la energía y la materia en las cuatro esferas del sistema de la Tierra.

T, U

technology
The use of scientific knowledge to solve problems or engineer new products, tools, or processes.

tecnología El uso de conocimientos científicos para resolver problemas o para diseñar nuevos productos, herramientas o procesos.

temperature
A measure of the average amount of kinetic energy of the particles in an object. (p. 105)

temperatura Una medida de la cantidad promedio de energía cinética de las partículas en un objeto.

theory
In science, a set of widely accepted explanations of observations and phenomena. A theory is a well-tested explanation that is consistent with all available evidence.

teoría En las ciencias, un conjunto de explicaciones de observaciones y fenómenos que es ampliamente aceptado. Una teoría es una explicación bien probada que es consecuente con la evidencia disponible.

thermal energy
The energy an object has due to the motion of its particles; the total amount of kinetic energy of particles in an object. (p. 111)

energía térmica La energía que tiene un objeto debido al movimiento de sus partículas; la cantidad total de energía cinética de las partículas en un objeto.

thermometer
A device for measuring temperature. (p. 107)

termómetro Un aparato para medir la temperatura.

R56

variable
Any factor that can change in a controlled experiment, observation, or model. (p. R30)

variable Cualquier factor que puede cambiar en un experimento controlado, en una observación o en un modelo.

volume
An amount of three-dimensional space, often used to describe the space that an object takes up. (p. 11)

volumen Una cantidad de espacio tridimensional; a menudo se usa este término para describir el espacio que ocupa un objeto.

W, X, Y, Z

weight
The force of gravity on an object. (p. 11)

peso La fuerza de la gravedad sobre un objeto.

R57

Index

Page numbers for definitions are printed in **boldface** type.
Page numbers for illustrations, maps, and charts are printed in *italics.*

A, B

C, D

E, F

G, H

INDEX

I, J

K, L

M

INDEX

N, O

INDEX

P, Q

R, S

T, U, V

W, X, Y, Z

Acknowledgments

Photography

Cover © Scott T. Smith/Corbis; **i** © Scott T. Smith/Corbis; **iii** *left (top to bottom)* Photograph of James Trefil by Evan Cantwell; Photograph of Rita Ann Calvo by Joseph Calvo; Photograph of Linda Carnine by Amilcar Cifuentes; Photograph of Sam Miller by Samuel Miller; *right (top to bottom)* Photograph of Kenneth Cutler by Kenneth A. Cutler; Photograph of Donald Steely by Marni Stamm; Photograph of Vicky Vachon by Redfern Photographics; **vi** © David Leahy/Getty Images; **vii** AP/Wide World Photos; **ix** Photographs by Sharon Hoogstraten; **xiv–xv** © Larry Hamill/age fotostock america, inc.; **xvi–xvii** © Fritz Poelking/age fotostock america, inc.; **xviii–xix** © Galen Rowell/Corbis; **xx–xxi** © Jack Affleck/SuperStock; **xxii** AP/Wide World Photos; **xxiii** © David Parker/IMI/University of Birmingham High, TC Consortium/Photo Researchers; **xxiv** *left* AP/Wide World Photos; *right Washington University Record;* **xxv** *top* © Kim Steele/Getty Images; *bottom* Reprinted with permission from S. Zhou et al., *SCIENCE* 291:1944–47. © 2001 AAAS; **xxvi–xxvii** © Mike Fiala/Getty Images; **xxvii** *left* © Derek Trask/Corbis; *right* AP/Wide World Photos; **xxxii** © The Chedd-Angier Production Company; **2–3, 3** Courtesy of NASA/JPL/Caltech; **4** *top* © Babakin Space Center, The Planetary Society; *bottom* © The Chedd-Angier Production Company; **6–7** © Steve Allen/Brand X Pictures; **7, 9** Photographs by Sharon Hoogstraten; **10** *left* © Antonio Mo/Getty Images; *right* © ImageState/Alamy; **11** © Tom Stewart/Corbis; **12, 13** Photographs by Sharon Hoogstraten; **14** *top* © Stewart Cohen/Getty Images; *bottom* Photograph by Sharon Hoogstraten; **14–15, 15** Photographs by Sharon Hoogstraten; **16** © Royalty-Free/Corbis; **17** Photograph by Sharon Hoogstraten; **18** © NatPhotos/Tony Sweet/Digital Vision; **19** © Jake Rajs/Getty Images; **20** Courtesy IBM Archives; **21** Photograph by Sharon Hoogstraten; **22** *left* © James L. Amos/Corbis; *right* © Omni Photo Communications, Inc./Index Stock; **23** © Richard Laird/Getty Images; **24** Photograph by Sharon Hoogstraten; **25** © Royalty-Free/Corbis; **26** © Nik Wheeler/Corbis; **27** Photograph by Sharon Hoogstraten; **30** © Robert F. Sisson/Getty Images; **31** Photograph by Sharon Hoogstraten; **34** *top* Photograph by Sharon Hoogstraten; *bottom left* © James L. Amos/Corbis; *bottom right* © Royalty-Free/Corbis; **36** Photographs by Sharon Hoogstraten; **38–39** © David Leahy/Getty Images; **39, 41** Photographs by Sharon Hoogstraten; **42** *left* Photograph by Sharon Hoogstraten; *right* © Dan Lim/Masterfile; **45** *top left* © Maryellen McGrath/Bruce Coleman Inc.; *top center* © Jean-Bernard Vernier/Corbis Sygma; *top right* © Angelo Cavalli/Getty Images; *bottom* © Garry Black/Masterfile; *inset* Photograph by Sharon Hoogstraten; **46** © Mark C. Burnett/Stock, Boston Inc./PictureQuest; **47** Photograph by Sharon Hoogstraten; **48** © J. Westrich/Masterfile; **49** *left* © Owen Franken/Corbis; *right* © Erich Lessing/Art Resource, New York; **50** © ImageState/Alamy; **51** *left* © Brand X Pictures; *right* © Peter Bowater/Alamy; **52** © Royalty-Free/Corbis; **53** © Winifred Wisniewski/Frank Lane Picture Agency/Corbis; **54** © A. Pasieka/Photo Researchers; **55** © Sean Ellis/Getty Images; **56** *top* © Royalty-Free/Corbis; *bottom* Photograph by Sharon Hoogstraten; **57, 58** Photographs by Sharon Hoogstraten; **59** © Lawrence Livermore National Laboratory/Photo Researchers; **60** *top left* © SPL/Photo Researchers; *top right* © Felix St. Clair Renard/Getty Images; *bottom* © David Young-Wolff/PhotoEdit; **61** Photograph by Sharon Hoogstraten; **62** © Alan Towse/Ecoscene/Corbis; **63** © Robert Essel NYC/Corbis; *inset* © The Cover Story/Corbis; **64** *top left* © Dan Lim/Masterfile; *top right* © Mark C. Burnett/Stock, Boston Inc./PictureQuest; *bottom* © David Young-Wolff/PhotoEdit; **66** © Winifred Wisniewski/Frank Lane Picture Agency/Corbis; **68–69** AP/Wide World Photos; **69, 71** Photographs by Sharon Hoogstraten; **72** © Alan Schein Photography/Corbis; **73** *top* © Patrick Ward/Corbis; *bottom* © NASA/Photo Researchers; **74** AP/Wide World Photos; **75** *top* © George H. H. Huey/Corbis; *bottom* Photograph by Sharon Hoogstraten; **76** *top* © Vladimir Pcholkin/Getty Images; *bottom* © Thomas Beach; **77** © Adam Gault/Digital Vision; **78** © Bill Aron/PhotoEdit; **79** © TempSport/Corbis; **80** © Robert Cameron/Getty Images; **81** *left* © Gunter Marx Photography/Corbis; *right* © Lester Lefkowitz/Corbis; **82** © Left Lane Productions/Corbis; **83** © Dorling Kindersley; **84** *top* © Grant Klotz/Alaska Stock Images/PictureQuest; *bottom* Photograph by Sharon Hoogstraten; **85, 86** Photographs by Sharon Hoogstraten; **87** *top left* © Royalty-Free/Corbis; *top right* Thinkstock, LLC; *bottom* AP/Wide World Photos; **88** © AFP/Corbis; *inset* © John Farmar; Cordaiy Photo Library Ltd./Corbis; **89** *top* © Sally A. Morgan; Ecoscene/Corbis; *bottom* Photograph by Sharon Hoogstraten; **90** © Joe Sohm/Visions of America, LLC/PictureQuest; **91** © Michael S. Lewis/Corbis; **92** *top* © Vladimir Pcholkin/Getty Images; *bottom* © AFP/Corbis; **96** © Don Farrall/Getty Images; **97** *top left* © Sheila Terry/Photo Researchers; *top center, top right* © Dorling Kindersley; *bottom* © SEF/Art Resource, New York; **98** *top left* Mary Evans Picture Library; *top right, bottom* © Dorling Kindersley; **99** © Mark Wiens/Masterfile; **100–101** © Steve Bloom/stevebloom.com; **101, 103** Photographs by Sharon Hoogstraten; **104** © Tracy Frankel/Getty Images; **105** Photographs by Sharon Hoogstraten; **106** © Daryl Benson/Masterfile; *inset* © Spencer Grant/PhotoEdit; **107** Photograph by Sharon Hoogstraten; **108** *top* © Steve Vidler/SuperStock; *bottom* © Chase Jarvis/Getty Images; **109** © FogStock/Alamy; *inset* © Gordon Wiltsie/Getty Images; **110** © David Bishop/Getty Images; **111** Thinkstock, LLC; **112** Photograph by Sharon Hoogstraten; **113** © Richard Bickel/Corbis; **115** *top left* © Jeremy Samuelson/FoodPix; *bottom left* © William Reavell-StockFood Munich/StockFood; *right* © Martin Jacobs/FoodPix; **116** Photograph by Sharon Hoogstraten; **117** © Brand X Pictures/Alamy; **119** © ImageState Royalty Free/Alamy; **120** *top left* E.C. Humphrey; *top right* Creatas®; *bottom* © Uwe Walz Gdt/age fotostock america, inc.; **122** *top* © Nancy Ney/Corbis; *bottom* Photograph by Sharon Hoogstraten; **123** Photograph by Sharon Hoogstraten; **124** *top* Photographs by Sharon Hoogstraten; *bottom* Thinkstock, LLC; **R28** © Photodisc/Getty Images.

Illustrations and Maps

Accurate Art, Inc. **127;** Ampersand Design Group **29, 115;** Stephen Durke **10, 11, 18, 20, 22, 30, 32, 33, 34, 81;** MapQuest.com, Inc. **114;** Dan Stuckenschneider **R11–R19, R22, R32**

Content Standards: 5–8

A. Science as Inquiry

As a result of activities in grades 5–8, all students should develop

Abilities Necessary to do Scientific Inquiry

A.1 Identify questions that can be answered through scientific investigations. Students should develop the ability to refine and refocus broad and ill-defined questions. An important aspect of this ability consists of students' ability to clarify questions and inquiries and direct them toward objects and phenomena that can be described, explained, or predicted by scientific investigations. Students should develop the ability to identify their questions with scientific ideas, concepts, and quantitative relationships that guide investigation.

A.2 Design and conduct a scientific investigation. Students should develop general abilities, such as systematic observation, making accurate measurements, and identifying and controlling variables. They should also develop the ability to clarify their ideas that are influencing and guiding the inquiry, and to understand how those ideas compare with current scientific knowledge. Students can learn to formulate questions, design investigations, execute investigations, interpret data, use evidence to generate explanations, propose alternative explanations, and critique explanations and procedures.

A.3 Use appropriate tools and techniques to gather, analyze, and interpret data. The use of tools and techniques, including mathematics, will be guided by the question asked and the investigations students design. The use of computers for the collection, summary, and display of evidence is part of this standard. Students should be able to access, gather, store, retrieve, and organize data, using hardware and software designed for these purposes.

A.4 Develop descriptions, explanations, predictions, and models using evidence. Students should base their explanation on what they observed, and as they develop cognitive skills, they should be able to differentiate explanation from description—providing causes for effects and establishing relationships based on evidence and logical argument. This standard requires a subject matter knowledge base so the students can effectively conduct investigations, because developing explanations establishes connections between the content of science and the contexts within which students develop new knowledge.

A.5 Think critically and logically to make the relationships between evidence and explanations. Thinking critically about evidence includes deciding what evidence should be used and accounting for anomalous data. Specifically, students should be able to review data from a simple experiment, summarize the data, and form a logical argument about the cause-and-effect relationships in the experiment. Students should begin to state some explanations in terms of the relationship between two or more variables.

A.6 Recognize and analyze alternative explanations and predictions. Students should develop the ability to listen to and respect the explanations proposed by other students. They should remain open to and acknowledge different ideas and explanations, be able to accept the skepticism of others, and consider alternative explanations.

A.7 Communicate scientific procedures and explanations. With practice, students should become competent at communicating experimental methods, following instructions, describing observations, summarizing the results of other groups, and telling other students about investigations and explanations.

A.8 Use mathematics in all aspects of scientific inquiry. Mathematics is essential to asking and answering questions about the natural world. Mathematics can be used to ask questions; to gather, organize, and present data; and to structure convincing explanations.

Understandings about Scientific Inquiry

A.9.a Different kinds of questions suggest different kinds of scientific investigations. Some investigations involve observing and describing objects, organisms, or events; some involve collecting specimens; some involve experiments; some involve seeking more information; some involve discovery of new objects and phenomena; and some involve making models.

A.9.b Current scientific knowledge and understanding guide scientific investigations. Different scientific domains employ different methods, core theories, and standards to advance scientific knowledge and understanding.

A.9.c Mathematics is important in all aspects of scientific inquiry.

A.9.d Technology used to gather data enhances accuracy and allows scientists to analyze and quantify results of investigations.

A.9.e Scientific explanations emphasize evidence, have logically consistent arguments, and use scientific principles, models, and theories. The scientific community accepts and uses such explanations until displaced by better scientific ones. When such displacement occurs, science advances.

A.9.f Science advances through legitimate skepticism. Asking questions and querying other scientists' explanations is part of scientific inquiry. Scientists evaluate the explanations proposed by other scientists by examining evidence, comparing evidence, identifying faulty reasoning, pointing out statements that go beyond the evidence, and suggesting alternative explanations for the same observations.

A.9.g Scientific investigations sometimes result in new ideas and phenomena for study, generate new methods or procedures for an investigation, or develop new technologies to improve the collection of data. All of these results can lead to new investigations.

B. Physical Science

As a result of their activities in grades 5–8, all students should develop an understanding of

Properties and Changes of Properties in Matter

B.1.a A substance has characteristic properties, such as density, a boiling point, and solubility, all of which are independent of the amount of the sample. A mixture of substances often can be separated into the original substances using one or more of the characteristic properties.

B.1.b Substances react chemically in characteristic ways with other substances to form new substances (compounds) with different characteristic properties. In chemical reactions, the total mass is conserved. Substances often are placed in categories or groups if they react in similar ways; metals is an example of such a group.

B.1.c Chemical elements do not break down during normal laboratory reactions involving such treatments as heating, exposure to electric current, or reaction with acids. There are more than 100 known elements that combine in a multitude of ways to produce compounds, which account for the living and nonliving substances that we encounter.

Motions and Forces

B.2.a The motion of an object can be described by its position, direction of motion, and speed. That motion can be measured and represented on a graph.

B.2.b An object that is not being subjected to a force will continue to move at a constant speed and in a straight line.

B.2.c If more than one force acts on an object along a straight line, then the forces will reinforce or cancel one another, depending on their direction and magnitude. Unbalanced forces will cause changes in the speed or direction of an object's motion.

Transfer of Energy

B.3.a Energy is a property of many substances and is associated with heat, light, electricity, mechanical motion, sound, nuclei, and the nature of a chemical. Energy is transferred in many ways.

B.3.b Heat moves in predictable ways, flowing from warmer objects to cooler ones, until both reach the same temperature.

B.3.c Light interacts with matter by transmission (including refraction), absorption, or scattering (including reflection). To see an object, light from that object—emitted by or scattered from it—must enter the eye.

B.3.d Electrical circuits provide a means of transferring electrical energy when heat, light, sound, and chemical changes are produced.

B.3.e In most chemical and nuclear reactions, energy is transferred into or out of a system. Heat, light, mechanical motion, or electricity might all be involved in such transfers.

B.3.f The sun is a major source of energy for changes on the earth's surface. The sun loses energy by emitting light. A tiny fraction of that light reaches the earth, transferring energy from the sun to the earth. The sun's energy arrives as light with a range of wavelengths, consisting of visible light, infrared, and ultraviolet radiation.

C. Life Science

As a result of their activities in grades 5–8, all students should develop understanding of

Structure and Function in Living Systems

C.1.a Living systems at all levels of organization demonstrate the complementary nature of structure and function. Important levels of organization for structure and function include cells, organs, tissues, organ systems, whole organisms, and ecosystems.

C.1.b All organisms are composed of cells—the fundamental unit of life. Most organisms are single cells; other organisms, including humans, are multicellular.

C.1.c Cells carry on the many functions needed to sustain life. They grow and divide, thereby producing more cells. This requires that they take in nutrients, which they use to provide energy for the work that cells do and to make the materials that a cell or an organism needs.

C.1.d Specialized cells perform specialized functions in multicellular organisms. Groups of specialized cells cooperate to form a tissue, such as a muscle. Different tissues are in turn grouped together to form larger functional units, called organs. Each type of cell, tissue, and organ has a distinct structure and set of functions that serve the organism as a whole.

C.1.e The human organism has systems for digestion, respiration, reproduction, circulation, excretion, movement, control, and coordination, and for protection from disease. These systems interact with one another.

C.1.f Disease is a breakdown in structures or functions of an organism. Some diseases are the result of intrinsic failures of the system. Others are the result of damage by infection by other organisms.

Reproduction and Heredity

C.2.a Reproduction is a characteristic of all living systems; because no individual organism lives forever, reproduction is essential to the continuation of every species. Some organisms reproduce asexually. Other organisms reproduce sexually.

C.2.b In many species, including humans, females produce eggs and males produce sperm. Plants also reproduce sexually—the egg and sperm are produced in the flowers of flowering plants. An egg and sperm unite to begin development of a new individual. That new individual receives genetic information from its mother (via the egg) and its father (via the sperm). Sexually produced offspring never are identical to either of their parents.

C.2.c Every organism requires a set of instructions for specifying its traits. Heredity is the passage of these instructions from one generation to another.

C.2.d Hereditary information is contained in genes, located in the chromosomes of each cell. Each gene carries a single unit of information. An inherited trait of an individual can be determined by one or by many genes, and a single gene can influence more than one trait. A human cell contains many thousands of different genes.

C.2.e The characteristics of an organism can be described in terms of a combination of traits. Some traits are inherited and others result from interactions with the environment.

Regulation and Behavior

C.3.a All organisms must be able to obtain and use resources, grow, reproduce, and maintain stable internal conditions while living in a constantly changing external environment.

C.3.b Regulation of an organism's internal environment involves sensing the internal environment and changing physiological activities to keep conditions within the range required to survive.

C.3.c Behavior is one kind of response an organism can make to an internal or environmental stimulus. A behavioral response requires coordination and communication at many levels, including cells, organ systems, and whole organisms. Behavioral response is a set of actions determined in part by heredity and in part from experience.

C.3.d An organism's behavior evolves through adaptation to its environment. How a species moves, obtains food, reproduces, and responds to danger are based in the species' evolutionary history.

Populations and Ecosystems

C.4.a A population consists of all individuals of a species that occur together at a given place and time. All populations living together and the physical factors with which they interact compose an ecosystem.

C.4.b Populations of organisms can be categorized by the function they serve in an ecosystem. Plants and some microorganisms are producers—they make their own food. All animals, including humans, are consumers, which obtain food by eating other organisms. Decomposers, primarily bacteria and fungi, are consumers that use waste materials and dead organisms for food. Food webs identify the relationships among producers, consumers, and decomposers in an ecosystem.

C.4.c For ecosystems, the major source of energy is sunlight. Energy entering ecosystems as sunlight is transferred by producers into chemical energy through photosynthesis. That energy then passes from organism to organism in food webs.

C.4.d The number of organisms an ecosystem can support depends on the resources available and abiotic factors, such as quantity of light and water, range of temperatures, and soil composition. Given adequate biotic and abiotic resources and no disease or predators, populations (including humans) increase at rapid rates. Lack of resources and other factors, such as predation and climate, limit the growth of populations in specific niches in the ecosystem.

Diversity and Adaptations of Organisms

C.5.a Millions of species of animals, plants, and microorganisms are alive today. Although different species might look dissimilar, the unity among organisms becomes apparent from an analysis of internal structures, the similarity of their chemical processes, and the evidence of common ancestry.

C.5.b Biological evolution accounts for the diversity of species developed through gradual processes over many generations. Species acquire many of their unique characteristics through biological adaptation, which involves the selection of naturally occurring variations in populations. Biological adaptations include changes in structures, behaviors, or physiology that enhance survival and reproductive success in a particular environment.

C.5.c Extinction of a species occurs when the environment changes and the adaptive characteristics of a species are insufficient to allow its survival. Fossils indicate that many organisms that lived long ago are extinct. Extinction of species is common; most of the species that have lived on the earth no longer exist.

D. Earth and Space Science

As a result of their activities in grades 5–8, all students should develop an understanding of

Structure of the Earth System

D.1.a The solid earth is layered with a lithosphere; hot, convecting mantle; and dense, metallic core.

D.1.b Lithospheric plates on the scales of continents and oceans constantly move at rates of centimeters per year in response to movements in the mantle. Major geological events, such as earthquakes, volcanic eruptions, and mountain building, result from these plate motions.

D.1.c Land forms are the result of a combination of constructive and destructive forces. Constructive forces include crustal deformation, volcanic eruption, and deposition of sediment, while destructive forces include weathering and erosion.

D.1.d Some changes in the solid earth can be described as the "rock cycle." Old rocks at the earth's surface weather, forming sediments that are buried, then compacted, heated, and often recrystallized into new rock. Eventually, those new rocks may be brought to the surface by the forces that drive plate motions, and the rock cycle continues.

D.1.e Soil consists of weathered rocks and decomposed organic material from dead plants, animals, and bacteria. Soils are often found in layers, with each having a different chemical composition and texture.

D.1.f Water, which covers the majority of the earth's surface, circulates through the crust, oceans, and atmosphere in what is known as the "water cycle." Water evaporates from the earth's surface, rises and cools as it moves to higher elevations, condenses as rain or snow, and falls to the surface where it collects in lakes, oceans, soil, and in rocks underground.

D.1.g Water is a solvent. As it passes through the water cycle it dissolves minerals and gases and carries them to the oceans.

D.1.h The atmosphere is a mixture of nitrogen, oxygen, and trace gases that include water vapor. The atmosphere has different properties at different elevations.

D.1.i Clouds, formed by the condensation of water vapor, affect weather and climate.

D.1.j Global patterns of atmospheric movement influence local weather. Oceans have a major effect on climate, because water in the oceans holds a large amount of heat.

D.1.k Living organisms have played many roles in the earth system, including affecting the composition of the atmosphere, producing some types of rocks, and contributing to the weathering of rocks.

Earth's History

D.2.a The earth processes we see today, including erosion, movement of lithospheric plates, and changes in atmospheric composition, are similar to those that occurred in the past. Earth history is also influenced by occasional catastrophes, such as the impact of an asteroid or comet.

D.2.b Fossils provide important evidence of how life and environmental conditions have changed.

Earth in the Solar System

D.3.a The earth is the third planet from the sun in a system that includes the moon, the sun, eight other planets and their moons, and smaller objects, such as asteroids and comets. The sun, an average star, is the central and largest body in the solar system.

D.3.b Most objects in the solar system are in regular and predictable motion. Those motions explain such phenomena as the day, the year, phases of the moon, and eclipses.

D.3.c Gravity is the force that keeps planets in orbit around the sun and governs the rest of the motion in the solar system. Gravity alone holds us to the earth's surface and explains the phenomena of the tides.

D.3.d The sun is the major source of energy for phenomena on the earth's surface, such as growth of plants, winds, ocean currents, and the water cycle. Seasons result from variations in the amount of the sun's energy hitting the surface, due to the tilt of the earth's rotation on its axis and the length of the day.

E. Science and Technology

As a result of activities in grades 5–8, all students should develop

Abilities of Technological Design

E.1 Identify appropriate problems for technological design. Students should develop their abilities by identifying a specified need, considering its various aspects, and talking to different potential users or beneficiaries. They should appreciate that for some needs, the cultural backgrounds and beliefs of different groups can affect the criteria for a suitable product.

E.2 Design a solution or product. Students should make and compare different proposals in the light of the criteria they have selected. They must consider constraints—such as cost, time, trade-offs, and materials needed—and communicate ideas with drawings and simple models.

E.3 Implement a proposed design. Students should organize materials and other resources, plan their work, make good use of group collaboration where appropriate, choose suitable tools and techniques, and work with appropriate measurement methods to ensure adequate accuracy.

E.4 Evaluate completed technological designs or products. Students should use criteria relevant to the original purpose or need, consider a variety of factors that might affect acceptability and suitability for intended users or beneficiaries, and develop measures of quality with respect to such criteria and factors; they should also suggest improvements and, for their own products, try proposed modifications.

E.5 Communicate the process of technological design. Students should review and describe any completed piece of work and identify the stages of problem identification, solution design, implementation, and evaluation.

Understandings about Science and Technology

E.6.a Scientific inquiry and technological design have similarities and differences. Scientists propose explanations for questions about the natural world, and engineers propose solutions relating to human problems, needs, and aspirations. Technological solutions are temporary; technologies exist within nature and so they cannot contravene physical or biological principles; technological solutions have side effects; and technologies cost, carry risks, and provide benefits.

E.6.b Many different people in different cultures have made and continue to make contributions to science and technology.

E.6.c Science and technology are reciprocal. Science helps drive technology, as it addresses questions that demand more sophisticated instruments and provides principles for better instrumentation and technique. Technology is essential to science, because it provides instruments and techniques that enable observations of objects and phenomena that are otherwise unobservable due to factors such as quantity, distance, location, size, and speed. Technology also provides tools for investigations, inquiry, and analysis.

E.6.d Perfectly designed solutions do not exist. All technological solutions have trade-offs, such as safety, cost, efficiency, and appearance. Engineers often build in back-up systems to provide safety. Risk is part of living in a highly technological world. Reducing risk often results in new technology.

E.6.e Technological designs have constraints. Some constraints are unavoidable, for example, properties of materials, or effects of weather and friction; other constraints limit choices in the design, for example, environmental protection, human safety, and aesthetics.

E.6.f Technological solutions have intended benefits and unintended consequences. Some consequences can be predicted, others cannot.

F. Science in Personal and Social Perspectives

As a result of activities in grades 5–8, all students should develop understanding of

Personal Health

F.1.a Regular exercise is important to the maintenance and improvement of health. The benefits of physical fitness include maintaining healthy weight, having energy and strength for routine activities, good muscle tone, bone strength, strong heart/lung systems, and improved mental health. Personal exercise, especially developing cardiovascular endurance, is the foundation of physical fitness.

F.1.b The potential for accidents and the existence of hazards imposes the need for injury prevention. Safe living involves the development and use of safety precautions and the recognition of risk in personal decisions. Injury prevention has personal and social dimensions.

F.1.c The use of tobacco increases the risk of illness. Students should understand the influence of short-term social and psychological factors that lead to tobacco use, and the possible long-term detrimental effects of smoking and chewing tobacco.

F.1.d Alcohol and other drugs are often abused substances. Such drugs change how the body functions and can lead to addiction.

F.1.e Food provides energy and nutrients for growth and development. Nutrition requirements vary with body weight, age, sex, activity, and body functioning.

F.1.f Sex drive is a natural human function that requires understanding. Sex is also a prominent means of transmitting diseases. The diseases can be prevented through a variety of precautions.

F.1.g Natural environments may contain substances (for example, radon and lead) that are harmful to human beings. Maintaining environmental health involves establishing or monitoring quality standards related to use of soil, water, and air.

Populations, Resources, and Environments

F.2.a When an area becomes overpopulated, the environment will become degraded due to the increased use of resources.

F.2.b Causes of environmental degradation and resource depletion vary from region to region and from country to country.

Natural Hazards

F.3.a Internal and external processes of the earth system cause natural hazards, events that change or destroy human and wildlife habitats, damage property, and harm or kill humans. Natural hazards include earthquakes, landslides, wildfires, volcanic eruptions, floods, storms, and even possible impacts of asteroids.

F.3.b Human activities also can induce hazards through resource acquisition, urban growth, land-use decisions, and waste disposal. Such activities can accelerate many natural changes.

F.3.c Natural hazards can present personal and societal challenges because misidentifying the change or incorrectly estimating the rate and scale of change may result in either too little attention and significant human costs or too much cost for unneeded preventive measures.

Risks and Benefits

F.4.a Risk analysis considers the type of hazard and estimates the number of people that might be exposed and the number likely to suffer consequences. The results are used to determine the options for reducing or eliminating risks.

F.4.b Students should understand the risks associated with natural hazards (fires, floods, tornadoes, hurricanes, earthquakes, and volcanic eruptions), with chemical hazards (pollutants in air, water, soil, and food), with biological hazards (pollen, viruses, bacterial, and parasites), social hazards (occupational safety and transportation), and with personal hazards (smoking, dieting, and drinking).

F.4.c Individuals can use a systematic approach to thinking critically about risks and benefits. Examples include applying probability estimates to risks and comparing them to estimated personal and social benefits.

F.4.d Important personal and social decisions are made based on perceptions of benefits and risks.

Science and Technology in Society

F.5.a Science influences society through its knowledge and world view. Scientific knowledge and the procedures used by scientists influence the way many individuals in society think about themselves, others, and the environment. The effect of science on society is neither entirely beneficial nor entirely detrimental.

F.5.b Societal challenges often inspire questions for scientific research, and social priorities often influence research priorities through the availability of funding for research.

F.5.c Technology influences society through its products and processes. Technology influences the quality of life and the ways people act and interact. Technological changes are often accompanied by social, political, and economic changes that can be beneficial or detrimental to individuals and to society. Social needs, attitudes, and values influence the direction of technological development.

F.5.d Science and technology have advanced through contributions of many different people, in different cultures, at different times in history. Science and technology have contributed enormously to economic growth and productivity among societies and groups within societies.

F.5.e Scientists and engineers work in many different settings, including colleges and universities, businesses and industries, specific research institutes, and government agencies.

F.5.f Scientists and engineers have ethical codes requiring that human subjects involved with research be fully informed about risks and benefits associated with the research before the individuals choose to participate. This ethic extends to potential risks to communities and property. In short, prior knowledge and consent are required for research involving human subjects or potential damage to property.

F.5.g Science cannot answer all questions and technology cannot solve all human problems or meet all human needs. Students should understand the difference between scientific and other questions. They should appreciate what science and technology can reasonably contribute to society and what they cannot do. For example, new technologies often will decrease some risks and increase others.

G. History and Nature of Science

As a result of activities in grades 5–8, all students should develop understanding of

Science as a Human Endeavor

G.1.a Women and men of various social and ethnic backgrounds—and with diverse interests, talents, qualities, and motivations—engage in the activities of science, engineering, and related fields such as the health professions. Some scientists work in teams, and some work alone, but all communicate extensively with others.

G.1.b Science requires different abilities, depending on such factors as the field of study and type of inquiry. Science is very much a human endeavor, and the work of science relies on basic human qualities, such as reasoning, insight, energy, skill, and creativity—as well as on scientific habits of mind, such as intellectual honesty, tolerance of ambiguity, skepticism, and openness to new ideas.

Nature of Science

G.2.a Scientists formulate and test their explanations of nature using observation, experiments, and theoretical and mathematical models. Although all scientific ideas are tentative and subject to change and improvement in principle, for most major ideas in science, there is much experimental and observational confirmation. Those ideas are not likely to change greatly in the future. Scientists do and have changed their ideas about nature when they encounter new experimental evidence that does not match their existing explanations.

G.2.b In areas where active research is being pursued and in which there is not a great deal of experimental or observational evidence and understanding, it is normal for scientists to differ with one another about the interpretation of the evidence or theory being considered. Different scientists might publish conflicting experimental results or might draw different conclusions from the same data. Ideally, scientists acknowledge such conflict and work towards finding evidence that will resolve their disagreement.

G.2.c It is part of scientific inquiry to evaluate the results of scientific investigations, experiments, observations, theoretical models, and the explanations proposed by other scientists. Evaluation includes reviewing the experimental procedures, examining the evidence, identifying faulty reasoning, pointing out statements that go beyond the evidence, and suggesting alternative explanations for the same observations. Although scientists may disagree about explanations of phenomena, about interpretations of data, or about the value of rival theories, they do agree that questioning, response to criticism, and open communication are integral to the process of science. As scientific knowledge evolves, major disagreements are eventually resolved through such interactions between scientists.

History of Science

G.3.a Many individuals have contributed to the traditions of science. Studying some of these individuals provides further understanding of scientific inquiry, science as a human endeavor, the nature of science, and the relationships between science and society.

G.3.b In historical perspective, science has been practiced by different individuals in different cultures. In looking at the history of many peoples, one finds that scientists and engineers of high achievement are considered to be among the most valued contributors to their culture.

G.3.c Tracing the history of science can show how difficult it was for scientific innovators to break through the accepted ideas of their time to reach the conclusions that we currently take for granted.

Project 2061 Benchmarks for Science Literacy

1. The Nature of Science

By the end of the 8th grade, students should know that

1.A The Scientific World View

1.A.1 When similar investigations give different results, the scientific challenge is to judge whether the differences are trivial or significant, and it often takes further studies to decide. Even with similar results, scientists may wait until an investigation has been repeated many times before accepting the results as correct.

1.A.2 Scientific knowledge is subject to modification as new information challenges prevailing theories and as a new theory leads to looking at old observations in a new way.

1.A.3 Some scientific knowledge is very old and yet is still applicable today.

1.A.4 Some matters cannot be examined usefully in a scientific way. Among them are matters that by their nature cannot be tested objectively and those that are essentially matters of morality. Science can sometimes be used to inform ethical decisions by identifying the likely consequences of particular actions but cannot be used to establish that some action is either moral or immoral.

1.B Scientific Inquiry

1.B.1 Scientists differ greatly in what phenomena they study and how they go about their work. Although there is no fixed set of steps that all scientists follow, scientific investigations usually involve the collection of relevant evidence, the use of logical reasoning, and the application of imagination in devising hypotheses and explanations to make sense of the collected evidence.

1.B.2 If more than one variable changes at the same time in an experiment, the outcome of the experiment may not be clearly attributable to any one of the variables. It may not always be possible to prevent outside variables from influencing the outcome of an investigation (or even to identify all of the variables), but collaboration among investigators can often lead to research designs that are able to deal with such situations.

1.B.3 What people expect to observe often affects what they actually do observe. Strong beliefs about what should happen in particular circumstances can prevent them from detecting other results. Scientists know about this danger to objectivity and take steps to try and avoid it when designing investigations and examining data. One safeguard is to have different investigators conduct independent studies of the same questions.

1.C The Scientific Enterprise

1.C.1 Important contributions to the advancement of science, mathematics, and technology have been made by different kinds of people, in different cultures, at different times.

1.C.2 Until recently, women and racial minorities, because of restrictions on their education and employment opportunities, were essentially left out of much of the formal work of the science establishment; the remarkable few who overcame those obstacles were even then likely to have their work disregarded by the science establishment.

1.C.3 No matter who does science and mathematics or invents things, or when or where they do it, the knowledge and technology that result can eventually become available to everyone in the world.

1.C.4 Scientists are employed by colleges and universities, business and industry, hospitals, and many government agencies. Their places of work include offices, classrooms, laboratories, farms, factories, and natural field settings ranging from space to the ocean floor.

1.C.5 In research involving human subjects, the ethics of science require that potential subjects be fully informed about the risks and benefits associated with the research and of their right to refuse to participate. Science ethics also demand that scientists must not knowingly subject coworkers, students, the neighborhood, or the community to health or property risks without their prior knowledge and consent. Because animals cannot make informed choices, special care must be taken in using them in scientific research.

1.C.6 Computers have become invaluable in science because they speed up and extend people's ability to collect, store, compile, and analyze data, prepare research reports, and share data and ideas with investigators all over the world.

1.C.7 Accurate record-keeping, openness, and replication are essential for maintaining an investigator's credibility with other scientists and society.

3. The Nature of Technology

By the end of the 8th grade, students should know that

3.A Technology and Science

3.A.1 In earlier times, the accumulated information and techniques of each generation of workers were taught on the job directly to the next generation of workers. Today, the knowledge base for technology can be found as well in libraries of print and electronic resources and is often taught in the classroom.

3.A.2 Technology is essential to science for such purposes as access to outer space and other remote locations, sample collection and treatment, measurement, data collection and storage, computation, and communication of information.

3.A.3 Engineers, architects, and others who engage in design and technology use scientific knowledge to solve practical problems. But they usually have to take human values and limitations into account as well.

3.B Design and Systems

3.B.1 Design usually requires taking constraints into account. Some constraints, such as gravity or the properties of the materials to be used, are unavoidable. Other constraints, including economic, political, social, ethical, and aesthetic ones, limit choices.

3.B.2 All technologies have effects other than those intended by the design, some of which may have been predictable and some not. In either case, these side effects may turn out to be unacceptable to some of the population and therefore lead to conflict between groups.

3.B.3 Almost all control systems have inputs, outputs, and feedback. The essence of control is comparing information about what is happening to what people want to happen and then making appropriate adjustments. This procedure requires sensing information, processing it, and making changes. In almost all modern machines, microprocessors serve as centers of performance control.

3.B.4 Systems fail because they have faulty or poorly matched parts, are used in ways that exceed what was intended by the design, or were poorly designed to begin with. The most common ways to prevent failure are pretesting parts and procedures, overdesign, and redundancy.

3.C Issues in Technology

3.C.1 The human ability to shape the future comes from a capacity for generating knowledge and developing new technologies—and for communicating ideas to others.

3.C.2 Technology cannot always provide successful solutions for problems or fulfill every human need.

3.C.3 Throughout history, people have carried out impressive technological feats, some of which would be hard to duplicate today even with modern tools. The purposes served by these achievements have sometimes been practical, sometimes ceremonial.

3.C.4 Technology has strongly influenced the course of history and continues to do so. It is largely responsible for the great revolutions in agriculture, manufacturing, sanitation and medicine, warfare, transportation, information processing, and communications that have radically changed how people live.

3.C.5 New technologies increase some risks and decrease others. Some of the same technologies that have improved the length and quality of life for many people have also brought new risks.

3.C.6 Rarely are technology issues simple and one-sided. Relevant facts alone, even when known and available, usually do not settle matters entirely in favor of one side or another. That is because the contending groups may have different values and priorities. They may stand to gain or lose in different degrees, or may make very different predictions about what the future consequences of the proposed action will be.

3.C.7 Societies influence what aspects of technology are developed and how these are used. People control technology (as well as science) and are responsible for its effects.

4. The Physical Setting

By the end of the 8th grade, students should know that

4.A The Universe

4.A.1 The sun is a medium-sized star located near the edge of a disk-shaped galaxy of stars, part of which can be seen as a glowing band of light that spans the sky on a very clear night. The universe contains many billions of galaxies, and each galaxy contains many billions of stars. To the naked eye, even the closest of these galaxies is no more than a dim, fuzzy spot.

4.A.2 The sun is many thousands of times closer to the earth than any other star. Light from the sun takes a few minutes to reach the earth, but light from the next nearest star takes a few years to arrive. The trip to that star would take the fastest rocket thousands of years. Some distant galaxies are so far away that their light takes several billion years to reach the earth. People on earth, therefore, see them as they were that long ago in the past.

4.A.3 Nine planets of very different size, composition, and surface features move around the sun in nearly circular orbits. Some planets have a great variety of moons and even flat rings of rock and ice particles orbiting around them. Some of these planets and moons show evidence of geologic activity. The earth is orbited by one moon, many artificial satellites, and debris.

4.A.4 Large numbers of chunks of rock orbit the sun. Some of those that the earth meets in its yearly orbit around the sun glow and disintegrate from friction as they plunge through the atmosphere—and sometimes impact the ground. Other chunks of rocks mixed with ice have long, off-center orbits that carry them close to the sun, where the sun's radiation (of light and particles) boils off frozen material from their surfaces and pushes it into a long, illuminated tail.

4.B The Earth

4.B.1 We live on a relatively small planet, the third from the sun in the only system of planets definitely known to exist (although other, similar systems may be discovered in the universe).

4.B.2 The earth is mostly rock. Three-fourths of its surface is covered by a relatively thin layer of water (some of it frozen), and the entire planet is surrounded by a relatively thin blanket of air. It is the only body in the solar system that appears able to support life. The other planets have compositions and conditions very different from the earth's.

4.B.3 Everything on or anywhere near the earth is pulled toward the earth's center by gravitational force.

4.B.4 Because the earth turns daily on an axis that is tilted relative to the plane of the earth's yearly orbit around the sun, sunlight falls more intensely on different parts of the earth during the year. The difference in heating of the earth's surface produces the planet's seasons and weather patterns.

4.B.5 The moon's orbit around the earth once in about 28 days changes what part of the moon is lighted by the sun and how much of that part can be seen from the earth—the phases of the moon.

4.B.6 Climates have sometimes changed abruptly in the past as a result of changes in the earth's crust, such as volcanic eruptions or impacts of huge rocks from space. Even relatively small changes in atmospheric or ocean content can have widespread effects on climate if the change lasts long enough.

4.B.7 The cycling of water in and out of the atmosphere plays an important role in determining climatic patterns. Water evaporates from the surface of the earth, rises and cools, condenses into rain or snow, and falls again to the surface. The water falling on land collects in rivers and lakes, soil, and porous layers of rock, and much of it flows back into the ocean.

4.B.8 Fresh water, limited in supply, is essential for life and also for most industrial processes. Rivers, lakes, and groundwater can be depleted or polluted, becoming unavailable or unsuitable for life.

4.B.9 Heat energy carried by ocean currents has a strong influence on climate around the world.

4.B.10 Some minerals are very rare and some exist in great quantities, but—for practical purposes—the ability to recover them is just as important as their abundance. As minerals are depleted, obtaining them becomes more difficult. Recycling and the development of substitutes can reduce the rate of depletion but may also be costly.

4.B.11 The benefits of the earth's resources—such as fresh water, air, soil, and trees—can be reduced by using them wastefully or by deliberately or inadvertently destroying them. The atmosphere and the oceans have a limited capacity to absorb wastes and recycle materials naturally. Cleaning up polluted air, water, or soil or restoring depleted soil, forests, or fishing grounds can be very difficult and costly.

4.C Processes that Shape the Earth

4.C.1 The interior of the earth is hot. Heat flow and movement of material within the earth cause earthquakes and volcanic eruptions and create mountains and ocean basins. Gas and dust from large volcanoes can change the atmosphere.

4.C.2 Some changes in the earth's surface are abrupt (such as earthquakes and volcanic eruptions) while other changes happen very slowly (such as uplift and wearing down of mountains). The earth's surface is shaped in part by the motion of water and wind over very long times, which act to level mountain ranges.

4.C.3 Sediments of sand and smaller particles (sometimes containing the remains of organisms) are gradually buried and are cemented together by dissolved minerals to form solid rock again.

4.C.4 Sedimentary rock buried deep enough may be reformed by pressure and heat, perhaps melting and recrystallizing into different kinds of rock. These re-formed rock layers may be forced up again to become land surface and even mountains. Subsequently, this new rock too will erode. Rock bears evidence of the minerals, temperatures, and forces that created it.

4.C.5 Thousands of layers of sedimentary rock confirm the long history of the changing surface of the earth and the changing life forms whose remains are found in successive layers. The youngest layers are not always found on top, because of folding, breaking, and uplift of layers.

4.C.6 Although weathered rock is the basic component of soil, the composition and texture of soil and its fertility and resistance to erosion are greatly influenced by plant roots and debris, bacteria, fungi, worms, insects, rodents, and other organisms.

4.C.7 Human activities, such as reducing the amount of forest cover, increasing the amount and variety of chemicals released into the atmosphere, and intensive farming, have changed the earth's land, oceans, and atmosphere. Some of these changes have decreased the capacity of the environment to support some life forms.

4.D Structure of Matter

4.D.1 All matter is made up of atoms, which are far too small to see directly through a microscope. The atoms of any element are alike but are different from atoms of other elements. Atoms may stick together in well-defined molecules or may be packed together in large arrays. Different arrangements of atoms into groups compose all substances.

4.D.2 Equal volumes of different substances usually have different weights.

4.D.3 Atoms and molecules are perpetually in motion. Increased temperature means greater average energy, so most substances expand when heated. In solids, the atoms are closely locked in position and can only vibrate. In liquids, the atoms or molecules have higher energy, are more loosely connected, and can slide past one another; some molecules may get enough energy to escape into a gas. In gases, the atoms or molecules have still more energy and are free of one another except during occasional collisions.

4.D.4 The temperature and acidity of a solution influence reaction rates. Many substances dissolve in water, which may greatly facilitate reactions between them.

4.D.5 Scientific ideas about elements were borrowed from some Greek philosophers of 2,000 years earlier, who believed that everything was made from four basic substances: air, earth, fire, and water. It was the combinations of these "elements" in different proportions that gave other substances their observable properties. The Greeks were wrong about those four, but now over 100 different elements have been identified, some rare and some plentiful, out of which everything is made. Because most elements tend to combine with others, few elements are found in their pure form.

4.D.6 There are groups of elements that have similar properties, including highly reactive metals, less-reactive metals, highly reactive nonmetals (such as chlorine, fluorine, and oxygen), and some almost completely nonreactive gases (such as helium and neon). An especially important kind of reaction between substances involves combination of oxygen with something else—as in burning or rusting. Some elements don't fit into any of the categories; among them are carbon and hydrogen, essential elements of living matter.

4.D.7 No matter how substances within a closed system interact with one another, or how they combine or break apart, the total weight of the system remains the same. The idea of atoms explains the conservation of matter: If the number of atoms stays the same no matter how they are rearranged, then their total mass stays the same.

4.E Energy Transformations

4.E.1 Energy cannot be created or destroyed, but only changed from one form into another.

4.E.2 Most of what goes on in the universe—from exploding stars and biological growth to the operation of machines and the motion of people—involves some form of energy being transformed into another. Energy in the form of heat is almost always one of the products of an energy transformation.

4.E.3 Heat can be transferred through materials by the collisions of atoms or across space by radiation. If the material is fluid, currents will be set up in it that aid the transfer of heat.

4.E.4 Energy appears in different forms. Heat energy is in the disorderly motion of molecules; chemical energy is in the arrangement of atoms; mechanical energy is in moving bodies or in elastically distorted shapes; gravitational energy is in the separation of mutually attracting masses.

4.F Motion

4.F.1 Light from the sun is made up of a mixture of many different colors of light, even though to the eye the light looks almost white. Other things that give off or reflect light have a different mix of colors.

4.F.2 Something can be "seen" when light waves emitted or reflected by it enter the eye—just as something can be "heard" when sound waves from it enter the ear.

4.F.3 An unbalanced force acting on an object changes its speed or direction of motion, or both. If the force acts toward a single center, the object's path may curve into an orbit around the center.

4.F.4 Vibrations in materials set up wavelike disturbances that spread away from the source. Sound and earthquake waves are examples. These and other waves move at different speeds in different materials.

4.F.5 Human eyes respond to only a narrow range of wavelengths of electromagnetic radiation—visible light. Differences of wavelength within that range are perceived as differences in color.

4.G Forces of Nature

4.G.1 Every object exerts gravitational force on every other object. The force depends on how much mass the objects have and on how far apart they are. The force is hard to detect unless at least one of the objects has a lot of mass.

4.G.2 The sun's gravitational pull holds the earth and other planets in their orbits, just as the planets' gravitational pull keeps their moons in orbit around them.

4.G.3 Electric currents and magnets can exert a force on each other.

5. The Living Environment

By the end of the 8th grade, students should know that

5.A Diversity of Life

5.A.1 One of the most general distinctions among organisms is between plants, which use sunlight to make their own food, and animals, which consume energy-rich foods. Some kinds of organisms, many of them microscopic, cannot be neatly classified as either plants or animals.

5.A.2 Animals and plants have a great variety of body plans and internal structures that contribute to their being able to make or find food and reproduce.

5.A.3 Similarities among organisms are found in internal anatomical features, which can be used to infer the degree of relatedness among organisms. In classifying organisms, biologists consider details of internal and external structures to be more important than behavior or general appearance.

5.A.4 For sexually reproducing organisms, a species comprises all organisms that can mate with one another to produce fertile offspring.

5.A.5 All organisms, including the human species, are part of and depend on two main interconnected global food webs. One includes microscopic ocean plants, the animals that feed on them, and finally the animals that feed on those animals. The other web includes land plants, the animals that feed on them, and so forth. The cycles continue indefinitely because organisms decompose after death to return food material to the environment.

5.B Heredity

5.B.1 In some kinds of organisms, all the genes come from a single parent, whereas in organisms that have sexes, typically half of the genes come from each parent.

5.B.2 In sexual reproduction, a single specialized cell from a female merges with a specialized cell from a male. As the fertilized egg, carrying genetic information from each parent, multiplies to form the complete organism with about a trillion cells, the same genetic information is copied in each cell.

5.B.3 New varieties of cultivated plants and domestic animals have resulted from selective breeding for particular traits.

5.C Cells

5.C.1 All living things are composed of cells, from just one to many millions, whose details usually are visible only through a microscope. Different body tissues and organs are made up of different kinds of cells. The cells in similar tissues and organs in other animals are similar to those in human beings but differ somewhat from cells found in plants.

5.C.2 Cells repeatedly divide to make more cells for growth and repair. Various organs and tissues function to serve the needs of cells for food, air, and waste removal.

5.C.3 Within cells, many of the basic functions of organisms—such as extracting energy from food and getting rid of waste—are carried out. The way in which cells function is similar in all living organisms.

5.C.4 About two-thirds of the weight of cells is accounted for by water, which gives cells many of their properties.

5.D Interdependence of Life

5.D.1 In all environments—freshwater, marine, forest, desert, grassland, mountain, and others—organisms with similar needs may compete with one another for resources, including food, space, water, air, and shelter. In any particular environment, the growth and survival of organisms depend on the physical conditions.

5.D.2 Two types of organisms may interact with one another in several ways: They may be in a producer/consumer, predator/prey, or parasite/host relationship. Or one organism may scavenge or decompose another. Relationships may be competitive or mutually beneficial. Some species have become so adapted to each other that neither could survive without the other.

5.E Flow of Matter and Energy

5.E.1 Food provides molecules that serve as fuel and building material for all organisms. Plants use the energy in light to make sugars out of carbon dioxide and water. This food can be used immediately for fuel or materials or it may be stored for later use. Organisms that eat plants break down the plant structures to produce the materials and energy they need to survive. Then they are consumed by other organisms.

5.E.2 Over a long time, matter is transferred from one organism to another repeatedly and between organisms and their physical environment. As in all material systems, the total amount of matter remains constant, even though its form and location change.

5.E.3 Energy can change from one form to another in living things. Animals get energy from oxidizing their food, releasing some of its energy as heat. Almost all food energy comes originally from sunlight.

5.F Evolution of Life

5.F.1 Small differences between parents and offspring can accumulate (through selective breeding) in successive generations so that descendants are very different from their ancestors.

5.F.2 Individual organisms with certain traits are more likely than others to survive and have offspring. Changes in environmental conditions can affect the survival of individual organisms and entire species.

5.F.3 Many thousands of layers of sedimentary rock provide evidence for the long history of the earth and for the long history of changing life forms whose remains are found in the rocks. More recently deposited rock layers are more likely to contain fossils resembling existing species.

6. The Human Organism

By the end of the 8th grade, students should know that

6.A Human Identity

6.A.1 Like other animals, human beings have body systems for obtaining and providing energy, defense, reproduction, and the coordination of body functions.

6.A.2 Human beings have many similarities and differences. The similarities make it possible for human beings to reproduce and to donate blood and organs to one another throughout the world. Their differences enable them to create diverse social and cultural arrangements and to solve problems in a variety of ways.

6.A.3 Fossil evidence is consistent with the idea that human beings evolved from earlier species.

6.A.4 Specialized roles of individuals within other species are genetically programmed, whereas human beings are able to invent and modify a wider range of social behavior.

6.A.5 Human beings use technology to match or excel many of the abilities of other species. Technology has helped people with disabilities survive and live more conventional lives.

6.A.6 Technologies having to do with food production, sanitation, and disease prevention have dramatically changed how people live and work and have resulted in rapid increases in the human population.

6.B Human Development

6.B.1 Fertilization occurs when sperm cells from a male's testes are deposited near an egg cell from the female ovary, and one of the sperm cells enters the egg cell. Most of the time, by chance or design, a sperm never arrives or an egg isn't available.

6.B.2 Contraception measures may incapacitate sperm, block their way to the egg, prevent the release of eggs, or prevent the fertilized egg from implanting successfully.

6.B.3 Following fertilization, cell division produces a small cluster of cells that then differentiate by appearance and function to form the basic tissues of an embryo. During the first three months of pregnancy, organs begin to form. During the second three months, all organs and body features develop. During the last three months, the organs and features mature enough to function well after birth. Patterns of human development are similar to those of other vertebrates.

6.B.4 The developing embryo—and later the newborn infant—encounters many risks from faults in its genes, its mother's inadequate diet, her cigarette smoking or use of alcohol or other drugs, or from infection. Inadequate child care may lead to lower physical and mental ability.

6.B.5 Various body changes occur as adults age. Muscles and joints become less flexible, bones and muscles lose mass, energy levels diminish, and the senses become less acute. Women stop releasing eggs and hence can no longer reproduce. The length and quality of human life are influenced by many factors, including sanitation, diet, medical care, sex, genes, environmental conditions, and personal health behaviors.

6.C Basic Functions

6.C.1 Organs and organ systems are composed of cells and help to provide all cells with basic needs.

6.C.2 For the body to use food for energy and building materials, the food must first be digested into molecules that are absorbed and transported to cells.

6.C.3 To burn food for the release of energy stored in it, oxygen must be supplied to cells, and carbon dioxide removed. Lungs take in oxygen for the combustion of food and they eliminate the carbon dioxide produced. The urinary system disposes of dissolved waste molecules, the intestinal tract removes solid wastes, and the skin and lungs rid the body of heat energy. The circulatory system moves all these substances to or from cells where they are needed or produced, responding to changing demands.

6.C.4 Specialized cells and the molecules they produce identify and destroy microbes that get inside the body.

6.C.5 Hormones are chemicals from glands that affect other body parts. They are involved in helping the body respond to danger and in regulating human growth, development, and reproduction.

6.C.6 Interactions among the senses, nerves, and brain make possible the learning that enables human beings to cope with changes in their environment.

6.D Learning

6.D.1 Some animal species are limited to a repertoire of genetically determined behaviors; others have more complex brains and can learn a wide variety of behaviors. All behavior is affected by both inheritance and experience.

6.D.2 The level of skill a person can reach in any particular activity depends on innate abilities, the amount of practice, and the use of appropriate learning technologies.

6.D.3 Human beings can detect a tremendous range of visual and olfactory stimuli. The strongest stimulus they can tolerate may be more than a trillion times as intense as the weakest they can detect. Still, there are many kinds of signals in the world that people cannot detect directly.

6.D.4 Attending closely to any one input of information usually reduces the ability to attend to others at the same time.

6.D.5 Learning often results from two perceptions or actions occurring at about the same time. The more often the same combination occurs, the stronger the mental connection between them is likely to be. Occasionally a single vivid experience will connect two things permanently in people's minds.

6.D.6 Language and tools enable human beings to learn complicated and varied things from others.

6.E Physical Health

6.E.1 The amount of food energy (calories) a person requires varies with body weight, age, sex, activity level, and natural body efficiency. Regular exercise is important to maintain a healthy heart/lung system, good muscle tone, and bone strength.

6.E.2 Toxic substances, some dietary habits, and personal behavior may be bad for one's health. Some effects show up right away, others may not show up for many years. Avoiding toxic substances, such as tobacco, and changing dietary habits to reduce the intake of such things as animal fat increases the chances of living longer.

6.E.3 Viruses, bacteria, fungi, and parasites may infect the human body and interfere with normal body functions. A person can catch a cold many times because there are many varieties of cold viruses that cause similar symptoms.

6.E.4 White blood cells engulf invaders or produce antibodies that attack them or mark them for killing by other white cells. The antibodies produced will remain and can fight off subsequent invaders of the same kind.

6.E.5 The environment may contain dangerous levels of substances that are harmful to human beings. Therefore, the good health of individuals requires monitoring the soil, air, and water and taking steps to keep them safe.

6.F Mental Health

6.F.1 Individuals differ greatly in their ability to cope with stressful situations. Both external and internal conditions (chemistry, personal history, values) influence how people behave.

6.F.2 Often people react to mental distress by denying that they have any problem. Sometimes they don't know why they feel the way they do, but with help they can sometimes uncover the reasons.

8. The Designed World

By the end of the 8th grade, students should know that

8.A Agriculture

8.A.1 Early in human history, there was an agricultural revolution in which people changed from hunting and gathering to farming. This allowed changes in the division of labor between men and women and between children and adults, and the development of new patterns of government.

8.A.2 People control the characteristics of plants and animals they raise by selective breeding and by preserving varieties of seeds (old and new) to use if growing conditions change.

8.A.3 In agriculture, as in all technologies, there are always trade-offs to be made. Getting food from many different places makes people less dependent on weather in any one place, yet more dependent on transportation and communication among far-flung markets. Specializing in one crop may risk disaster if changes in weather or increases in pest populations wipe out that crop. Also, the soil may be exhausted of some nutrients, which can be replenished by rotating the right crops.

8.A.4 Many people work to bring food, fiber, and fuel to U.S. markets. With improved technology, only a small fraction of workers in the United States actually plant and harvest the products that people use. Most workers are engaged in processing, packaging, transporting, and selling what is produced.

8.B Materials and Manufacturing

8.B.1 The choice of materials for a job depends on their properties and on how they interact with other materials. Similarly, the usefulness of some manufactured parts of an object depends on how well they fit together with the other parts.

8.B.2 Manufacturing usually involves a series of steps, such as designing a product, obtaining and preparing raw materials, processing the materials mechanically or chemically, and assembling, testing, inspecting, and packaging. The sequence of these steps is also often important.

8.B.3 Modern technology reduces manufacturing costs, produces more uniform products, and creates new synthetic materials that can help reduce the depletion of some natural resources.

8.B.4 Automation, including the use of robots, has changed the nature of work in most fields, including manufacturing. As a result, high-skill, high-knowledge jobs in engineering, computer programming, quality control, supervision, and maintenance are replacing many routine, manual-labor jobs. Workers therefore need better learning skills and flexibility to take on new and rapidly changing jobs.

8.C Energy Sources and Use

8.C.1 Energy can change from one form to another, although in the process some energy is always converted to heat. Some systems transform energy with less loss of heat than others.

8.C.2 Different ways of obtaining, transforming, and distributing energy have different environmental consequences.

8.C.3 In many instances, manufacturing and other technological activities are performed at a site close to an energy source. Some forms of energy are transported easily, others are not.

8.C.4 Electrical energy can be produced from a variety of energy sources and can be transformed into almost any other form of energy. Moreover, electricity is used to distribute energy quickly and conveniently to distant locations.

8.C.5 Energy from the sun (and the wind and water energy derived from it) is available indefinitely. Because the flow of energy is weak and variable, very large collection systems are needed. Other sources don't renew or renew only slowly.

8.C.6 Different parts of the world have different amounts and kinds of energy resources to use and use them for different purposes.

8.D Communication

8.D.1 Errors can occur in coding, transmitting, or decoding information, and some means of checking for accuracy is needed. Repeating the message is a frequently used method.

8.D.2 Information can be carried by many media, including sound, light, and objects. In this century, the ability to code information as electric currents in wires, electromagnetic waves in space, and light in glass fibers has made communication millions of times faster than is possible by mail or sound.

8.E Information Processing

8.E.1 Most computers use digital codes containing only two symbols, 0 and 1, to perform all operations. Continuous signals (analog) must be transformed into digital codes before they can be processed by a computer.

8.E.2 What use can be made of a large collection of information depends upon how it is organized. One of the values of computers is that they are able, on command, to reorganize information in a variety of ways, thereby enabling people to make more and better uses of the collection.

8.E.3 Computer control of mechanical systems can be much quicker than human control. In situations where events happen faster than people can react, there is little choice but to rely on computers. Most complex systems still require human oversight, however, to make certain kinds of judgments about the readiness of the parts of the system (including the computers) and the system as a whole to operate properly, to react to unexpected failures, and to evaluate how well the system is serving its intended purposes.

8.E.4 An increasing number of people work at jobs that involve processing or distributing information. Because computers can do these tasks faster and more reliably, they have become standard tools both in the workplace and at home.

8.F Health Technology

8.F.1 Sanitation measures such as the use of sewers, landfills, quarantines, and safe food handling are important in controlling the spread of organisms that cause disease. Improving sanitation to prevent disease has contributed more to saving human life than any advance in medical treatment.

8.F.2 The ability to measure the level of substances in body fluids has made it possible for physicians to make comparisons with normal levels, make very sophisticated diagnoses, and monitor the effects of the treatments they prescribe.

8.F.3 It is becoming increasingly possible to manufacture chemical substances such as insulin and hormones that are normally found in the body. They can be used by individuals whose own bodies cannot produce the amounts required for good health.

9. The Mathematical World

By the end of the 8th grade, students should know that

9.A Numbers

9.A.1 There have been systems for writing numbers other than the Arabic system of place values based on tens. The very old Roman numerals are now used only for dates, clock faces, or ordering chapters in a book. Numbers based on 60 are still used for describing time and angles.

9.A.2 A number line can be extended on the other side of zero to represent negative numbers. Negative numbers allow subtraction of a bigger number from a smaller number to make sense, and are often used when something can be measured on either side of some reference point (time, ground level, temperature, budget).

9.A.3 Numbers can be written in different forms, depending on how they are being used. How fractions or decimals based on measured quantities should be written depends on how precise the measurements are and how precise an answer is needed.

9.A.4 The operations + and – are inverses of each other—one undoes what the other does; likewise x and ÷ .

9.A.5 The expression *a/b* can mean different things: *a* parts of size *1/b* each, *a* divided by *b,* or *a* compared to *b.*

9.A.6 Numbers can be represented by using sequences of only two symbols (such as 1 and 0, on and off); computers work this way.

9.A.7 Computations (as on calculators) can give more digits than make sense or are useful.

9.B Symbolic Relationships

9.B.1 An equation containing a variable may be true for just one value of the variable.

9.B.2 Mathematical statements can be used to describe how one quantity changes when another changes. Rates of change can be computed from differences in magnitudes and vice versa.

9.B.3 Graphs can show a variety of possible relationships between two variables. As one variable increases uniformly, the other may do one of the following: increase or decrease steadily, increase or decrease faster and faster, get closer and closer to some limiting value, reach some intermediate maximum or minimum, alternately increase and decrease indefinitely, increase or decrease in steps, or do something different from any of these.

9.C Shapes

9.C.1 Some shapes have special properties: triangular shapes tend to make structures rigid, and round shapes give the least possible boundary for a given amount of interior area. Shapes can match exactly or have the same shape in different sizes.

9.C.2 Lines can be parallel, perpendicular, or oblique.

9.C.3 Shapes on a sphere like the earth cannot be depicted on a flat surface without some distortion.

9.C.4 The graphic display of numbers may help to show patterns such as trends, varying rates of change, gaps, or clusters. Such patterns sometimes can be used to make predictions about the phenomena being graphed.

9.C.5 It takes two numbers to locate a point on a map or any other flat surface. The numbers may be two perpendicular distances from a point, or an angle and a distance from a point.

9.C.6 The scale chosen for a graph or drawing makes a big difference in how useful it is.

9.D Uncertainty

9.D.1 How probability is estimated depends on what is known about the situation. Estimates can be based on data from similar conditions in the past or on the assumption that all the possibilities are known.

9.D.2 Probabilities are ratios and can be expressed as fractions, percentages, or odds.

9.D.3 The mean, median, and mode tell different things about the middle of a data set.

9.D.4 Comparison of data from two groups should involve comparing both their middles and the spreads around them.

9.D.5 The larger a well-chosen sample is, the more accurately it is likely to represent the whole. But there are many ways of choosing a sample that can make it unrepresentative of the whole.

9.D.6 Events can be described in terms of being more or less likely, impossible, or certain.

9.E Reasoning

9.E.1 Some aspects of reasoning have fairly rigid rules for what makes sense; other aspects don't. If people have rules that always hold, and good information about a particular situation, then logic can help them to figure out what is true about it. This kind of reasoning requires care in the use of key words such as if, and, not, or, all, and some. Reasoning by similarities can suggest ideas but can't prove them one way or the other.

9.E.2 Practical reasoning, such as diagnosing or troubleshooting almost anything, may require many-step, branching logic. Because computers can keep track of complicated logic, as well as a lot of information, they are useful in a lot of problem-solving situations.

9.E.3 Sometimes people invent a general rule to explain how something works by summarizing observations. But people tend to overgeneralize, imagining general rules on the basis of only a few observations.

9.E.4 People are using incorrect logic when they make a statement such as "If *A* is true, then *B* is true; but *A* isn't true, therefore *B* isn't true either."

9.E.5 A single example can never prove that something is always true, but sometimes a single example can prove that something is not always true.

9.E.6 An analogy has some likenesses to but also some differences from the real thing.

10. Historical Perspectives

By the end of the 8th grade, students should know that

10.A Displacing the Earth from the Center of the Universe

10.A.1 The motion of an object is always judged with respect to some other object or point and so the idea of absolute motion or rest is misleading.

10.A.2 Telescopes reveal that there are many more stars in the night sky than are evident to the unaided eye, the surface of the moon has many craters and mountains, the sun has dark spots, and Jupiter and some other planets have their own moons.

10.F Understanding Fire

10.F.1 From the earliest times until now, people have believed that even though millions of different kinds of material seem to exist in the world, most things must be made up of combinations of just a few basic kinds of things. There has not always been agreement, however, on what those basic kinds of things are. One theory long ago was that the basic substances were earth, water, air, and fire. Scientists now know that these are not the basic substances. But the old theory seemed to explain many observations about the world.

10.F.2 Today, scientists are still working out the details of what the basic kinds of matter are and of how they combine, or can be made to combine, to make other substances.

10.F.3 Experimental and theoretical work done by French scientist Antoine Lavoisier in the decade between the American and French revolutions led to the modern science of chemistry.

10.F.4 Lavoisier's work was based on the idea that when materials react with each other many changes can take place but that in every case the total amount of matter afterward is the same as before. He successfully tested the concept of conservation of matter by conducting a series of experiments in which he carefully measured all the substances involved in burning, including the gases used and those given off.

10.F.5 Alchemy was chiefly an effort to change base metals like lead into gold and to produce an elixir that would enable people to live forever. It failed to do that or to create much knowledge of how substances react with each other. The more scientific study of chemistry that began in Lavoisier's time has gone far beyond alchemy in understanding reactions and producing new materials.

10.G Splitting the Atom

10.G.1 The accidental discovery that minerals containing uranium darken photographic film, as light does, led to the idea of radioactivity.

10.G.2 In their laboratory in France, Marie Curie and her husband, Pierre Curie, isolated two new elements that caused most of the radioactivity of the uranium mineral. They named one radium because it gave off powerful, invisible rays, and the other polonium in honor of Madame Curie's country of birth. Marie Curie was the first scientist ever to win the Nobel prize in two different fields—in physics, shared with her husband, and later in chemistry.

10.I Discovering Germs

10.I.1 Throughout history, people have created explanations for disease. Some have held that disease has spiritual causes, but the most persistent biological theory over the centuries was that illness resulted from an imbalance in the body fluids. The introduction of germ theory by Louis Pasteur and others in the 19th century led to the modern belief that many diseases are caused by microorganisms—bacteria, viruses, yeasts, and parasites.

10.I.2 Pasteur wanted to find out what causes milk and wine to spoil. He demonstrated that spoilage and fermentation occur when microorganisms enter from the air, multiply rapidly, and produce waste products. After showing that spoilage could be avoided by keeping germs out or by destroying them with heat, he investigated animal diseases and showed that microorganisms were involved. Other investigators later showed that specific kinds of germs caused specific diseases.

10.I.3 Pasteur found that infection by disease organisms—germs—caused the body to build up an immunity against subsequent infection by the same organisms. He then demonstrated that it was possible to produce vaccines that would induce the body to build immunity to a disease without actually causing the disease itself.

10.I.4 Changes in health practices have resulted from the acceptance of the germ theory of disease. Before germ theory, illness was treated by appeals to supernatural powers or by trying to adjust body fluids through induced vomiting, bleeding, or purging. The modern approach emphasizes sanitation, the safe handling of food and water, the pasteurization of milk, quarantine, and aseptic surgical techniques to keep germs out of the body; vaccinations to strengthen the body's immune system against subsequent infection by the same kind of microorganisms; and antibiotics and other chemicals and processes to destroy microorganisms.

10.I.5 In medicine, as in other fields of science, discoveries are sometimes made unexpectedly, even by accident. But knowledge and creative insight are usually required to recognize the meaning of the unexpected.

10.J Harnessing Power

10.J.1 Until the 1800s, most manufacturing was done in homes, using small, handmade machines that were powered by muscle, wind, or running water. New machinery and steam engines to drive them made it possible to replace craftsmanship with factories, using fuels as a source of energy. In the factory system, workers, materials, and energy could be brought together efficiently.

10.J.2 The invention of the steam engine was at the center of the Industrial Revolution. It converted the chemical energy stored in wood and coal, which were plentiful, into mechanical work. The steam engine was invented to solve the urgent problem of pumping water out of coal mines. As improved by James Watt, it was soon used to move coal, drive manufacturing machinery, and power locomotives, ships, and even the first automobiles.

11. Common Themes

By the end of the 8th grade, students should know that

11.A Systems

11.A.1 A system can include processes as well as things.

11.A.2 Thinking about things as systems means looking for how every part relates to others. The output from one part of a system (which can include material, energy, or information) can become the input to other parts. Such feedback can serve to control what goes on in the system as a whole.

11.A.3 Any system is usually connected to other systems, both internally and externally. Thus a system may be thought of as containing subsystems and as being a subsystem of a larger system.

11.B Models

11.B.1 Models are often used to think about processes that happen too slowly, too quickly, or on too small a scale to observe directly, or that are too vast to be changed deliberately, or that are potentially dangerous.

11.B.2 Mathematical models can be displayed on a computer and then modified to see what happens.

11.B.3 Different models can be used to represent the same thing. What kind of a model to use and how complex it should be depends on its purpose. The usefulness of a model may be limited if it is too simple or if it is needlessly complicated. Choosing a useful model is one of the instances in which intuition and creativity come into play in science, mathematics, and engineering.

11.C Constancy and Change

11.C.1 Physical and biological systems tend to change until they become stable and then remain that way unless their surroundings change.

11.C.2 A system may stay the same because nothing is happening or because things are happening but exactly counterbalance one another.

11.C.3 Many systems contain feedback mechanisms that serve to keep changes within specified limits.

11.C.4 Symbolic equations can be used to summarize how the quantity of something changes over time or in response to other changes.

11.C.5 Symmetry (or the lack of it) may determine properties of many objects, from molecules and crystals to organisms and designed structures.

11.C.6 Cycles, such as the seasons or body temperature, can be described by their cycle length or frequency, what their highest and lowest values are, and when these values occur. Different cycles range from many thousands of years down to less than a billionth of a second.

11.D Scale

11.D.1 Properties of systems that depend on volume, such as capacity and weight, change out of proportion to properties that depend on area, such as strength or surface processes.

11.D.2 As the complexity of any system increases, gaining an understanding of it depends increasingly on summaries, such as averages and ranges, and on descriptions of typical examples of that system.

12. Habits of Mind

By the end of the 8th grade, students should know that

12.A Values and Attitudes

12.A.1 Know why it is important in science to keep honest, clear, and accurate records.

12.A.2 Know that hypotheses are valuable, even if they turn out not to be true, if they lead to fruitful investigations.

12.A.3 Know that often different explanations can be given for the same evidence, and it is not always possible to tell which one is correct.

12.B Computation and Estimation

12.B.1 Find what percentage one number is of another and figure any percentage of any number.

12.B.2 Use, interpret, and compare numbers in several equivalent forms such as integers, fractions, decimals, and percents.

12.B.3 Calculate the circumferences and areas of rectangles, triangles, and circles, and the volumes of rectangular solids.

12.B.4 Find the mean and median of a set of data.

12.B.5 Estimate distances and travel times from maps and the actual size of objects from scale drawings.

12.B.6 Insert instructions into computer spreadsheet cells to program arithmetic calculations.

12.B.7 Determine what unit (such as seconds, square inches, or dollars per tankful) an answer should be expressed in from the units of the inputs to the calculation, and be able to convert compound units (such as yen per dollar into dollar per yen, or miles per hour into feet per second).

12.B.8 Decide what degree of precision is adequate and round off the result of calculator operations to enough significant figures to reasonably reflect those of the inputs.

12.B.9 Express numbers like 100, 1,000, and 1,000,000 as powers of 10.

12.B.10 Estimate probabilities of outcomes in familiar situations, on the basis of history or the number of possible outcomes.

11.D Scale

11.D.1 Properties of systems that depend on volume, such as capacity and weight, change out of proportion to properties that depend on area, such as strength or surface processes.

11.D.2 As the complexity of any system increases, gaining an understanding of it depends increasingly on summaries, such as averages and ranges, and on descriptions of typical examples of that system.

12. Habits of Mind

By the end of the 8th grade, students should know that

12.A Values and Attitudes

12.A.1 Know why it is important in science to keep honest, clear, and accurate records.

12.A.2 Know that hypotheses are valuable, even if they turn out not to be true, if they lead to fruitful investigations.

12.A.3 Know that often different explanations can be given for the same evidence, and it is not always possible to tell which one is correct.

12.B Computation and Estimation

12.B.1 Find what percentage one number is of another and figure any percentage of any number.

12.B.2 Use, interpret, and compare numbers in several equivalent forms such as integers, fractions, decimals, and percents.

12.B.3 Calculate the circumferences and areas of rectangles, triangles, and circles, and the volumes of rectangular solids.

12.B.4 Find the mean and median of a set of data.

12.B.5 Estimate distances and travel times from maps and the actual size of objects from scale drawings.

12.B.6 Insert instructions into computer spreadsheet cells to program arithmetic calculations.

12.B.7 Determine what unit (such as seconds, square inches, or dollars per tankful) an answer should be expressed in from the units of the inputs to the calculation, and be able to convert compound units (such as yen per dollar into dollar per yen, or miles per hour into feet per second).

12.B.8 Decide what degree of precision is adequate and round off the result of calculator operations to enough significant figures to reasonably reflect those of the inputs.

12.B.9 Express numbers like 100, 1,000, and 1,000,000 as powers of 10.

12.B.10 Estimate probabilities of outcomes in familiar situations, on the basis of history or the number of possible outcomes.

12.C Manipulation and Observation

12.C.1 Use calculators to compare amounts proportionally.

12.C.2 Use computers to store and retrieve information in topical, alphabetical, numerical, and key-word files, and create simple files of their own devising.

12.C.3 Read analog and digital meters on instruments used to make direct measurements of length, volume, weight, elapsed time, rates, and temperature, and choose appropriate units for reporting various magnitudes.

12.C.4 Use cameras and tape recorders for capturing information.

12.C.5 Inspect, disassemble, and reassemble simple mechanical devices and describe what the various parts are for; estimate what the effect that making a change in one part of a system is likely to have on the system as a whole.

12.D Communication Skills

12.D.1 Organize information in simple tables and graphs and identify relationships they reveal.

12.D.2 Read simple tables and graphs produced by others and describe in words what they show.

12.D.3 Locate information in reference books, back issues of newspapers and magazines, compact disks, and computer databases.

12.D.4 Understand writing that incorporates circle charts, bar and line graphs, two-way data tables, diagrams, and symbols.

12.D.5 Find and describe locations on maps with rectangular and polar coordinates.

12.E Critical-Response Skills

12.E.1 Question claims based on vague attributions (such as "Leading doctors say...") or on statements made by celebrities or others outside the area of their particular expertise.

12.E.2 Compare consumer products and consider reasonable personal trade-offs among them on the basis of features, performance, durability, and cost.

12.E.3 Be skeptical of arguments based on very small samples of data, biased samples, or samples for which there was no control sample.

12.E.4 Be aware that there may be more than one good way to interpret a given set of findings.

12.E.5 Notice and criticize the reasoning in arguments in which (1) fact and opinion are intermingled or the conclusions do not follow logically from the evidence given, (2) an analogy is not apt, (3) no mention is made of whether the control groups are very much like the experimental group, or (4) all members of a group (such as teenagers or chemists) are implied to have nearly identical characteristics that differ from those of other groups.